AF835482

Laser Surface Texturing for Engineering Materials

Laser surface texturing (LST) is a process that uses lasers to modify or prepare a material's surface for improved performance. This book examines laser surface texturing techniques for various materials and applications. It also explores the microstructural, tribological, anti-corrosion, and mechanical behaviors of laser-textured component surfaces to understand the mechanisms for improving lifecycle and reducing frictional losses.

This book:

- Considers LST of metals, ceramics, and polymers
- Offers a unique mix of research and commercials aspects
- Includes process parameters, mechanical properties results, and research outcomes
- Discusses adhesive strength, wettability, microstructural properties, mechanical and tribological behavior, surface roughness, coefficient of friction, and wear rate
- Features case studies
- Describes the challenges, applications, material properties, and bonding strength of various materials

Offering a concise approach to a complex technology with minimal mathematical equations, this work provides an overview of surface texturing techniques, their applications, analysis, and future advancements for materials, mechanical, and manufacturing engineers.

Jeyaprakash Natarajan is an associate professor in the Department of Mechanical Engineering, Global College of Engineering and Technology, Muscat, Oman. He received his PhD in production engineering from the National Institute of Technology, Tiruchirappalli, India.

Raj Kumar Durairaj is a professor in the Department of Mechanical Engineering, Sri Ramakrishna College of Engineering, Perambalur, India. He obtained his PhD in mechanical engineering from Anna University, India.

Ibrahim Abdullah Alnaser is the director of the Center of Excellence for Research in Engineering Materials at King Saud University, Riyadh, Saudi Arabia. He earned his PhD in mechanical engineering from the University of Tulsa, USA.

Laser Surface Texturing for Engineering Materials

Edited by Jeyaprakash Natarajan,
Raj Kumar Durairaj, and Ibrahim Abdullah Alnaser

CRC Press
Taylor & Francis Group
Boca Raton London New York

CRC Press is an imprint of the
Taylor & Francis Group, an **informa** business

Designed cover image: © 2027 Shutterstock

First edition published 2027
by CRC Press
2385 NW Executive Center Drive, Suite 320, Boca Raton FL 33431

and by CRC Press
4 Park Square, Milton Park, Abingdon, Oxon, OX14 4RN

CRC Press is an imprint of Taylor & Francis Group, LLC

© 2027 selection and editorial matter, Jeyaprakash Natarajan, Raj Kumar Durairaj, and Ibrahim Abdullah Alnaser; individual chapters, the contributors

Reasonable efforts have been made to publish reliable data and information, but the author and publisher cannot assume responsibility for the validity of all materials or the consequences of their use. The authors and publishers have attempted to trace the copyright holders of all material reproduced in this publication and apologize to copyright holders if permission to publish in this form has not been obtained. If any copyright material has not been acknowledged please write and let us know so we may rectify in any future reprint.

Except as permitted under U.S. Copyright Law, no part of this book may be reprinted, reproduced, transmitted, or utilized in any form by any electronic, mechanical, or other means, now known or hereafter invented, including photocopying, microfilming, and recording, or in any information storage or retrieval system, without written permission from the publishers.

For permission to photocopy or use material electronically from this work, access www.copyright.com or contact the Copyright Clearance Center, Inc. (CCC), 222 Rosewood Drive, Danvers, MA 01923, 978-750-8400. For works that are not available on CCC please contact mpkbookspermissions@tandf.co.uk

Trademark notice: Product or corporate names may be trademarks or registered trademarks and are used only for identification and explanation without intent to infringe.

For Product Safety Concerns and Information please contact our EU representative GPSR@taylorandfrancis.com. Taylor & Francis Verlag GmbH, Kaufingerstraße 24, 80331 München, Germany.

ISBN: 978-1-032-89278-8 (hbk)
ISBN: 978-1-032-89279-5 (pbk)
ISBN: 978-1-003-54208-7 (ebk)

DOI: 10.1201/9781003542087

Typeset in Times
by Apex CoVantage, LLC

Contents

Preface

Surface engineering has become a decisive enabler for improving the performance, reliability, and sustainability of modern engineering systems. Among the wide range of surface modification techniques, laser surface texturing (LST) has emerged as a powerful, flexible, and precise approach for tailoring surface functionalities without altering the bulk properties of materials. *Laser Surface Texturing for Engineering Materials* is conceived to provide a comprehensive, systematic, and application-oriented understanding of LST, addressing both fundamental principles and advanced engineering practices across diverse material systems and industrial domains.

The motivation for this book arises from the rapid expansion of laser-based surface engineering in sectors such as manufacturing, tribology, biomedical engineering, energy systems, and additive manufacturing. Laser surface texturing enables controlled micro- and nano-scale features that significantly influence mechanical, tribological, thermal, wetting, and corrosion behaviors. Despite the growing body of research, a unified reference that integrates process fundamentals, simulation, surface characterization, and real-world applications remains limited. This book aims to bridge that gap by offering an interdisciplinary perspective that combines materials science, laser–matter interaction, mechanical engineering, and surface physics.

The opening chapter introduces the fundamentals of surface texturing techniques, emphasizing the evolution, classification, and advantages of laser-based approaches over conventional methods. This foundation is extended in Chapter 2, which focuses on the simulation and systematic study of laser surface texturing parameters, highlighting the role of modeling in predicting surface morphology, thermal fields, and process optimization. Recognizing that surface performance is strongly influenced by auxiliary treatments, Chapter 3 examines pre-, during-, and post-processing strategies and their impact on surface integrity and functional outcomes.

Subsequent chapters explore the mechanical, tribological, and thermal implications of laser surface texturing. Chapters 4 and 7 provide in-depth analyses of textured coatings and bulk materials—including metals, ceramics, and polymers—while Chapter 5 addresses surface heat transfer enhancement, an area of increasing relevance for energy and thermal management applications. The importance of tooling and manufacturing efficiency is emphasized in Chapter 6, which discusses LST on tool surfaces for both conventional and unconventional machining processes.

Functional surface behaviors such as hydrophobicity, corrosion resistance, and biocompatibility are systematically addressed in Chapters 8, 10, and 12. These chapters highlight how laser-induced textures can be engineered to meet stringent requirements in marine, biomedical, and chemical environments. Chapter 9 presents a focused industrial case study on piston top surface texturing, demonstrating the practical benefits of LST in reducing residual stresses and improving surface quality in engine components. The growing relevance of advanced manufacturing is reflected in Chapter 11, which examines the integration of LST with additively manufactured parts to overcome inherent surface and performance limitations.

This book is intended for researchers, graduate students, practicing engineers, and industry professionals seeking both theoretical insight and practical guidance. By combining fundamental understanding with application-driven case studies, *Laser Surface Texturing for Engineering Materials* aims to serve as a valuable reference and a catalyst for future innovation in surface engineering.

Jeyaprakash Natarajan,
Raj Kumar Durairaj,
and Ibrahim Abdullah Alnaser

Contributors

A. Riyaz Ahmed
Department of Projects
National Institute of Electronics and Information Technology
Chennai, Tamil Nadu, India

Habtamu Alemayehu
Mechanical Engineering
Haramaya Institute of Technology
Haramaya University
Haramaya, Dire Dawa, Ethiopia

Kannan Ganesa Balamurugan
Department of Mechanical Engineering
IFET College of Engineering
Villupuram, Tamil Nadu, India

V. M. Basanthy
Mechanical Engineering Department
Global College of Engineering and Technology
Muscat, Oman

Firomsa Dadi
Mechanical Engineering
Haramaya Institute of Technology
Haramaya University
Haramaya, Dire Dawa, Ethiopia

Suresh Goka
Department of Mechanical Engineering
National Institute of Technology
Warangal, Telangana, India

Monaf Hodhod
Department of Mechanical Engineering
College of Engineering
King Faisal University
Al-Ahsa, Kingdom of Saudi Arabia

K. Kamalakannan
Department of Mechanical Engineering
IFET College of Engineering
Villupuram, Tamil Nadu, India

Kumaran M.
School of Mechanical Engineering
Faculty of Engineering and Technology
SRM Institute of Science and Technology
Trichy, Tamil Nadu, India

Durairaj Manikandan
Department of Mechanical Engineering
SRM TRP Engineering College
Trichy, Tamil Nadu, India

S. Mohan
Department of Mechanical Engineering
Dhanalakshmi Srinivasan University
Trichy, Tamil Nadu, India

Syed Quadir Moinuddin
Department of Mechanical Engineering
College of Engineering
King Faisal University
Al-Ahsa, Kingdom of Saudi Arabia

V. Mugendiran
Department of Production Technology
Madras Institute of Technology
Anna University
Chennai, India

Jayanth N.
Centre of Excellence in Product Design and Smart Manufacturing
Maulana Azad National Institute of Technology
Bhopal, Madhya Pradesh, India

Jeyaprakash Natarajan
Department of Mechanical Engineering
Global College of Engineering and Technology
Muscat, Oman

Saad Patel
Department of Mechanical Engineering
Vellore Institute of Technology
Vellore, Tamil Nadu, India

G. Prabu
Graduate Institute of Manufacturing Technology
National Taipei University of Technology
Taipei, Taiwan

Durairaj Rajkumar
Department of Mechanical Engineering
Sri Ramakrishna College of Engineering
Perambalur, Tamil Nadu, India

S. Ravi
Department of Production Engineering
National Institute of Technology
Trichy, Tamil Nadu, India

T. Sathies
Department of Mechanical Engineering
Amrita School of Engineering
Amrita Vishwa Vidyapeetham
Andhra Pradesh, India

S. Sathish
Department of Production Technology
Madras Institute of Technology
Anna University
Chennai, India

Lakshmanan Vellaichamy
Department of Mechanical Engineering
SRM TRP Engineering College
Trichy, Tamil Nadu, India

Govindarajan Venkatesan
Mechanical Engineering
Haramaya Institute of Technology
Haramaya University
Haramaya, Dire Dawa, Ethiopia

S. Vijayakumar
Department of Production Technology
Madras Institute of Technology
Anna University
Chennai, India

Che-Hua Yang
Graduate Institute of Manufacturing Technology
National Taipei University of Technology
Taipei, Taiwan

1 Introduction to Surface Texturing Techniques and Its Applications

Govindarajan Venkatesan, Habtamu Alemayehu, Jeyaprakash Natarajan, and Firomsa Dadi

1.1 FUNDAMENTALS AND DEFINITIONS

Surface texturing is an advanced engineering technique that modifies the surface characteristics of materials to enhance their performance for specific applications [1]. This process involves creating highly controlled patterns, structures, or features at micro to nanoscale levels, significantly affecting a material's physical, chemical, and mechanical properties [2]. It plays a pivotal role in optimizing critical features like friction control, wear resistance, adhesion, and optical performance, making it indispensable across various industries, including automotive, aerospace, electronics, and biomedical engineering [3].

One of the key goals of surface texturing is to precisely manipulate interactions between materials and their environments, such as friction, heat generation, corrosion, or chemical bonding [4]. By altering surface topography, engineers can fine-tune these interactions, improving material performance to meet specific operational needs [5]. This capability is crucial for advancing technologies, improving energy efficiency, and solving complex engineering challenges [6].

Surface texturing finds numerous applications, particularly in reducing friction in mechanical systems [7]. Friction, a natural byproduct of surface contact, causes energy loss and wear [8]. Textured surfaces minimize these frictional forces, extending machinery lifespan and improving energy efficiency. For example, in engine components and bearings, reduced friction lowers energy consumption and wear, contributing to sustainability and reliability [9].

The customization allowed by surface texturing also plays a key role in improving adhesion, which is vital in industries like automotive, aerospace, and electronics. Textured surfaces enhance microscopic mechanical interlocking, leading to stronger bonds between materials. This property is particularly valuable in applications requiring durable bonding, such as construction materials and protective coatings designed to withstand corrosion in extreme environments [10, 11].

Beyond mechanical and adhesive improvements, surface texturing has a transformative impact on optics and renewable energy technologies. In solar energy, textured surfaces maximize light absorption and reduce reflectivity, leading to more efficient

DOI: 10.1201/9781003542087-1

energy conversion in solar panels. Similarly, optical devices such as lenses, mirrors, and display screens benefit from surface textures that optimize light manipulation, enhancing performance and versatility [12, 13].

In the automotive industry, surface texturing improves engine performance, fuel efficiency, and component lifespan by reducing friction and wear. Textured tires also provide better grip and water drainage under wet conditions, enhancing safety and performance [14]. In biomedical engineering, surface texturing has revolutionized the functionality of implants and medical devices. By promoting better biocompatibility and osseointegration (the process where bone cells grow into the implant surface), textured surfaces improve the stability and longevity of devices like dental prosthetics, joint replacements, and stents [15, 16]

Recent advancements in manufacturing technologies, such as laser ablation, chemical etching, and additive manufacturing (3D printing), have expanded the possibilities of surface texturing. These techniques enable the creation of complex, high-precision surface patterns, driving innovations in self-cleaning surfaces, anti-corrosion coatings, and more. For instance, self-cleaning surfaces are now used in construction, electronics, and consumer goods, reducing maintenance and extending material durability [17].

Looking to the future, the potential applications of surface texturing are vast. Research is exploring multifunctional textures that combine properties like friction reduction with antibacterial or antifungal capabilities, opening up new possibilities in healthcare, food processing, and construction [18]. "Smart" surfaces that respond to environmental stimuli such as temperature, pressure, or light could revolutionize industries like aerospace and wearable electronics [19].

Surface texturing is reshaping material science and engineering, offering solutions to some of today's most pressing technological challenges. As manufacturing technologies continue to advance, its applications are expected to expand, further empowering engineers to unlock new possibilities across industries [20].

1.2 HISTORICAL OVERVIEW OF SURFACE TEXTURING TECHNIQUES

Ancient beginnings: Surface texturing has a long history, dating back to prehistoric times, when early humans modified surfaces to meet practical needs. These foundational techniques laid the groundwork for future advancements in material manipulation. Early civilizations quickly realized that altering surface properties improved the functionality and durability of tools, weapons, and everyday objects [21].

Stone Age methods: During the Stone Age, humans developed rudimentary methods such as chipping to create sharp edges on stones, greatly enhancing their cutting and hunting capabilities. Polishing further refined these tools, producing sharper edges and increasing their effectiveness. These early surface texturing techniques demonstrated an understanding of material performance, setting the stage for future technological advancements [22].

Metalworking in antiquity: The advent of metallurgy marked a significant turning point in surface texturing. As early craftsmen began working with materials like copper, bronze, and iron, they developed new techniques such as hammering to shape and strengthen metals. Processes like quenching (rapid cooling) hardened surfaces, while engraving served both decorative and practical purposes, as seen in weapons, jewelry, and artifacts. This period demonstrated how surface texturing could blend functionality with aesthetics, meeting the growing complexity of societal needs [23]

Medieval advances (specialized applications): By the medieval era, surface texturing techniques had been refined to serve both industrial and artistic purposes. Armor and weapons were textured to enhance durability and resilience in battle, while Gothic architecture featured intricate stone carvings that added both beauty and structural strength. This era showcased the adaptability of surface texturing, with its applications expanding from tools to art and architecture [24].

The Industrial Revolution (mechanization and mass production): The Industrial Revolution brought mechanized processes that revolutionized surface texturing. Machines such as lathes and milling devices enabled precise and scalable modifications, while techniques like shot peening and sandblasting became widely used to strengthen and texture materials. These innovations transformed industries such as automotive, aerospace, and heavy machinery, allowing for mass production with consistent quality [25].

Twentieth-century advancements (precision engineering): The 20th century marked a new era of precision engineering, driven by advancements in micro-texturing and advanced coating technologies. Micro-texturing, which reduced friction and enhanced wear resistance, became crucial in automotive and aerospace industries. Additionally, coating methods like electroplating and chemical vapor deposition improved properties such as corrosion resistance and biocompatibility. These developments reflected a deeper understanding of materials at the microscopic level, revolutionizing surface engineering [26].

Twenty-first-century innovations (nanotechnology and sustainability): In the 21st century, nanotechnology and sustainable practices have reshaped surface texturing. Nano-texturing has enabled the creation of surfaces with unique properties, including self-cleaning capabilities, anti-reflective coatings, and enhanced biocompatibility for medical applications. Sustainable methods, such as waterjet texturing, have minimized environmental impact while still achieving intricate and high-performance designs. These innovations highlight the growing emphasis on eco-friendly solutions and cutting-edge performance across industries.

From its prehistoric origins to the precision engineering of today, surface texturing has evolved in complexity and application, transforming materials to meet ever-changing functional, aesthetic, and environmental needs [27].

1.3 SURFACE TEXTURING TECHNIQUES

The variety of surface texturing techniques such as mechanical abrasion, thermal treatments, and laser patterning offers customizable solutions tailored to meet the specific performance and design needs of different industries [28].

1.3.1 Mechanical Abrasion Based Surface Texturing Technique

Mechanical abrasion is a widely used surface texturing method that involves physically altering or removing material through friction using abrasive tools or grit, resulting in a roughened surface. Common techniques such as grinding, sandblasting, polishing, buffing, and shot peening improve friction, prepare surfaces for coatings, and enhance electrical performance [29]. This cost-effective and versatile process is commonly applied in industries like automotive, manufacturing, and electronics [30]. Figure 1.1 provides a schematic 3D representation of the mechanical abrasion process for surface texturing. Despite its effectiveness, mechanical abrasion has limitations in precision, generates debris, and can lead to material wastage.

1.3.2 Chemical Etching Based Surface Texturing Technique

Chemical etching, or chemical milling, is a precise surface texturing method that uses chemical reactions to selectively remove material, creating intricate, high-precision patterns [31]. The process begins by selecting a material with the required mechanical and chemical properties, followed by thorough cleaning to remove contaminants. A protective photoresist layer is then applied, and the desired pattern is transferred onto it using UV light [32]. During the development phase, unexposed areas of the photoresist are dissolved, exposing the underlying material, which is then chemically etched to form the textured pattern. The photoresist is subsequently stripped away, and the surface is inspected to ensure the etched patterns meet design specifications [33]. Finishing treatments are applied to enhance functionality or appearance [34].

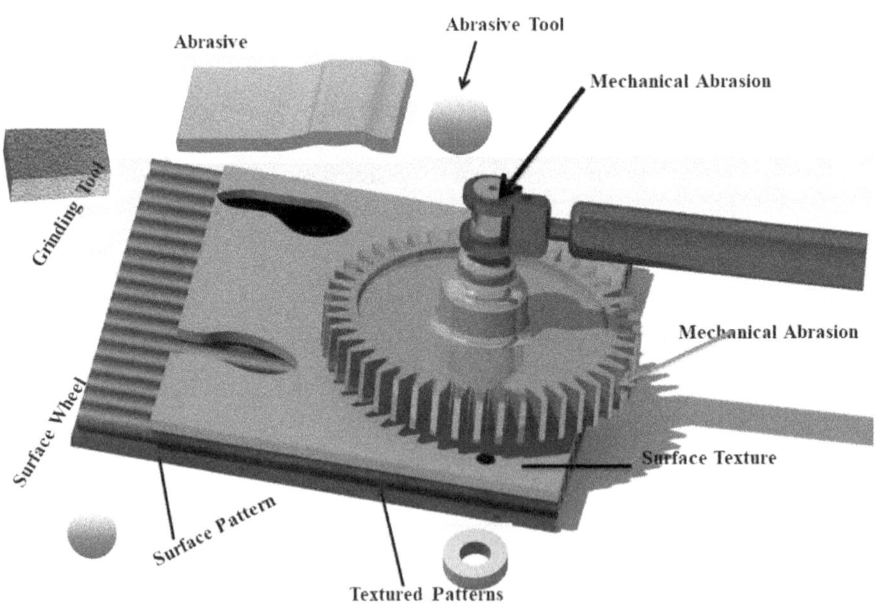

FIGURE 1.1 3D representation of the mechanical abrasion process for surface texturing.

This highly accurate technique, widely used in microfabrication, decorative designs, and functional surface engineering, offers key advantages, including high precision, uniform material removal, and the ability to create fine details without mechanical stress. Techniques such as wet etching, dry etching, and photochemical etching are commonly employed in industries like electronics, aerospace, and decorative applications [35]. However, the process involves handling chemicals that must be managed carefully and presents environmental challenges, such as waste disposal. Figure 1.2 illustrates the chemical etching process for surface texturing.

1.3.3 Thermal Treatments-Based Surface Texturing Technique

Thermal treatments involve the controlled heating and cooling of materials to alter their surface properties, such as hardness, roughness, and microstructure, making them particularly effective in improving wear resistance and creating textured surfaces [36]. Techniques like flame spraying, plasma treatments, annealing, and thermal spraying are widely used in industries such as aerospace, automotive, and medical, where components are exposed to extreme conditions [37]. These treatments are beneficial for creating durable, wear-resistant surfaces but require specialized equipment and expertise [38]. While effective, thermal processes can be energy intensive and may cause thermal distortion in materials sensitive to high temperatures. Figure 1.3 illustrates the thermal treatment process for surface texturing.

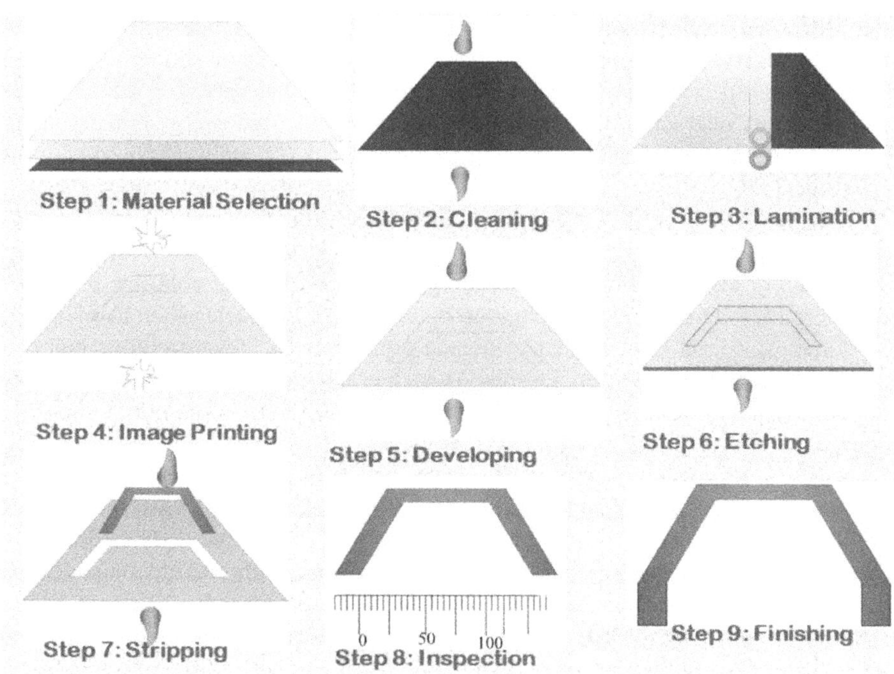

FIGURE 1.2 Schematic of chemical etching process for surface texturing.

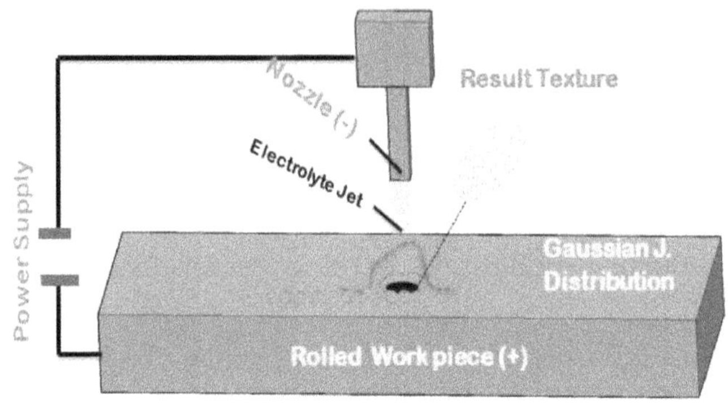

FIGURE 1.3 Schematic of thermal treatment process.

1.3.4 Laser Patterning-Based Surface Texturing Technique

Laser patterning is an advanced surface texturing technique that uses high-energy laser beams to create micro and nano-scale patterns with exceptional precision. The process works by directing a laser onto the material surface, where concentrated energy removes material to form intricate designs [39]. It begins with a computer controlling the laser's parameters, such as power, speed, and path, ensuring accuracy and repeatability. The laser passes through an aperture and beam splitter to optimize intensity and uniformity before being focused into a precise spot size, allowing fine control over its interaction with the surface [40]. This focused energy induces localized thermal, photochemical, or photophysical effects to create the desired patterns or textures [41].

Laser patterning is widely used for applications that improve surface properties, enhance functionality, or create micro/nano-scale textures. Key methods include direct laser interference patterning, femtosecond laser ablation, and laser engraving [42]. This versatile technique is utilized in industries such as biomedical, automotive, and electronics to enhance properties like friction reduction, cell adhesion, and light absorption. Despite its high precision, laser patterning presents challenges, including high setup costs and the need for skilled operators to ensure accuracy.

Laser patterning, along with other surface texturing techniques like mechanical abrasion, chemical etching, and thermal treatments, offers unique advantages that can be tailored to meet specific industrial needs [43]. These methods significantly improve wear resistance, corrosion resistance, fatigue strength, and biocompatibility, optimizing the performance and longevity of materials. As industries continue to evolve, these texturing methods will further advance, providing increasingly efficient and sustainable solutions for demanding applications in sectors such as aerospace, automotive, medical, and manufacturing [43]. Figure 1.4 illustrates the laser patterning process for surface texturing.

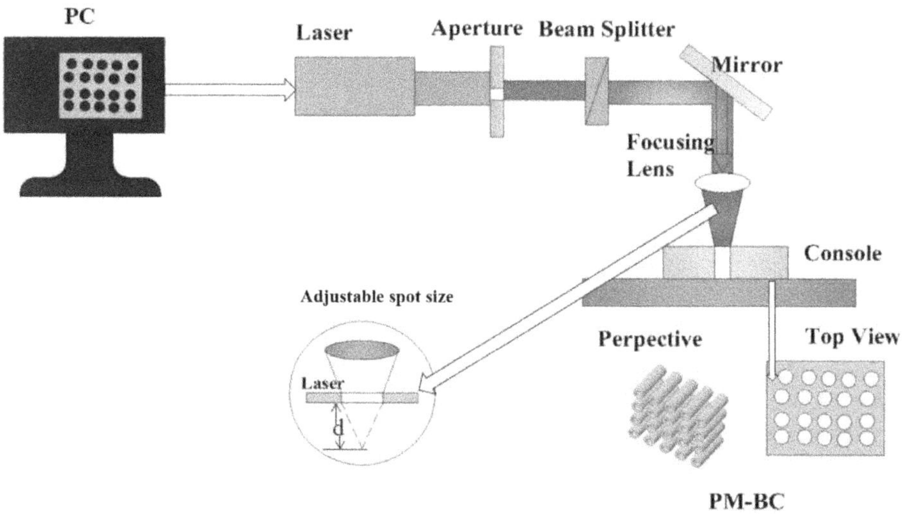

FIGURE 1.4 Schematic of laser patterning process.

1.4 EMERGING TECHNOLOGIES

Recent advancements in artificial intelligence (AI) and 3D printing have revolutionized surface texturing, pushing the boundaries of design and production. AI-driven tools optimize surface textures by simulating performance under various conditions, enabling highly customized solutions. Meanwhile, 3D printing allows for the direct integration of complex textures into components during production, offering unprecedented precision and customization for industries like aerospace, healthcare, and electronics. The evolution of surface texturing is a testament to humanity's continuous pursuit of innovation, with each era introducing techniques that address contemporary challenges and lay the foundation for future progress. Today, the integration of AI, 3D printing, and nanotechnology continues to drive engineering advancements, unlocking transformative possibilities across diverse fields [44]

1.5 EMERGING DEVELOPMENTS AND INNOVATIONS IN SURFACE TEXTURING

Surface texturing technology has undergone significant advancements in recent years, evolving from its initial focus on improving material properties like friction, wear resistance, and adhesion, to becoming a transformative force across various high-performance industries [45]. As new materials, manufacturing techniques, and design methodologies emerge, surface texturing is rapidly adapting to meet the demands of specialized applications, particularly in aerospace, automotive, biomedical engineering, and electronics. Engineers continue to push the boundaries of material science, expanding the potential applications of surface texturing, making it one of the most transformative technologies in modern engineering.

One key development is *laser-assisted surface texturing (LAST)*, which enables precise creation of micro and nano-scale textures on materials [46]. This technique offers exceptional control over surface properties, allowing for the production of complex structures previously unattainable. With high-speed laser systems, engineers can now tailor the surface characteristics of components for specific applications, such as reducing friction in mechanical parts or improving heat dissipation in electronic devices. LAST has gained widespread adoption in industries where performance and precision are critical, including aerospace, biomedical engineering, and electronics.

Another significant trend is the rise of *bioinspired textures*, which replicate natural surface properties like the water-repellent qualities of lotus leaves, the anti-fouling features of sharkskin, and the optical properties of butterfly wings. These bioinspired designs are being applied to create self-cleaning surfaces and anti-reflective coatings and improve the performance of renewable energy technologies [47]. For example, sharkskin-inspired textures are used in aerospace and marine industries to reduce drag, while lotus-leaf-inspired textures are employed for self-cleaning solar panels and windows [48]. By mimicking nature's efficient designs, engineers are improving sustainability, functionality, and performance across diverse sectors.

The integration of *additive manufacturing (3D printing)* with surface texturing is also transforming the field. This combination allows for the production of customized, complex components that were once impossible to create using traditional methods. The ability to control both the geometry and the texture of components enables the design of optimized parts that reduce weight, enhance performance, and increase durability. This innovation is particularly valuable in industries like aerospace, automotive, and healthcare, where lightweight, high-performance components are critical [49].

Another exciting development is *dynamic surface textures*, which respond to external stimuli such as heat, light, moisture, or mechanical stress. These responsive textures can adjust their properties to optimize performance under varying conditions. For example, in the automotive and aerospace industries, dynamic textures can improve fuel efficiency and reduce wear by adapting to changes in temperature and pressure [50]. In wearable technology, responsive surfaces can adjust to body temperature, enhancing both comfort and functionality.

As sustainability becomes an increasingly important focus, *eco-friendly surface texturing methods* are gaining traction. Traditional surface texturing techniques often rely on harmful chemicals or energy-intensive processes, which can negatively impact the environment [51]. To address this, researchers are exploring green alternatives such as chemical-free etching and sustainable coatings that reduce chemical usage and energy consumption [52]. Surface texturing is also being applied to enhance the efficiency of renewable energy systems, including solar panels and wind turbines, further promoting environmental sustainability.

The integration of *artificial intelligence (AI)* and *machine learning* into surface texturing processes is also accelerating innovation [53]. AI algorithms can optimize the design and application of surface textures by analyzing large datasets and predicting the ideal texture patterns for specific applications [54]. Machine learning models improve the precision and cost-effectiveness of surface texturing, allowing for faster production cycles and more efficient manufacturing processes [55]. This combination of AI and machine learning is driving the development of new materials and applications, opening new frontiers for surface texturing technologies.

With these advancements, ranging from laser-assisted texturing and bioinspired designs to dynamic textures and sustainable practices, the potential for surface texturing across industries is vast. As industries continue to demand higher efficiency, performance, and sustainability, surface texturing will remain a crucial technology, enabling the creation of high-performance, environmentally friendly materials and products that meet the evolving demands of the 21st century.

1.6 IMPORTANCE AND APPLICATIONS OF SURFACE TEXTURING

Surface texturing is a transformative technique that modifies material surfaces to enhance functionality, efficiency, and durability across various industries. By altering surface properties, it improves performance in areas such as stress resistance, friction reduction, biocompatibility, and energy efficiency, contributing significantly to the automotive, aerospace, healthcare, electronics, and energy sectors. This widespread application improves material performance, longevity, and sustainability [45].

1.6.1 Automotive Sectors

The following are some of the applications of surface texturing in the automotive sector [45].

- Surface texturing enhances engine performance, safety, and design by minimizing friction in components like pistons, cylinders, and bearings, improving lubrication and fuel efficiency. This reduction in friction also lowers energy consumption, contributing to lower fuel costs and helping meet environmental regulations.
- Textured components increase wear resistance, extending service life and reducing the need for frequent maintenance.
- Brake systems benefit from textured discs and pads that effectively channel heat away, preventing overheating and ensuring consistent braking performance.
- Micro-textures on brake surfaces reduce noise and vibration, improving safety and comfort.
- On vehicle exteriors, anti-glare surfaces improve visibility, while hydrophobic textures, inspired by natural surfaces like lotus leaves, repel water and dirt, reducing maintenance.

1.6.2 Aerospace Sectors

In the aerospace industry, surface texturing is crucial for components that are exposed to extreme conditions. Some of the key applications of surface texturing in aerospace include the following [56]:

- Surface texturing improves the efficiency and durability of high-performance components, such as turbine blades, which operate under extreme temperatures and aerodynamic forces.

- Riblet-like textures, inspired by shark skin, reduce drag and optimize airflow, resulting in better fuel efficiency and improved engine performance.
- Micro-texturing enhances heat dissipation, preventing thermal buildup that could degrade materials.
- Surface texturing is applied to aircraft wings and fuselages to reduce aerodynamic drag, improving fuel efficiency and supporting sustainability efforts.
- Lightweight textured coatings help achieve weight reduction goals without sacrificing structural integrity.
- In spacecraft, textured heat shields and coatings manage thermal flux, protect against space radiation, and enhance adhesion during re-entry.

1.6.3 Electronics Sectors

For electronics, surface texturing improves the performance and durability of components such as sensors, connectors, and micro-electromechanical systems. Some of the key applications of surface texturing in aerospace include the following [57]:

- Surface texturing plays an essential role in improving thermal management, reducing heat buildup, and extending the lifespan of microchips and semiconductors.
- Textured surfaces help prevent overheating, ensuring optimal device performance and
- Micro-textures also enhance the electrical properties of semiconductor materials, boosting processing speeds and energy efficiency.
- In optical components, anti-reflective coatings with precisely engineered textures reduce glare and improve lens clarity.
- Nano-textured solar panels reduce light reflection, enhancing energy absorption and overall efficiency.
- Printed circuit boards benefit from improved solderability, reliable electrical connections, and enhanced corrosion resistance in harsh environments.

1.6.4 Healthcare Sectors

In healthcare, surface texturing significantly boosts the functionality of medical implants and devices. Textured surfaces improve Osseo integration in orthopedic and dental implants, increasing stability and longevity. Some of the key applications of surface texturing in healthcare include the following [58]:

- Surface texturing is transforming the development of implants, medical devices, and drug delivery systems.
- In orthopedic and dental implants, textured surfaces improve adhesion, accelerating healing and reducing the risk of rejection.
- Textured surfaces also offer greater wear resistance, making them ideal for joint replacements, such as hip and knee prostheses.

- For surgical tools, surface texturing reduces microbial adhesion, enhancing sterility and safety.
- Textured surfaces increase the durability of medical devices, ensuring performance over repeated procedures.
- In drug delivery systems, surface texturing regulates the release of medication for consistent and precise therapeutic outcomes.
- Microfluidic devices benefit from textured surfaces that facilitate smoother fluid flow, improving diagnostic accuracy and treatment effectiveness.

1.6.5 Energy Sector

In the energy sector, surface texturing plays a crucial role in renewable energy technologies. Textured photovoltaic cells reduce light reflection and increase energy conversion efficiency, making solar power systems more cost effective. Similarly, textured turbine blades in wind energy systems reduce drag, optimize performance at lower wind speeds, and enhance energy production, all of which contribute to lower operational costs and greater sustainability. In oil and gas, surface treatments such as shot peening and sandblasting increase the wear resistance of components like drill bits and valves, reducing maintenance needs and extending their operational life [59].

Overall, surface texturing is a vital technology driving innovation across multiple industries. As research advances, it continues to offer advanced solutions to meet the evolving challenges of modern industry, enhancing sustainability and improving the performance, durability, and efficiency of materials and components across a wide range of applications [60].

1.7 BENEFITS AND ADVANTAGES OFFERED BY SURFACE TEXTURING

Surface texturing is a critical process in modern engineering, enhancing the performance, functionality, and aesthetics of materials across various industries [61]. By modifying surface properties, techniques like shot peening, abrasive blasting, laser treatment, and chemical etching improve key attributes such as wear resistance, corrosion resistance, fatigue strength, lubrication efficiency, and biocompatibility [45]. These improvements contribute to the durability, longevity, and cost-effectiveness of products, reducing maintenance and extending lifespans crucial for sectors like aerospace, automotive, and manufacturing, where components are exposed to harsh conditions.

Wear resistance is particularly important for components subject to high friction, such as gears, turbine blades, and machine tools, and is enhanced through processes like shot peening and abrasive blasting, which alter surface topography to withstand abrasive forces. Corrosion resistance is also improved by creating protective coatings that shield materials from environmental factors like moisture and chemicals. Techniques like anodizing and electroplating are widely used in aerospace and marine applications, resulting in more durable engine parts and ship hulls [62].

Furthermore, surface texturing improves fatigue strength, which is vital for materials subjected to repetitive loading cycles. Shot peening, for example, introduces compressive residual stresses that reduce the likelihood of crack formation, thereby enhancing resistance to failure under cyclic stress [63]. In medical applications, surface texturing boosts the biocompatibility of implants, promoting tissue integration and reducing the risk of rejection. It also contributes to improved lubrication efficiency in mechanical systems by reducing friction between contacting surfaces, resulting in energy savings and longer-lasting moving parts.

Beyond functional benefits, surface texturing allows manufacturers to create visually appealing products with unique patterns or finishes, adding aesthetic value to consumer goods, automotive interiors, and electronics [64].

As technology advances, the applications of surface texturing are expected to expand, offering further improvements in performance, sustainability, and efficiency. This makes it an indispensable tool in modern engineering, addressing both practical and design needs across multiple sectors.

1.8 CHALLENGES AND LIMITATIONS OF SURFACE TEXTURING TECHNIQUES

Despite significant advancements in surface texturing across various industries, several challenges and limitations must be addressed to maximize its broader application and efficiency.

One of the key challenges is the complexity of producing precise micro- and nanoscale textures, particularly with advanced techniques like LAST. While LAST offers exceptional control over surface properties, achieving consistent texture uniformity remains difficult, especially when dealing with irregular or intricate geometries. Additionally, the precision required for these techniques demands costly equipment and specialized expertise, making it less accessible for smaller enterprises or industries with limited resources.

Another significant hurdle is the compatibility of materials with various texturing methods. Not all materials can be effectively processed with every texturing technique, which limits the versatility of certain approaches [65]. For example, laser-assisted texturing may not be suitable for all metals, polymers, or ceramics, as their inherent physical properties may interfere with the process [66]. Similarly, bioinspired textures, although efficient, require careful attention to material properties and surface treatments to ensure long-term performance. These material compatibility issues restrict the applicability of surface texturing, particularly in industries that handle a broad range of materials [45].

Scalability also poses a challenge for the widespread adoption of surface texturing. Techniques such as LAST and additive manufacturing are often more suited for small-scale or specialized production rather than large-scale manufacturing. This becomes problematic in sectors like automotive and consumer electronics, where cost-effective mass production is critical. The time and cost involved in producing custom components with complex textures may outweigh the performance benefits, impacting the economic feasibility of surface texturing for mass production.

Dynamic surface textures, which hold promise for adaptive systems, also face technical challenges. These textures, designed to respond to external factors such as temperature, light, or mechanical stress, must maintain consistent performance over time. However, environmental wear and exposure can degrade their adaptability, diminishing their functionality. Additionally, integrating dynamic textures into existing systems often requires substantial design adjustments, increasing development time and costs, which can slow their adoption in certain industries.

Finally, the environmental impact of traditional surface texturing methods remains a critical issue. Many conventional techniques rely on energy-intensive processes or harmful chemicals, contributing to environmental degradation. Although more sustainable alternatives, such as chemical-free etching and eco-friendly coatings, are being explored, they are not yet widely available or cost-effective for all applications. To mitigate the environmental footprint of surface texturing, it is crucial to develop more sustainable and energy-efficient methods. Until these alternatives become more scalable and cost effective, industries will face challenges in balancing the need for high-performance surfaces with environmental considerations.

These challenges underscore the importance of continued research and innovation to unlock the full potential of surface texturing techniques and overcome the barriers to their broader application across industries.

1.9 THE IMPACT OF SURFACE TEXTURING ON TECHNOLOGICAL ADVANCEMENT AND INNOVATION

Surface texturing has become a critical driver of technological advancement across various industries by enhancing the performance, durability, and functionality of materials. By modifying surface properties such as friction, adhesion, and wear resistance, surface texturing enables the creation of high-performing components, leading to significant breakthroughs in aerospace, automotive, electronics, healthcare, and renewable energy sectors. As the demand for efficient and high-performance products grows, surface texturing plays a key role in overcoming challenges once thought insurmountable.

In the aerospace industry, surface texturing has advanced critical technologies by improving the performance of components like turbine blades and heat exchangers. Techniques such as LAST create micro-scale patterns that reduce friction, enhance heat dissipation, and extend the lifespan of aerospace systems, contributing to more sustainable and cost-effective solutions. Similarly, in healthcare, surface texturing enhances the biocompatibility of medical devices, including implants and prosthetics, promoting better integration with human tissue. Bioinspired textures, like those mimicking sharkskin or lotus leaves, improve self-cleaning properties and reduce bacterial growth, which lowers infection risks and accelerates recovery.

The electronics industry has also greatly benefited from surface texturing, particularly in improving heat dissipation and prolonging the lifespan of devices like smartphones, laptops, and semiconductors. Textured surfaces enhance thermal management, ensuring efficient operation without overheating, thus reducing maintenance costs and improving user experience. In addition, surface texturing plays a crucial role in the development of environmentally friendly solutions. By improving the efficiency of renewable energy systems such as solar panels and wind turbines, surface texturing enhances energy capture and system sustainability, while eco-friendly methods like chemical-free etching align with the growing demand for greener manufacturing processes.

In sum, surface texturing is essential not only for improving performance across a range of industries but also for advancing sustainability, highlighting its transformative potential in the future of technology.

1.10 FUTURE PERSPECTIVES AND RECOMMENDATIONS

To fully realize the potential of surface texturing techniques and overcome the existing challenges, several key recommendations should be pursued in future research and development:

- Efforts should continue to improve the precision of micro- and nano-scale texturing, especially for complex geometries. Innovations in LAST and other advanced methods should focus on enhancing uniformity and reliability, particularly for irregular shapes. Additionally, research into automation and AI-guided precision could reduce human error and improve consistency.
- Future research should focus on developing more versatile surface texturing methods that can handle a wider range of materials, including metals, polymers, and ceramics, without compromising performance. Studies on the interactions between different materials and surface textures could lead to more efficient treatments and customized solutions for specific applications.
- To overcome scalability challenges, efforts should focus on adapting texturing technologies for mass production, particularly in automotive and electronics. Simplifying processes for methods like LAST and additive manufacturing can reduce costs for high-volume manufacturing. Exploring hybrid methods that combine traditional and advanced techniques may offer a cost-effective solution for scaling.
- Future research on dynamic surface textures should focus on enhancing durability and long-term performance, particularly through improved materials and coatings. Innovations to increase wear resistance and adaptability to environmental changes are key. Additionally, simplifying integration into existing systems will support broader adoption in industries that benefit from adaptive features.
- Efforts should focus on developing eco-friendly surface texturing methods, including chemical-free etching and energy-efficient processes. Adopting circular economy principles, like reusing waste materials and minimizing resource use, will also help reduce the environmental impact.
- Collaboration among industry, academia, and material scientists is key to advancing surface texturing techniques and materials. Joint efforts can foster innovation, standardization, and improve the efficiency and applicability of these methods across industries.

By focusing on these recommendations, the surface texturing industry can overcome its current limitations, leading to broader application, improved performance, and more sustainable practices across industries. Continued innovation and investment in research will be pivotal to unlocking the full potential of surface texturing techniques in the future.

REFERENCES

[1] Coblas, D. G., Fatu, A., Maoui, A., & Hajjam, M. (2014). Manufacturing Textured Surfaces: State of Art and Recent Developments. Proceedings of the Institution of Mechanical Engineers, Part J: Journal of Engineering Tribology, 229(1), 3–29. https://doi.org/10.1177/1350650114542242

[2] Jeon, H. J., Simon, C. G., Jr., & Kim, G. H. (2024). A Mini-Review: Cell Response to Microscale, Nanoscale, and Hierarchical Patterning of Surface Structure. Journal of Biomedical Materials Research Part B: Applied Biomaterials. https://doi.org/10.1002/jbm.b.33158

[3] Gao, K., Zhang, Y., Yi, J., Dong, F., & Chen, P. (2024). Overview of Surface Modification Techniques for Titanium Alloys in Modern Material Science: A Comprehensive Analysis. Coatings, 14, 148. https://doi.org/10.3390/coatings14010148

[4] Nsilani Kouediatouka, A., Ma, Q., Liu, Q., Mawignon, F. J., Rafique, F., & Dong, G. (2022). Design Methodology and Application of Surface Texture: A Review. Coatings, 12, 1015. https://doi.org/10.3390/coatings12071015

[5] Pereira, A. C., Nayak, V. V., Coelho, P. G., & Witek, L. (2024). Integrative Modeling and Experimental Insights into 3D and 4D Printing Technologies. Polymers, 16, 2686. https://doi.org/10.3390/polym16192686

[6] Arnold, F. (2001). Combinatorial and Computational Challenges for Biocatalyst Design. Nature, 409, 253–257. https://doi.org/10.1038/35051731

[7] Arslan, A., Masjuki, H. H., Kalam, M. A., Varman, M., Mufti, R. A., Mosarof, M. H., . . . Quazi, M. M. (2016). Surface Texture Manufacturing Techniques and Tribological Effect of Surface Texturing on Cutting Tool Performance: A Review. Critical Reviews in Solid State and Materials Sciences, 41(6), 447–481. https://doi.org/10.1080/10408436.2016.1186597

[8] Soleimani, H., & Moavenian, M. (2017). Tribological Aspects of Wheel–Rail Contact: A Review of Wear Mechanisms and Effective Factors on Rolling Contact Fatigue. Urban Rail Transit, 3, 227–237. https://doi.org/10.1007/s40864-017-0072-2

[9] A, A. S., Onu, P., A, A. A., & Samuel, A. (2024). Green Tribology, Sustainable Lubrication Practices and Environmental Impact: A Mini Review. *2024 International Conference on Science, Engineering and Business for Driving Sustainable Development Goals (SEB4SDG)*, Omu-Aran, Nigeria, pp. 1–9. https://doi.org/10.1109/SEB4SDG60871.2024.10630171. keywords: {Renewable energy sources;Lubricants;Friction;Biomimetics;Green products;Air pollution;Lubrication;green tribology;Sustainable lubrication;friction;Environmental Impact}.

[10] Thakur, A., & Kumar, A. (2024). Challenges and Opportunities in the Development of Nano-Hybrid Smart Coatings. https://doi.org/10.1021/bk-2024-1469.ch015.

[11] Samantaa, A., Wanga, Q., Shawb, S. K., & Ding, H. (2020). Roles of Chemistry Modification for Laser Textured Metal Alloys to Achieve Extreme Surface Wetting Behaviors. Materials and Design, 192, 108744.

[12] Riveiro, A., Maçon, A. L. B., del Val, J., Comesaña, R., & Pou, J. (2018). Laser Surface Texturing of Polymers for Biomedical Applications. Frontiers Physics, 6, 16. https://doi.org/10.3389/fphy.2018.00016

[13] Kim, M. S., Lee, J. H., & Kwak, M. K. (2020). Review: Surface Texturing Methods for Solar Cell Efficiency Enhancement. International Journal of Precision Engineering Manufacturing, 21, 1389–1398. https://doi.org/10.1007/s12541-020-00337-5

[14] Zhang, L., Wang, R., Zhou, H., & Wang, G. (2022). Estimation of the Friction Behaviour of Rubber on Wet Rough Road, and Its Application to Tyre Wet Skid Resistance, Using Numerical Simulation. Symmetry, 14, 2541. https://doi.org/10.3390/sym14122541
[15] Li, J., Zhou, P., Attarilar, S., & Shi, H. (2021). Innovative Surface Modification Procedures to Achieve Micro/Nano-Graded Ti-Based Biomedical Alloys and Implants. Coatings, 11, 647. https://doi.org/10.3390/coatings11060647
[16] Dong, H., Liu, H., Zhou, N., Li, Q., Yang, G., Chen, L., & Mou, Y. (2020). Surface Modified Techniques and Emerging Functional Coating of Dental Implants. Coatings, 10, 1012. https://doi.org/10.3390/coatings10111012
[17] Ahmed, N., Darwish, S., & Alahmari, A. M. (2016). Laser Ablation and Laser-Hybrid Ablation Processes: A Review. Materials and Manufacturing Processes, 31(9), 1121–1142. https://doi.org/10.1080/10426914.2015.1048359
[18] Staroń, A., & Długosz, O. (2021). Antimicrobial Properties of Nanoparticles in the Context of Advantages and Potential Risks of Their Use. Journal of Environmental Science and Health, Part A, 56(6), 680–693. https://doi.org/10.1080/10934529.2021.1917936
[19] Biswas, M. C., Chakraborty, S., Bhattacharjee, A., & Mohammed, Z. (2021). 4D Printing of Shape Memory Materials for Textiles: Mechanism, Mathematical Modeling, and Challenges. Advanced Functional Materials. https://doi.org/10.1002/adfm.202100257
[20] Isahak, W. N. R. W., & Al-Amiery, A. (2024). Catalysts Driving Efficiency and Innovation in Thermal Reactions: A Comprehensive Review. Green Technologies and Sustainability, 2(2), May, 100078.
[21] David Kingery, W., Vandiver, P. B., & Prickett, M. (1988). The Beginnings of Pyrotechnology, Part II: Production and Use of Lime and Gypsum Plaster in the Pre-Pottery Neolithic Near East. Journal of Field Archaeology, 15(2), 219–243. https://doi.org/10.1179/009346988791974501
[22] Greenfield, H. J. (2006). Slicing Cut Marks on Animal Bones: Diagnostics for Identifying Stone Tool Type and Raw Material. Journal of Field Archaeology, 31(2), 147–163. https://doi.org/10.1179/009346906791071972
[23] Smith, C. S. (1972). Penrose Memorial Lecture. Metallurgical Footnotes to the History of Art, Proceedings of the American Philosophical Society, 116(2), April 17, pp. 97–135 (39 pages) Published By: University of Pennsylvania Press.
[24] Tsirogiannis, E. C., Daskalakis, E., & Vogiatzis, C. (2024). Foivos Psarommatis c, Paulo Bartolo, Advanced Composite Armor Protection Systems for Military Vehicles: Design Methodology, Ballistic Testing, and Comparison. Composites Science and Technology, 251, 26 May, 110486.
[25] Aguilar, A., Roman-Flores, A., & Huegel, J. C. (2013). Design, Refinement, Implementation and Prototype Testing of a Reconfigurable Lathe-Mill. Journal of Manufacturing Systems, 32(2), April, 364–371.
[26] Singh, D., & Nagdev, S. (2024). Novel Biomaterials Based Strategies for Neurodegeneration. Recent Advancements and Future Prospects, 21(8).
[27] Malik, S., Muhammad, K., & Waheed, Y. (2023). Nanotechnology: A Revolution in Modern Industry. Molecules, 28, 661. https://doi.org/10.3390/molecules28020661
[28] Wu, Z., Bao, H., Xing, Y. et al. (2021). Tribological Characteristics and Advanced Processing Methods of Textured Surfaces: A Review. International Journal of Advanced Manufacturing Technology, 114, 1241–1277. https://doi.org/10.1007/s00170-021-06954-2
[29] Rathnam, G. S. (2022). The Effect of Different Surface Treatments to Reduce Surface Roughness before PVD Coatings. International Journal of Mechanics of Solids, 3(1), 20–26. https://doi.org/10.22271/2707806X.2022.v3.i1a.8
[30] Salifu, S., Desai, D., Ogunbiyi, O. et al. (2022). Recent Development in the Additive Manufacturing of Polymer-Based Composites for Automotive Structures: A Review.

International Journal of Advanced Manufacturing Technology, 119, 6877–6891. https://doi.org/10.1007/s00170-021-08569-z
[31] Arslan, A., Masjuki, H. H., Kalam, M. A., Varman, M., Mufti, R. A., Mosarof, M. H., . . . Quazi, M. M. (2016). Surface Texture Manufacturing Techniques and Tribological Effect of Surface Texturing on Cutting Tool Performance: A Review. Critical Reviews in Solid State and Materials Sciences, 41(6), 447–481. https://doi.org/10.1080/10408436.2016.1186597
[32] Khalkhal, F., Chaney, K. H., & Muller, S. J. (2016). Optimization and Application of Dry Film Photoresist for Rapid Fabrication of High-Aspect-Ratio Microfluidic Devices. Microfluid Nanofluid, 20, 153. https://doi.org/10.1007/s10404-016-1817-z
[33] Rengarajan, V., Geng, J., & Huang, Y. (2020). Fabrication of Tapered 3D Microstructure Arrays Using Dual-Exposure Lithography (DEL). Micromachines, 11, 903. https://doi.org/10.3390/mi11100903
[34] Arslan, A., Masjuki, H. H., Kalam, M. A., Varman, M., Mufti, R. A., Mosarof, M. H., . . . Quazi, M. M. (2016). Surface Texture Manufacturing Techniques and Tribological Effect of Surface Texturing on Cutting Tool Performance: A Review. Critical Reviews in Solid State and Materials Sciences, 41(6), 447–481. https://doi.org/10.1080/10408436.2016.1186597
[35] Zantye, P. B., Kumar, A., & Sikder, A. K. (2004). Chemical Mechanical Planarization for Microelectronics Applications. Materials Science and Engineering: R: Reports, 45(3–6), 15 October, 89–220.
[36] Domanski, M., & Webb, J. (2007). A Review of Heat Treatment Research. Lithic Technology, 32(2), 153–194. https://doi.org/10.1080/01977261.2007.11721052
[37] Fauchais, P. L., Heberlein, J. V. R., & Boulos, M. I. (2014). Industrial Applications of Thermal Spraying Technology. In: Thermal Spray Fundamentals. Springer, Boston, MA. https://doi.org/10.1007/978-0-387-68991-3_18
[38] Qadir, D., Sharif, R., Nasir, R., Awad, A., & Mannan, H. A. (2024, January). A Review on Coatings Through Thermal Spraying. Chemical Papers, 78(1), 71–91.
[39] Khonina, S. N., Butt, M. A., & Kazanskiy, N. L. (2024, May). A Review on Reconfigurable Metalenses Revolutionizing Flat Optics. Advanced Optical Materials, 12(14), 2302794.
[40] Feng, D., Yan, Y., Jin, G., & Fan, S. (2004). Design and Fabrication of Continuous-Profile Diffractive Micro-Optical Elements as a Beam Splitter. Applied Optics, 43(29), 5476–5480. https://doi.org/10.1364/AO.43.005476
[41] Lei, S., Zhao, X., Yu, X., Hu, A., Vukelic, S., Jun, M. B., Joe, H. E., Yao, Y. L., & Shin, Y. C. (2020, March). Ultrafast Laser Applications in Manufacturing Processes: A State-of-the-Art Review. Journal of Manufacturing Science and Engineering, 42(3), 031005.
[42] Barbero, C. A., & Acevedo, D. F. (2022). Manufacturing Functional Polymer Surfaces by Direct Laser Interference Patterning (DLIP): A Polymer Science View. Nanomanufacturing, 2, 229–264. https://doi.org/10.3390/nanomanufacturing2040015
[43] Alkunte, S., Fidan, I., Naikwadi, V., Gudavasov, S., Ali, M. A., Mahmudov, M., Hasanov, S., & Cheepu, M. (2024). Advancements and Challenges in Additively Manufactured Functionally Graded Materials: A Comprehensive Review. Journal Manufacturing and Materials Processing, 8, 23. https://doi.org/10.3390/jmmp8010023
[44] Bulavko, G. (2024). Organic Photovoltaics: A Journey through Time, Advancements, and Future Opportunities. History of Science and Technology, 14(1), 10–32. https://doi.org/10.32703/2415-7422-2024-14-1-10-32
[45] Arslan, A., Masjuki, H. H., Kalam, M. A., Varman, M., Mufti, R. A., Mosarof, M. H., . . . Quazi, M. M. (2016). Surface Texture Manufacturing Techniques and Tribological Effect of Surface Texturing on Cutting Tool Performance: A Review. Critical Reviews in Solid State and Materials Sciences, 41(6), 447–481. https://doi.org/10.1080/10408436.2016.1186597

[46] Arulvel, S., Dsilva, W. R. D., Jain, A., Kandasamy, J., & Singhal, M. (2023). Laser Processing Techniques for Surface Property Enhancement: Focus on Material Advancement. Surfaces and Interfaces, 42(Part A), November, 103293.
[47] Vijayan, P. P., & Puglia, D. (2019). Biomimetic Multifunctional Materials: A Review. Emergent Mater, 2, 391–415. https://doi.org/10.1007/s42247-019-00051-7
[48] Harun-Ur-Rashid, M., Jahan, I., Foyez, T., & Imran, A. B. (2023). Bio-Inspired Nanomaterials for Micro/Nanodevices: A New Era in Biomedical Applications. Micromachines, 14, 1786. https://doi.org/10.3390/mi14091786
[49] Wawryniuk, Z., Brancewicz-Steinmetz, E., & Sawicki, J. (2024). Revolutionizing Transportation: An Overview of 3D Printing in Aviation, Automotive, and Space Industries. International Journal of Advanced Manufacturing Technology, 134, 3083–3105. https://doi.org/10.1007/s00170-024-14226-y
[50] Zhai, W., Bai, L., Zhou, R., Fan, X., Kang, G., Liu, Y., & Zhou, K. (2021, June). Recent Progress on Wear-Resistant Materials: Designs, Properties, and Applications. Advanced Science, 8(11), 2003739.
[51] Boye, J. I., & Arcand, Y. (2013). Current Trends in Green Technologies in Food Production and Processing. Food Engineering Reviews, 5, 1–17. https://doi.org/10.1007/s12393-012-9062-z
[52] Yang, D., Lee, Y., Kang, H. et al. (2024). Green Manufacturing of Electrically-Tunable Smart Light-Weight Planar Optics: A Review. International Journal of Precision Engineering and Manufacturing-Green Technology, 11, 1029–1051. https://doi.org/10.1007/s40684-024-00621-z
[53] Manjunath, K., Tewary, S., Khatri, N., & Cheng, K. (2021). Monitoring and Predicting the Surface Generation and Surface Roughness in Ultraprecision Machining: A Critical Review. Machines, 9, 369. https://doi.org/10.3390/machines9120369
[54] Ghosh, S., Knoblauch, R., El Mansori, M. et al. (2025). Towards AI Driven Surface Roughness Evaluation in Manufacturing: A Prospective Study. Journal of Intelligent Manufacturing, 36, 4519–4548. https://doi.org/10.1007/s10845-024-02493-1
[55] Babu, S. S., Mourad, A. H. I., Harib, K. H., & Vijayavenkataraman, S. (2022). Recent Developments in the Application of Machine-Learning Towards Accelerated Predictive Multiscale Design and Additive Manufacturing. Virtual and Physical Prototyping, 18(1). https://doi.org/10.1080/17452759.2022.2141653
[56] Vijayakumar, P., Raja, S., Rusho, M. A. et al. (2024). Investigations on Microstructure, Crystallographic Texture Evolution, Residual Stress and Mechanical Properties of Additive Manufactured Nickel-Based Superalloy for Aerospace Applications: Role of Industrial Ageing Heat Treatment. Journal of the Brazilian Society of Mechanical Science Engineering, 46, 356. https://doi.org/10.1007/s40430-024-04940-9
[57] Kim, N. (2017). Recent Progress of Functional Coating Materials and Technologies for Polycarbonate. Journal of Coatings Technology and Research, 14, 21–34. https://doi.org/10.1007/s11998-016-9837-x
[58] Caló, E., & Khutoryanskiy, V. V. (2015). Biomedical Applications of Hydrogels: A Review of Patents and Commercial Products. European Polymer Journal, 65, April, 252–267.
[59] Kim, M. S., Lee, J. H., & Kwak, M. K. (2020). Review: Surface Texturing Methods for Solar Cell Efficiency Enhancement. International Journal of Precision Engineering Manufacturing, 21, 1389–1398. https://doi.org/10.1007/s12541-020-00337-5
[60] Tofail, S. A. M., Koumoulos, E. P., Bandyopadhyay, A., Bose, S., O'Donoghue, L., & Charitidis, C. (2018). Additive Manufacturing: Scientific and Technological Challenges, Market Uptake and Opportunities. Materials Today, 21(1), January–February, 22–37.

[61] Obilor, A. F., Pacella, M., Wilson, A. et al. (2022). Micro-Texturing of Polymer Surfaces Using Lasers: A Review. International Journal of Advanced Manufacturing Technology, 120, 103–135. https://doi.org/10.1007/s00170-022-08731-1
[62] Xu, J., Lu, H., Cai, L., Liao, Y., & Lian, J. (2023). Surface Protection Technology for Metallic Materials in Marine Environments. Materials, 16, 6822. https://doi.org/10.3390/ma16206822
[63] R, S., P, G., Gupta, R. K. et al. (2019). Laser Shock Peening and Its Applications: A Review. Lasers Manuf. Material Processing, 6, 424–463. https://doi.org/10.1007/s40516-019-00098-8
[64] Zuo, H., Jones, M., Hope, T., & Jones, R. (2016). Sensory Perception of Material Texture in Consumer Products. The Design Journal, 19(3), 405–427. https://doi.org/10.1080/14606925.2016.1149318
[65] Ali, F., Kalva, S. N., & Koc, M. (2024). Advancements in 3D Printing Techniques for Biomedical Applications: A Comprehensive Review of Materials Consideration, Post Processing, Applications, and Challenges. Discover Materials, 4, 53. https://doi.org/10.1007/s43939-024-00115-4
[66] De Zanet, A., Casalegno, V., & Salvo, M. (2021). Laser Surface Texturing of Ceramics and Ceramic Composite Materials: A Review. Ceramics International, 47(6), 15 March, 7307–7320.

2 Simulation and Study of Laser Surface Texturing Parameters

Jeyaprakash Natarajan and T. Sathies

2.1 INTRODUCTION

Surface treatment techniques are needed in various industries for improving the surface characteristics of components. Surface treatment techniques generate artificial structures on the surface of material to improve their corrosion resistance, wear resistance, adhesion, and appearance. Industries like automotive, aerospace, electronics, construction, and medical devices depend on surface treatment methods for improving the functionality of the components. One of the primary reasons for deploying surface treatment in industries is to protect materials against corrosion and wear [1,2]. Several metals, when exposed to moisture and air, are prone to oxidation and rust, which can significantly reduce their lifespan. Techniques such as galvanizing, anodizing, and electroplating are commonly used to apply protective coatings to metals, for improving corrosion resistance. Galvanizing involves coating steel or iron with a layer of zinc, which acts as a sacrificial layer, and protects the base metal. Anodizing is used for creating a thick, protective oxide layer on the surface of the aluminum components. In several industries, wear resistance is also very important because materials are subjected to friction and abrasion. Carburizing, nitriding, and induction hardening are all surface hardening methods used to improve hardness and wear resistance of metal. When carbon is added to the surface layer of steel through carburizing, the outer layer becomes harder while the substrate retains its toughness. Nitriding improves the hardness and wear resistance without involving quenching. During nitriding process, nitrogen is diffused into the surface to improve hardness. Induction hardening heats the metal's surface with electromagnetic induction and then cools it quickly, because of which the surface becomes harder. Surface treatments also used to improve bonding with other components in adhesive bonding applications. In industries such as electronics and automotive manufacturing, a strong adhesion between surfaces is required to assemble the components. Techniques like surface roughening, plasma treatment, and chemical etching are employed to modify the surface textures of components and enhance the bonding of adhesives, coatings, or paints on base components [3,4].

Laser-based surface treatment methods are widely used in many fields because they are precise, efficient, and flexible. These methods use high-power laser beams to change the surface properties of materials. Laser-based surface texturing method

 DOI: 10.1201/9781003542087-2

can work on a specific area which makes it possible to change the surfaces of materials without changing their bulk properties. Laser hardening, laser cladding, and laser texturing are the laser-based surface treatment methods used to improve wear resistance, corrosion resistance, and surface roughness [5,6]. Laser hardening is used to make metals harder on the outside, while retaining the inner functional characteristics. Laser cladding adds a layer of material on top of a substrate and this protective coating makes the surface more resistant to wear and corrosion. Laser-based methods are also environment friendly because laser-based processes are noncontact methods, reducing waste and not harmful for surface processing. These methods can also be used on a wide range of materials, such as metals, ceramics, and polymers, which makes them useful in many different industries. Laser-based surface treatments are needed in manufacturing, aerospace, and automotive industries for improving the properties of materials while maintaining accuracy and efficiency [7,8].

Laser surface texturing (LST) is an important laser-based surface treatment technique because it offers several advantages for industrial applications. While other laser-based treatments, such as laser hardening, laser cladding, and laser ablation, help to alter surface's chemical composition or adding protective layers, LST improves the physical texture of surfaces. This modification helps to alter friction, lubrication, and adhesion characters of components. As a result, LST is widely used in automotive, aerospace, biomedical engineering, and manufacturing. LST can create highly controlled micro- and nano-scale patterns on material surfaces. These patterns (Figure 2.1) can be tailored to meet specific functional requirements, such as reducing friction in mechanical components and improving the adhesion of coatings and paints in automotives. In contrast to other laser-based techniques, LST focuses on optimizing the surface topography. Precise control over surface roughness and patterning can lead to significant performance improvements in various applications [9].

For example, in the automotive industry, LST finds application in piston rings and cylinder liners for reducing friction. LST-created micro textures retain lubrication in

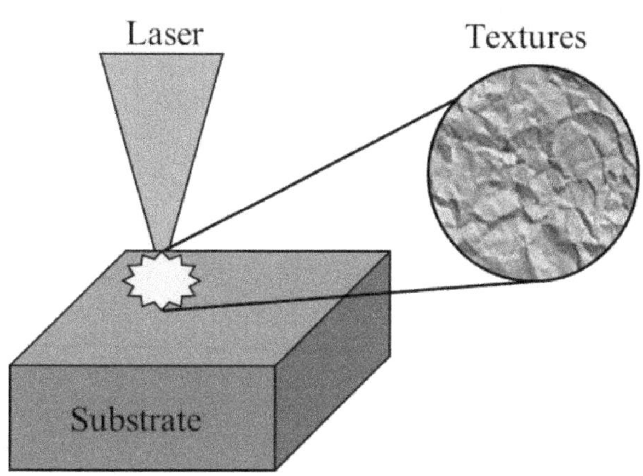

FIGURE 2.1 Schematic of laser surface texturing.

the contact area, reduce wear, and improve the efficiency of the engine components. In biomedical engineering, LST is used to improve the implant's biocompatibility and functionality. LST can create specific surface textures for improving the integration of implant with surrounding tissue. LST is distinct from laser cladding, because laser cladding develops biocompatible coating instead of optimizing the surface texture for better tissue integration [10]. Moreover, LST is a noncontact process, and it minimizes contamination or damage to the materials. Since LST is capable of creating intricate, repeatable patterns with high precision and consistency, it is useful in applications where surface topography is essential. In order to obtain better surface textures for improving the mechanical and tribological characteristics, the process parameters used during laser surface texturing must be optimized. Figure 2.2 represents the process parameters used in LST process. The optimization of these process parameters varies from material to material depending upon the material type and its properties. This chapter discusses the effect of different LST parameters on various properties.

2.2 LASER POWER

Laser power is an important process parameter in LST and affects the properties of textured surface. Laser power influences the amount of energy applied on the material's surface, and impacts the various characteristics of textured features. It is necessary to optimize the laser power for getting desired performance, durability, and functionality.

2.2.1 Influence on Surface Morphology

The laser power in LST affects the morphology of the surface texture. When a laser beam interacts with a material, the energy absorbed by the surface causes localized

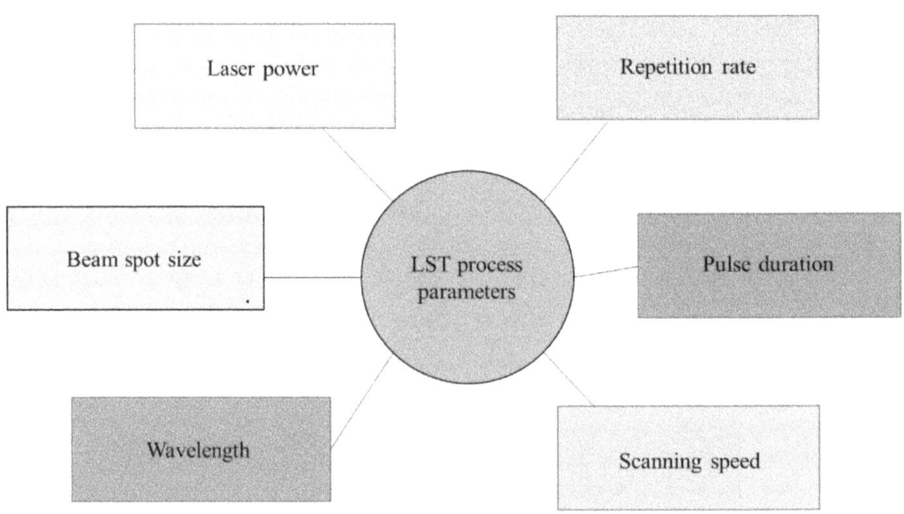

FIGURE 2.2 Process parameters used in LST.

heating, melting, vaporization, or ablation [11,12]. Higher laser power delivers more energy to the surface of the substrate and generates deeper and wider grooves. Higher energy develops pronounced textures, and it is beneficial for applications requiring higher surface roughness, such as improving adhesion, friction, or surface area for coatings. However, excessive laser power may lead to excessive material removal, melting, and the formation of large heat-affected zones (HAZ). These zones alter the microstructure and mechanical properties of the textured material and develop issues like reduced hardness, strength, or wear resistance. Therefore, careful selection of laser power is necessary for getting desired surface features.

2.2.2 Role in Controlling Heat-Affected Zones

The size and extent of the HAZ developed around the textured area depends on the laser power. In HAZ, materials experiences change in its microstructure due to the heat generated during the laser process. A higher laser power results in a larger HAZ and affects phase transformations, grain growth, and residual stresses. These changes can affect the material's performance [13]. To minimize the HAZ, the laser power must be optimized. Using lower laser power reduces the thermal input to the material and reduces the size of the HAZ and preserves the original microstructure and mechanical properties of the material. When texturing materials like ceramics, polymers, or composite materials, excessive heat generates cracking, delamination, and other damage.

2.2.3 Impact on Processing Efficiency

Laser power also influences the processing efficiency of LST. Higher laser power increases the material removal rate, and creates deeper textures in a shorter duration [14]. This is required in applications where higher productivity is needed. However, a balance between speed and surface quality must be maintained. While higher power makes LST faster, it may also lead to rougher surfaces, and generate inconsistent patterns. Lower laser power might produce finer and more uniform textures, but the processing speed decreases. Therefore, optimized laser power must achieve efficient production and develop high-quality surface textures. The laser power depends on the application and the desired characteristics of the textured surface.

2.2.4 Influence on Material Properties

The properties of textured surface depends on the laser power. The surface hardness and wear resistance are important in automotive and cutting tool industries and the laser power used for texturing should not affect these properties. Controlled laser power can induce desirable microstructural changes, such as fine martensitic structure formation in steel [15,16]. The usage of excessive laser power may lead to formation of cracks, voids, and oxidation in materials like titanium and aluminum alloys. These defects can degrade the material's performance, and affect corrosion resistance and mechanical strength. Thus, laser power must be carefully chosen to tailor the material's properties based on the application.

2.3 PULSE DURATION

Pulse duration in LST influences the quality of the textured surfaces. It refers to the duration for which the laser emits energy during each pulse, and it determines the interaction of laser with the material [17].

2.3.1 Impact on Material Interaction

Since LST influences the interaction between the laser and the material being processed, shorter pulse durations (femtoseconds to picoseconds) result in minimal heat diffusion into the material. As a result, precise material removal happens and well-defined microstructures are created. The short pulses deliver energy in shorter duration and reduces the thermal effects. Conversely, longer pulse durations lead to increased heat diffusion, and generates broader HAZ. Higher pulse duration might cause defects such as cracks, warping, and alter the mechanical properties. Usage of higher pulse duration affects polymers and thin films significantly. Hence optimal pulse duration must be selected for getting desired surface texture [18].

2.3.2 Control Over Surface Texture

Pulse duration affects the characteristics of the textured surface. When shorter pulse durations are used, the energy is concentrated for shorter duration and it allows control over the ablation process. As a result, intricate and complex surface features are created at the micro- and nano-scale, and it helps to improve properties like adhesion, lubrication, and friction. Such features are required in biomedical implants, automotive components, and microelectromechanical systems. Longer pulse durations might lead to less defined and rougher surface textures [19]. When laser energy is applied for more duration, larger molten pools are created and results in more material removal.

2.3.3 Minimization of Thermal Effects

The pulse duration used during LST significantly influences thermal effects. Short pulse durations reduce heat accumulation in the material, and minimizes thermal damage. When processing low-melting-point material, pulse duration plays a major role [20]. The usage of ultrafast lasers with short pulse durations helps researchers to create textures with high quality and material's properties. In the aerospace industry, it is necessary to create high-quality textures with minimal thermal effects. Short pulse durations can fabricate components with enhanced performance.

2.3.4 Influence on Processing Speed

Pulse duration has a major role on the processing speed of LST. Shorter pulse leads to slower scanning strategies and it helps to achieve the desired texture. In contrast, longer pulse durations improve the processing speed and the energy is applied for more period, because of which faster material removal happens at longer pulse durations [21]. Hence, the relationship between the pulse duration and processing speed must be optimized to get textures surface with desired properties. Faster processing may reduce production time, but may affect the final product's performance. Hence a balance between speed and quality must be maintained when selecting pulse duration for LST.

2.3.5 Effects on Material Properties

The pulse duration also influences the material properties. Shorter pulse durations can increase hardness, improve wear resistance, and avoid thermal defects. In contrast, longer pulse durations may introduce residual stresses and phase changes and affect the material properties. For instance, higher pulse duration may result in phase transformations in metals, and affect the mechanical properties. Hence, pulse duration must be carefully chosen for getting desired mechanical properties.

2.4 REPETITION RATE

Repetition rate, also known as pulse frequency, is an important parameter in LST and can impact the quality and characteristics of the textured surface. It refers to the number of laser pulses delivered to a material surface per second. Repetition rate controls interaction between laser and material, and influences the properties of textured surfaces. Hence it is necessary to understand the effect of repetition rate to optimize LST for various applications [22,23].

2.4.1 Influence on Material Interaction

The repetition rate influences the interaction between laser and material's surface. When a high repetition rate is employed, the laser delivers faster pulses in subsequent cycles and increases the energy accumulation in the texturing region. As a result, thermal effects on the material get enhanced, and the material removal rate is higher. However, the energy accumulation depends on the laser power and pulse duration. When repetition rate is high, it can lead to excessive heating and results in over melting and warping. On the other hand, a lower repetition rate provides more cooling duration between pulses, and maintains the properties of heat-sensitive materials. When the duration between pulses increases, thermal distortion can be reduced and finer textures with minimal HAZ can be achieved. Therefore, the repetition rate must be controlled in combination with other parameters for getting desired texture [24,25].

2.4.2 Control Over Surface Features

Higher repetition rates generate denser and uniform textures, because laser pulse overlapping provides better control over the surface features. Lower repetition rates may result in larger surface features and it is used in applications where a rougher surface is required for improving friction and wear resistance [26]. For example, in automotive applications, a textured surface with larger features can retain the lubricants in engine components, and improve the performance of engine.

2.4.3 Impact on Processing Efficiency

The repetition rate has a control on the efficiency of the laser surface texturing process. Higher repetition rates increase the processing speed because laser pulses are applied in a shorter period, and it results in faster material removal and texture formation. By optimizing the repetition rate, production cost of laser texturing can be reduced while maintaining the quality of the textured surface [27–29]. Along with the repetition rate, factors such as material type, laser power, and pulse duration must be considered for getting better result from textured surface. Improper selection of process parameter may lead to poor surface quality, and higher thermal effects.

2.4.4 Effects on Surface Properties

Higher pulse frequency improves the hardness and wear resistance of textured surface by modifying the microstructure. Higher repetition rates refine the microstructures and improve strength and toughness [30]. If the repetition rate is too high, it might cause the material to overheat and change phases, and affects the performance of textured surface.

2.4.5 Thermal Management Considerations

Managing thermal effects is an important part of the laser surface texturing process, and the repetition rate plays a major role. When the repetition rate is high, there is a greater chance of thermal accumulation, which can cause undesired thermal effects such as changes in microstructure, phase changes, and even over melting of the textured material. So, to control the thermal profile of the material while it is being laser textured, it is necessary to manage the repetition rate. By lowering the repetition rate, heat may move better between pulses, which lowers the chance of thermal damage and retains the material's characteristics. This is important when working with low-melting-point materials.

2.5 SCANNING SPEED

The speed of the laser source significantly influences the quality, efficiency, and characteristics of the textured surface. Scanning speed is the speed at which the laser beam passes across the surface of the material while it is being textured. For getting the desired surface qualities with minimum defect, it's important to optimize scanning speed. This is an important factor to consider for several industrial applications [31].

2.5.1 Effect on the Shape of the Surface

The shape of the texture changes directly with the scanning speed. When the laser pulse moves fast, it might not give the material enough energy, which can lead to shallow or poorly defined textures. This can be a problem in applications where certain surface properties, such as more roughness or better adherence, are needed. On the other hand, slower scanning speeds allow more laser energy to be applied into a smaller region, which makes grooves or craters deeper and surface features more pronounced. This ability to regulate surface morphology means that scanning speed is an important aspect in customizing textures for certain functional needs [32,33].

2.5.2 Effect on How Heat Is Handled

The scanning speed of the laser beam is very important for controlling the thermal effects during the laser texturing process. When the laser beam moves slowly, the laser stays on a location for longer duration, which causes more heat to build up. This might create thermal effects, such as melting or deformation of the material, especially on substrates that are sensitive to heat. Conversely, a faster scanning speed helps to spread out heat more evenly, which lowers the risk of thermal damage and keeps the material's structure intact. This balance is very important in situations where exact surface features are needed without changing the properties of the substrate material.

2.5.3 Processing Efficiency

The scanning speed also impacts the overall processing efficiency of LST. In high-throughput production settings where efficiency is key, faster scanning rates reduce the processing time a lot. But a trade-off between speed and surface quality must be maintained. While faster scanning increases the speed of texturing, it can affect the quality of the textured surface. So, it's important to identify the best scanning speed for the laser texturing process to get the best results in terms of both speed and quality [34].

2.5.4 Effects on Material Properties

The scanning speed can influence the properties of textured surface. A slower scanning speed can generate desirable microstructure and enhance the hardness and wear resistance of the material. In contrast, excessively high scanning speed can lead to insufficient energy transfer, and affect the mechanical properties of textured surface [35]. Therefore, it is necessary to adjust the scanning speed in combination with other parameters, such as laser power and pulse duration, for producing textures with better properties.

2.6 LASER BEAM SPOT SIZE AND WAVELENGTH

Beam spot size and wavelength are the two important process parameters in LST. It is necessary to understand the effect of these parameters for optimizing the overall LST process.

2.6.1 Importance of Beam Spot Size

2.6.1.1 Definition and Influence on Resolution

When the laser beam irradiates the surface to be textured, the beam spot size indicates the beam's diameter. This parameter affects the resolution and precision of the developed textures in LST. A smaller beam spot size allows the laser energy to be delivered very precisely, which makes it possible to develop detailed and complex surface features at the micro- and nano-scale. This level of accurate texture is useful for biomedical implants, microelectromechanical systems, and advanced electronic components. High-resolution surface textures developed by LST can enhance properties like adhesion, wettability, and biocompatibility [36].

2.6.1.2 Surface Morphology Control

It is possible to control surface morphology by adjusting beam spot size. Sharper and more defined textures can be created by smaller spot size and it is useful in applications involving high surface area or specific frictional properties. For example, in tribological applications, finely textured surfaces produced by LST can reduce friction and improve wear resistance, and operational life of component. On the other hand, a bigger beam spot size can make energy distributions wider and can be used develop deeper grooves or wider features but the resolution of the textures created at higher beam spot size may be lesser [37].

2.6.1.3 Trade-Offs in Texturing

While smaller beam spot sizes allow for finer details, they reduce the processing speeds; the laser beam has to be moved precisely and slowly. A trade-off between speed and quality is necessary for applications where efficiency is very important. In such cases, it is necessary to optimize the beam spot size for achieving a balance between processing speed and the desired surface quality.

2.6.2 Impact of Wavelength

2.6.2.1 Wavelength and Material Interaction

The wavelength of the laser light is another important that affects the interaction laser and the substrate material. Different materials have different absorption characteristics and it affects the efficiency of the texturing process. For instance, shorter wavelengths produced by ultraviolet (UV) lasers can develop textures on polymers and certain ceramics. Because of this absorption property, UV lasers are good for precisely texturing heat-sensitive materials because they can produce the desired texture with minimal thermal effects [38].

In contrast, longer wavelengths, generated by carbon dioxide (CO_2) lasers, are used to create textures on metals and thicker materials. When the laser wavelengths increase, the interaction depth increases and deeper material penetration is possible but it may induce thermal effects. Higher wavelength is used for applications that require deeper textures or coatings, but it is not suitable to work on substrates that are sensitive to heat.

2.6.2.2 Material-Specific Considerations

When selecting laser wavelength for LST, it is necessary to understand the properties of material being processed. For example, metals tend to reflect shorter wavelengths, because of which longer wavelength is suitable for processing metal. By carefully selecting the laser wavelength based on material type, it is possible to apply appropriate laser energy on material surface and generate suitable textures. This property is needed in applications such as surface hardening, where specific surface properties like hardness and wear resistance are desired.

2.6.2.3 Combined Effects on Surface Quality

The interaction between beam spot size and wavelength affects the quality of the textured surface. A suitable LST parameter combination can deliver optimal energy and it allows the production of high-resolution textures with lesser thermal damage. For instance, using a small beam spot size with an appropriate wavelength can generate high-quality textures and improve the adhesion characteristics and wear resistance.

If the deployed beam spot size and wavelength do not meet the properties of material to be textured, it may lead to suboptimal surface quality. For example, excessive heat generation can result in melting or warping of the material, while insufficient energy may produce undesirable textures. Hence, it is essential to understand the relationship between wavelength, spot size, and material properties for achieving better textures on surface through LST [39].

2.6.2.4 Applications and Industry Impact

The importance of beam spot size and wavelength extends across various industries, including aerospace, automotive, biomedical, and electronics. In aerospace, precise surface textures can enhance aerodynamic properties and reduce friction, contributing to improved fuel efficiency. In the automotive sector, laser texturing is employed to optimize surface characteristics in engine components and brake systems, leading to enhanced performance and longevity. In biomedical applications, the ability to create finely textured surfaces can improve the integration of implants with surrounding tissues, promoting faster healing and reducing the risk of rejection. In electronics, precise laser texturing can enhance surface properties for better adhesion of coatings and improve the performance of microelectronic devices [40].

2.7 IMPORTANCE OF OPTIMIZATION OF PROCESS PARAMETERS USED IN LST

The process parameters of LST must be optimized for producing textures with desired dimension and properties. The impact of process parameters is shown in Figure 2.3. The properties of textured surface depend on LST parameters like laser power, pulse duration, repetition rate, scanning speed, beam spot size, and wavelength. When the parameters are optimized, sufficient energy will be delivered by the laser source and it will result in suitable textures without thermal damage [41–43]. The depth and morphology of the developed texture can be controlled by optimizing laser power and scanning speed, because of which the adhesion characteristics, wear resistance,

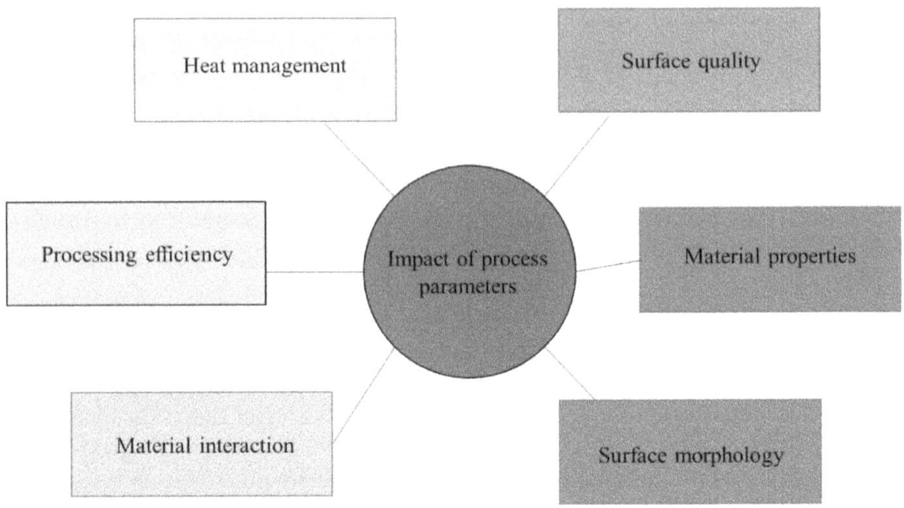

FIGURE 2.3 Impact of process parameters.

and overall properties of the textured component can be improved. Additionally, optimizing process parameters can lead to increased efficiency and reduced production costs. By fine-tuning these parameters, manufacturers can improve the processing speed without compromising the quality of the textured surface. Moreover, optimized parameters reduce cracks and overheating and improve the durability and functionality of the treated surfaces. Since several industries are working on LST, it is necessary to optimize the LST parameters for various materials before using the LST for functional components.

2.8 SIMULATION IN LST

Before performing actual LST, simulations are conducted to predict and optimize the surface characteristics. Even though LST is used in applications like tribology, adhesion, and wettability, a complex laser and material interaction including heat transfer, phase changes, and material removal happens in LST. As a result, it is difficult to achieve desired surface textures and material properties through experimental methods alone. Simulations help the researchers to model the effects of pulse duration, energy, and wavelength on the surface characteristics of textured surface. Through simulation, it is possible to predict the surface roughness, texture patterns, and the defects that can happen in texturing. The need for trial-and-error experiments can be reduced through simulation [43–46].

Moreover, simulation helps to understand the principles behind the formation of microstructures and the thermal effects on the material. By understanding these principles, it is possible to achieve desired textured surfaces that meets application requirement. Thus, simulations in laser surface texturing helps to enhance efficiency, reduce costs, and improve the accuracy of LST. The different role of simulation in LST is provided in Figure 2.4.

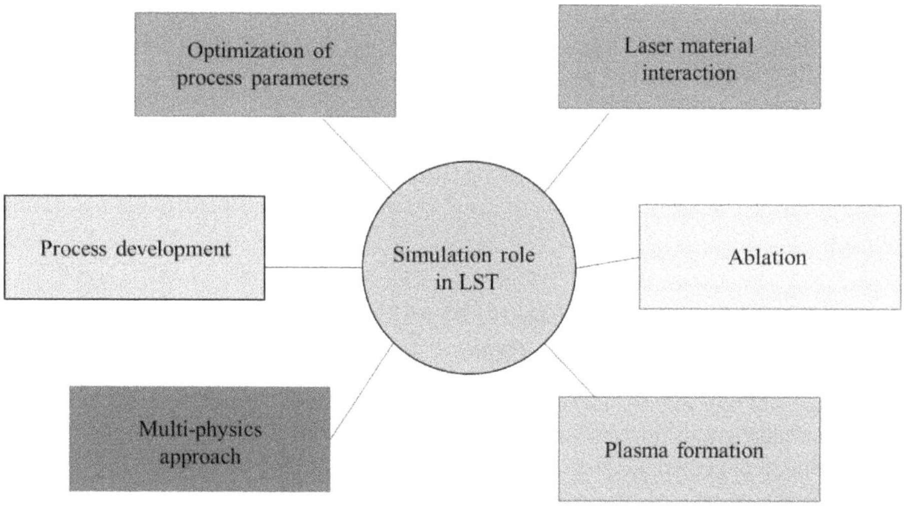

FIGURE 2.4 Simulation role in LST.

2.9 ROLE OF SIMULATION IN LASER MATERIAL INTERACTION

The process of LST involves applying the laser beams directly onto a material's surface, leading to localized heating, melting, vaporization, and even plasma formation. These physical phenomena happen in a fraction of a second and it depends on the material properties, laser parameters (such as wavelength, pulse duration, energy, and repetition rate), and environmental conditions. The interaction between the laser and the material affects the final geometry and properties of created textures. Since a complex interaction happens between the thermal, mechanical, and optical phenomena, it is difficult to predict the outcomes of LST processes through experiments alone [47]. Experimental methods are time consuming and costly, and require numerous trial experiments to find the optimal processing parameters. Further, LST is a rapid process and it is difficult to capture all the details using experimental techniques alone.

2.10 ROLE OF SIMULATION IN MATERIAL REMOVAL AND ABLATION

Material removal is an important phenomenon in LST, because it influences the texture and morphology of the surface. During laser irradiation, material removal happens through different principles like vaporization, spallation, and phase explosion. The specific mechanism depends on the laser parameters and the material's properties [48]. It is necessary to understand the material removal process for achieving the desired surface characteristics. Simulation of material removal and ablation processes involves modeling the interaction between the laser beam and the material at the micro and nanoscales. These simulations can predict the depth and shape of the microstructures formed on the surface, and calculate the amount of material removed during each laser pulse. By simulating different laser parameters, such as pulse duration, energy density, and beam spot size, it is possible to optimize the material removal process. Through simulation, the parameters required for developing hydrophobic or hydrophilic surfaces can be identified. Additionally, ablation simulations provide information about recast layers that can affect the surface's functional properties. The principle behind recast layers formation must be known for developing durable and high-performance textured surface.

2.11 ROLE OF SIMULATION IN PLASMA AND SHOCKWAVE FORMATION

The usage of high-energy laser pulses in LST generates plasma and shockwaves. In plasma state, the ionization of material happens, and the materials in that state exhibit different properties in comparison to their solid or liquid phases. Because of the formation of plasma, the energy absorption by the material during LST and the corresponding material removal mechanism are changed, and it may not be possible to develop desired textures on the material surface. For simulating plasma formation and its effects on laser–material interaction, researchers used to model the ionization processes and electromagnetic field generation, and then interaction between the developed plasma and the material are investigated in detail. Researchers usually perform these simulations with the help of particle-in-cell or hydrodynamic models, because they help to capture the dynamics propagation of plasma [49]. In addition to plasma, shockwaves are also generated within the material because of the rapid irradiation of laser. These shockwaves may induce mechanical deformation, spallation, and generate microcracks during LST process. By simulating shockwaves, it is possible to understand their effect on material properties and it can help in optimizing the LST process to minimize damage.

2.12 ROLE OF SIMULATION IN MICROSTRUCTURAL EVOLUTION

The microstructure developed during LST affects the hardness, wear resistance, and corrosion resistance. During LST, laser irradiation generates the rapid heating and cooling cycles and it can alter the material's microstructure. Changes like grain refinement, phase transformations, and new phase formation are possible in LST and they can influence the surface's mechanical and chemical properties. Simulating microstructure evolution during LST involves modeling the thermodynamic and kinetic processes, which can govern phase transformations and grain growth. These simulations are performed using phase-field models or cellular automata, and it helps to capture interactions between heat transfer, diffusion, and phase changes [50]. By simulating microstructure evolution, researchers predict the effect of different laser parameters on the material's final microstructure. In applications requiring high wear resistance, simulations help in the selection of laser parameters, which can promote grain refinement and the formation of hard phases. In applications requiring high toughness, simulations can help to avoid the formation of brittle phases.

2.13 MULTI-SCALE AND MULTI-PHYSICS SIMULATIONS

The interaction between the applied laser and material surface during LST is a multi-scale and multi-physics problem. Phenomena like ionization and phase transitions, heat transfer, and mechanical deformation happen in microscale to nanoscale during LST and affect the properties of resultant structure. The thermal, optical, mechanical, and electromagnetic properties of the material affect the phenomena happening during LST. To understand in detail about laser and material interaction during LST, multi-scale and multi-physics simulations are proposed by researchers. These simulations combine different computational models by involving different physical phenomena [51]. For example, a multi-physics simulation of LST combine thermal models (e.g., FEM), optical models (e.g., ray tracing), and mechanical models (e.g., finite element analysis) to identify the principle behind laser material interaction. Multi-scale and multi-physics simulations are used to understand and predict the properties at the atomic level that affect the performance of the textured surface. By integrating atomistic simulations, heat transfer simulations, and mechanical deformation, researchers understand the effect of laser parameters on the final surface texture and properties.

2.14 ROLE OF SIMULATION IN VIRTUAL PROTOTYPING AND PROCESS DEVELOPMENT

Simulations in LST helps to develop process and perform virtual prototyping. Digital models of the LST process are developed in virtual prototyping and they help researchers to test the properties of texture patterns before creating them on the actual surface. By simulating laser irradiation, material removal, and microstructure evolution, virtual prototyping allows the researchers to identify the issues that can happen in LST. Further, optimization is possible in a virtual environment. By using these techniques, companies reduce their product development time and achieve faster time to market of their product. Through virtual prototyping, it is possible to develop new products and processes in shorter duration and time involved in development of physical prototypes are reduced [52]. With the help of virtual reality tool, the LST parameters required for developing textures with new laser sources (e.g., ultrafast lasers) and on new materials (e.g., composites, high entropy glass, high-entropy alloys) can be identified and used for testing purpose.

2.15 ROLE OF SIMULATION IN OPTIMIZATION OF PROCESS PARAMETERS

It is possible to perform LST parameter optimization through simulation. Since several parameters are involved in LST (e.g., pulse duration, energy density, wavelength, repetition rate), it is difficult to identify optimum parameters for producing textures with desired properties. Through simulations, the parameter space of LST can be systematically explored and optimal parameter conditions can be identified. Techniques like gradient-based methods, genetic algorithms, and machine learning are used to perform optimization. By integrating simulations and optimization algorithms, researchers identify the optimum laser parameters easily, and reduce the time and cost associated with experimental trials. The performance of the textured surface depends on process parameters of LST [53]. These parameters govern the laser energy applied to the surface, their interaction with the material, and the resultant surface morphology of the developed texture. Laser power and pulse duration control the energy density, and it can affect the depth of material ablation, microstructures evolution, and thermal effects. Scanning speed and beam overlap determine the texture pattern and uniformity. But a complex relationship exists between these parameters. Hence a systematic investigation is required to identify the process parameters that can develop textures with required properties at minimal cost and time. Simulation helps to model and predict the outcomes of different parameter combinations, and reduce the dependency on experiments for parameter selection. Inverse modeling helps simulation to achieve desired surface characteristics. In inverse modeling, desired surface properties (e.g., roughness, texture pattern, microstructure) are specified and then simulations are performed to determine the laser parameters that help in getting the mentioned properties. This approach is useful in applications like microfluidic devices or biomedical implants, where precise control over surface features is required.

2.16 CHALLENGES DURING SIMULATION

LST is widely investigated for modifying the materials' surfaces and functional properties. Even though simulations help to understand and optimize the LST processes, several challenges are involved in simulation. This section discusses the challenges encountered during the simulation of laser surface texturing, including complexity in modeling, computational limitations, validation and integration with experimental data, and the need for multi-scale approaches.

2.16.1 Complexity in Modeling Laser–Material Interactions

It is difficult to accurately model the complex interactions between laser radiation and the material. The interaction between laser radiation and material involves various physical phenomena, including the following:

1. **Absorption and scattering:** The absorption of laser radiation by the material depends on its optical properties, and it can vary based on the wavelength of applied laser, surface condition, and material composition of sample. For accurately modeling these interactions, full details about sample material and the laser's characteristics are needed [54].
2. **Heat transfer:** During LST, localized heating happens and it results in phase changes such as melting, vaporization, and ablation. It is necessary to predict the temperature distribution and heat flow within the material for understanding the resulting microstructure. However, the transient nature of LST makes the thermal modeling complex because the temperature gradients change continuously during the LST process.

3. **Material removal mechanisms:** Material removal during LST depends on thermal ablation, phase explosion, and melt ejection. Factors such as energy density, pulse duration, and material properties influence the mentioned mechanism. It is difficult to capture these mechanisms accurately and perform simulation.

2.16.2 Computational Limitations

For simulating LST processes, complex mathematical models are required [55]. This section presents the challenges involved in computational modelling:

1. **High computational demand:** High-resolution spatial and temporal discretization is needed for simulating the laser and material interaction. Fine mesh sizes and short time steps are involved in such models, which makes excessive computational demands. Hence, high-performance computers are required to conduct such simulations.
2. **Time-consuming calculations:** Simulations involve multiple physical phenomena, including thermal, mechanical, and fluid dynamics interactions. Hence, numerous calculations are involved in simulation and it increases the simulation duration.
3. **Scalability issues:** As the model complexity increases, it is difficult to expand the application of simulation. It may not be possible to include detailed interactions on standard computational platforms. Hence, a balance between model complexity and computational feasibility is required.

2.16.3 Validation and Integration with Experimental Data

Furthermore, it is difficult to validate the results of simulation through experiment. Following are the challenges involved in this stage:

1. **Lack of comprehensive experimental data:** Even though experiments are needed to validate simulation results, comprehensive datasets involving a range of laser parameters and material conditions for performing machine learning in LST are not available. Because of the absence of standard datasets, it is difficult to validate models and the confidence of simulation results gets reduced.
2. **Difficulties in experimental replication:** It is difficult to identify all the parameters involved in experimentation including material properties, laser system performance, and environmental factors. Because of a lack of data for simulation, the predictions made by simulations are poor in certain cases and it makes the validation process difficult.
3. **Integration challenges:** It is also difficult to integrate simulation and experimental results to refine models and improve accuracy. Effective integration requires robust methodologies that can assimilate experimental findings into simulation frameworks, and adjust parameters and boundary conditions based on actual experiment.

2.16.4 Need for Multi-Scale Approaches

During LST, interaction happens from atomic to macroscopic levels [56]. However, capturing these interactions in a single simulation is difficult:

1. **Multi-scale modeling complexity:** During LST, different physical phenomena operate at different scales. Molecular dynamics simulations provide information about atomic level interactions, while continuum models are used to predict macroscopic behavior. Integrating these different methods becomes complex and requires advanced methodologies.
2. **Computational resources for multi-scale simulations:** Multi-scale simulations require huge computational resources because they involve large datasets.

2.16.5 Material and Process Variability

The variability of materials and laser processes introduces additional challenges in simulation:

1. **Diverse material properties:** Different materials exhibit different responses to laser irradiation based on their thermal, optical, and mechanical properties. Predicting these properties in simulations needs several characterizations and detailed data about the material. The simulation becomes complicated when the material composition, microstructure, and surface conditions differ between each trial.
2. **Process parameter sensitivity:** LST is sensitive to laser power, pulse duration, and scanning speed. Variations in these parameters affect the resultant surface texture and quality. Simulating these variations is difficult because it needs detailed exploration of the parameter space and advanced optimization techniques [57].

2.17 FUTURE DIRECTIONS IN SIMULATION

Despite these challenges, advancements in computational methods and technologies help to improve the simulation of LST. Some potential future directions include the following:

1. **Enhanced computational techniques:** By developing efficient algorithms and numerical methods, the processing speed and accuracy of simulations can be improved. Techniques such as adaptive mesh refinement and parallel computing can enhance the simulation capabilities [58].
2. **Integration of machine learning:** Machine learning techniques can be used analyze simulation data and identify patterns, and predict outcomes based on input parameters. Incorporating machine learning into the simulation process can improve optimization and reduce the time required for simulations [59].

3. **Collaborative platforms for data sharing:** Collaborative platforms for sharing experimental data and simulation results can be developed to enhance validation and improve model accuracy. Such platforms can help the researchers to collaborate and exchange their findings on simulations [60].

2.18 CONCLUSION

Studies on LST process parameters optimization and simulations are needed for advancing surface engineering and optimizing material properties for various applications. Through simulation techniques, researchers can model the complex interactions between laser radiation and materials, and understand the thermal effects, material removal mechanisms, and microstructural changes. These simulations help to identify optimal process parameters, such as laser power, pulse duration, and scanning speed, for achieving desired surface characteristics. By leveraging methods such as thermal simulations, material removal models, and multi-physics approaches, researchers can explore a wide range of parameter combinations efficiently. This systematic investigation can reduce the requirements on experimental trials and refine the laser texturing techniques. By integrating machine learning and artificial intelligence into LST simulation, the predictive capabilities of simulations can be enhanced and optimal conditions required for specific applications can be identified easily. As the demand for high-performance materials continues to grow across industries, the importance of accurately simulating and studying process parameters in LST will increase. Continued advancements in computational methods and data-driven approaches promise to further enhance our ability to tailor surface properties to meet specific requirements, driving innovation in fields such as tribology, biomedical engineering, and optical applications. Ultimately, the synergy between simulation and experimental validation will lead to more efficient and effective laser surface texturing processes.

2.19 ACKNOWLEDGEMENT

The authors thank the editors of this book for providing an opportunity to make this contribution.

2.20 CONFLICTS OF INTEREST

The authors declare no conflicts of interest.

REFERENCES

[1] Etsion I. State of the art in laser surface texturing. Journal of Tribology. 2005 Jan 1;127(1):248–253.

[2] Mao B, Siddaiah A, Liao Y, Menezes PL. Laser surface texturing and related techniques for enhancing tribological performance of engineering materials: A review. Journal of Manufacturing Processes. 2020 May 1;53:153–173.

[3] Karuppasamy SS, Yang CH. Adapting the time-domain synthetic aperture focusing technique (T-SAFT) to laser ultrasonics for imaging the subsurface defects. Sensors. 2023 Sep 22;23(19):8036.
[4] Shivakoti I, Kibria G, Cep R, Pradhan BB, Sharma A. Laser surface texturing for biomedical applications: A review. Coatings. 2021 Jan 22;11(2):124.
[5] Wahab JA, Ghazali MJ, Yusoff WM, Sajuri Z. Enhancing material performance through laser surface texturing: A review. Transactions of the IMF. 2016 Jul 3;94(4):193–198.
[6] Natarajan J, Yang CH, Karuppasamy SS. Investigation on microstructure, nanohardness and corrosion response of laser cladded colmonoy-6 particles on 316L steel substrate. Materials. 2021 Oct 18;14(20):6183.
[7] Kumar V, Verma R, Kango S, Sharma VS. Recent progresses and applications in laser-based surface texturing systems. Materials Today Communications. 2021 Mar 1;26:101736.
[8] Vincent C, Monteil G, Barriere T, Gelin JC. Control of the quality of laser surface texturing. Microsystem Technologies. 2008 Oct;14(9):1553–1557.
[9] Jeyaprakash N, Yang CH, Karuppasamy SS, Rajendran DK. Correlation of microstructural with corrosion behaviour of Ti-6Al-4V specimens developed through selective laser melting technique. Proceedings of the Institution of Mechanical Engineers, Part E: Journal of Process Mechanical Engineering. 2022 Oct;236(5):2240–2251.
[10] Vilhena LM, Sedlaček MM, Podgornik B, Vižintin J, Babnik A, Možina J. Surface texturing by pulsed Nd: YAG laser. Tribology International. 2009 Oct 1;42(10):1496–1504.
[11] Karuppasamy SS, Jeyaprakash N, Yang CH. Microstructure, nanoindentation and corrosion behaviour of colmonoy-5 deposition on SS410 substrate using laser cladding process. Arabian Journal for Science and Engineering. 2022 Jul;47(7):8751–8767.
[12] Voevodin AA, Zabinski JS. Laser surface texturing for adaptive solid lubrication. Wear. 2006 Dec 20;261(11–12):1285–1292.
[13] Etsion I. Laser surface texturing and applications. Recent Developments in Wear Prevention, Friction and Lubrication. 2010:137–158.
[14] Jeyaprakash N, Yang CH, Karuppasamy SS, Dhineshkumar SR. Evaluation of microstructure, nanoindentation and corrosion behavior of laser cladded Stellite-6 alloy on Inconel-625 substrate. Materials Today Communications. 2022 Jun 1;31:103370.
[15] Shivakoti I, Kibria G, Das S, Sharma A, Pradhan BB, Chatterjee S. Laser surface texturing on Ti-6Al-4V. Materials and Manufacturing Processes. 2021 May 19;36(7):858–867.
[16] Shamsul Baharin AF, Ghazali MJ, Wahab AJ. Laser surface texturing and its contribution to friction and wear reduction: A brief review. Industrial Lubrication and Tribology. 2016 Feb 8;68(1):57–66.
[17] Jeyaprakash N, Yang CH, Susila P, Karuppasamy SS. Laser cladding of NiCrMoFeNbTa particles on Inconel 625 alloy: Microstructure and corrosion resistance. Transactions of the Indian Institute of Metals. 2023 Feb;76(2):599–612.
[18] Liu Z, Niu T, Lei Y, Luo Y. Metal surface wettability modification by nanosecond laser surface texturing: A review. Biosurface and Biotribology. 2022 Jun;8(2):95–120.
[19] Dunn A, Wlodarczyk KL, Carstensen JV, Hansen EB, Gabzdyl J, Harrison PM, Shephard JD, Hand DP. Laser surface texturing for high friction contacts. Applied Surface Science. 2015 Dec 1;357:2313–2319.
[20] Murugesan P, Satheeshkumar V, Jeyaprakash N, Yang CH, Karuppasamy SS. Effect of α-Al and Si precipitates on microstructural evaluation and corrosion behavior of laser powder bed fusion printed AlSi10Mg plates in seawater environment. Metals and Materials International. 2023 Sep;29(9):2515–2532.

[21] Brizmer V, Kligerman Y, Etsion I. A laser surface textured parallel thrust bearing. Tribology Transactions. 2003 Jan 1;46(3):397–403.
[22] Seid Ahmed Y, DePaiva MJ, Amorim LF, Torres DR, de Rossi W, Veldhuis CS. Laser surface texturing and characterization of austenitic stainless steel for the improvement of its surface properties. The International Journal of Advanced Manufacturing Technology. 2021 Jul;115(5):1795–1808.
[23] Jeyaprakash N, Yang CH, Karuppasamy SS, Duraiselvam M. Stellite 6 cladding on AISI type 316L stainless steel: Microstructure, nanohardness and corrosion resistance. Transactions of the Indian Institute of Metals. 2023 Feb;76(2):491–503.
[24] Bonse J, Kirner SV, Griepentrog M, Spaltmann D, Krüger J. Femtosecond laser texturing of surfaces for tribological applications. Materials. 2018 May 15;11(5):801.
[25] De Zanet A, Casalegno V, Salvo M. Laser surface texturing of ceramics and ceramic composite materials: A review. Ceramics International. 2021 Mar 15;47(6):7307–7320.
[26] Jeyaprakash N, Alnaser IA, Cheng Y, Karuppasamy SS. Effect of heat treatment on microstructure and mechanical characteristics of Laser Powder Bed Fusion (LPBF) produced CuCrZr alloy. Tribology International. 2024 May 26:109826.
[27] Kovalchenko A, Ajayi O, Erdemir A, Fenske G, Etsion I. The effect of laser surface texturing on transitions in lubrication regimes during unidirectional sliding contact. Tribology International. 2005 Mar 1;38(3):219–225.
[28] Karuppasamy SS, Jeyaprakash N, Yang CH. Application of corrosion-resistant laser claddings. In Handbook of Laser-Based Sustainable Surface Modification and Manufacturing Techniques 2023 Jul 5 (pp. 27–56). CRC Press.
[29] Etsion I, Sher EJ. Improving fuel efficiency with laser surface textured piston rings. Tribology International. 2009 Apr 1;42(4):542–547.
[30] Jeyaprakash N, Yang CH, Karuppasamy SS. Laser cladding of Colmonoy 6 particles on Inconel 625 substrate: Microstructure and corrosion resistance. Surface Review and Letters. 2022 Aug 15;29(8):2250102.
[31] Han J, Zhang F, Van Meerbeek B, Vleugels J, Braem A, Castagne S. Laser surface texturing of zirconia-based ceramics for dental applications: A review. Materials Science and Engineering: C. 2021 Apr 1;123:112034.
[32] Brizmer V, Kligerman Y. A laser surface textured journal bearing. Journal of Tribology. 2012 July;134(3):031702.
[33] Schille J, Schneider L, Mauersberger S, Szokup S, Höhn S, Pötschke J, Reiß F, Leidich E, Löschner U. High-rate laser surface texturing for advanced tribological functionality. Lubricants. 2020 Mar 20;8(3):33.
[34] Verma PK, Vasudev H, Jeyaprakash N. Thermal barrier coatings and claddings for aerospace applications in extreme environment. In Aerospace Engineering Materials 2026; (pp. 110–133). CRC Press.
[35] Etsion I, Halperin G. A laser surface textured hydrostatic mechanical seal. Tribology Transactions. 2002 Jan 1;45(3):430–434.
[36] Jeyaprakash N, Yang CH. Microstructure and wear behaviour of SS420 micron layers on Ti–6Al–4V substrate using laser cladding process. Transactions of the Indian Institute of Metals. 2020 Jun;73(6):1527–1533.
[37] Ryk G, Kligerman Y, Etsion I. Experimental investigation of laser surface texturing for reciprocating automotive components. Tribology Transactions. 2002 Jan 1;45(4):444–449.
[38] Karuppasamy SS, Yang CH. Laser ultrasonic imaging based on synthetic aperture focusing technique algorithm. International Journal of Civil and Environmental Engineering. 2014 Jan 1;15(1):1–5.

[39] Pfleging W, Kumari R, Besser H, Scharnweber T, Majumdar JD. Laser surface textured titanium alloy (Ti–6Al–4V): Part 1–Surface characterization. Applied Surface Science. 2015 Nov 15;355:104–111.
[40] Wang S, Yan F, Chen A. Tribological effects of laser surface texturing and residual stress. Industrial Lubrication and Tribology. 2018 Jan 8;70(1):126–132.
[41] Jeyaprakash N, Karuppasamy SS, Yang CH. Interlayers in FSSW process: An overview. Friction Stir Spot Welding.:21–38.
[42] Braun D, Greiner C, Schneider J, Gumbsch P. Efficiency of laser surface texturing in the reduction of friction under mixed lubrication. Tribology International. 2014 Sep 1;77:142–147.
[43] Mehta A, Vasudev H, Jeyaprakash N. Oxide-based composite coatings for gas turbines to combat corrosion. In Thermal Spray Coatings for High-Temperature Conditions 2025, (pp. 135–147). CRC Press.
[44] Wan Y, Xiong DS. The effect of laser surface texturing on frictional performance of face seal. Journal of Materials Processing Technology. 2008 Feb 1;197(1–3):96–100.
[45] Jeyaprakash N, Duraiselvam M, Aditya SV. Numerical modeling of WC-12% Co laser alloyed cast iron in high temperature sliding wear condition using response surface methodology. Surface Review and Letters. 2018;25(07):1950009.
[46] Liu S, Sai Q, Wang S, Williams J. Effects of laser surface texturing and lubrication on the vibrational and tribological performance of sliding contact. Lubricants. 2022 Jan 11;10(1):10.
[47] Jeyaprakash N, Yang CH, Duraiselvam M, Prabu G, Tseng SP, Kumar DR. Investigation of high temperature wear performance on laser processed nodular iron using optimization technique. Results in Physics. 2019 Dec 1;15:102585.
[48] Gualtieri E, Borghi A, Calabri L, Pugno N, Valeri S. Increasing nanohardness and reducing friction of nitride steel by laser surface texturing. Tribology International. 2009 May 1;42(5):699–705.
[49] Jeyaprakash N, Yang CH, Karuppasamy SS, Radhika N. Microstructural evolution and anti-corrosion characteristics of as-cast and LPBF printed Inconel plates in seawater environment. Surface Review and Letters (SRL). 2024;31(4):1–5.
[50] Artur K, Norbert R, Łukasz O, Jacek P, Marcin S, Jozef B. Laser surface texturing: characteristics and applications. In System Safety: Human-Technical Facility-Environment 2023 Dec 29 (Vol. 5, No. 1, pp. 240–248). Quality and Production Managers Association.
[51] Zhang JJ, Zhang JG, Rosenkranz A, Zhao XL, Song YL. Surface textures fabricated by laser surface texturing and diamond cutting–influence of texture depth on friction and wear. Advanced Engineering Materials. 2018 Apr;20(4):1700995.
[52] He D, Zheng S, Pu J, Zhang G, Hu L. Improving tribological properties of titanium alloys by combining laser surface texturing and diamond-like carbon film. Tribology International. 2015 Feb 1;82:20–27.
[53] Lamraoui A, Costil S, Langlade C, Coddet C. Laser surface texturing (LST) treatment before thermal spraying: A new process to improve the substrate-coating adherence. Surface and Coatings Technology. 2010 Dec 25;205:S164–S167.
[54] Caiwang TA, Jianhui SU, Ziwei FE, Yifan LI, Bo CH, Xiaoguo SO. Laser joining of CFRTP to titanium alloy via laser surface texturing. Chinese Journal of Aeronautics. 2021 May 1;34(5):103–114.
[55] Pou P, Riveiro A, Del Val J, Comesaña R, Penide J, Arias-González F, Soto R, Lusquiños F, Pou J. Laser surface texturing of Titanium for bioengineering applications. Procedia Manufacturing. 2017 Jan 1;13:694–701.

[56] Bathe R, Sai Krishna V, Nikumb SK, Padmanabham GJ. Laser surface texturing of gray cast iron for improving tribological behavior. Applied Physics A. 2014 Oct;117:117–123.
[57] Amanov A, Tsuboi R, Oe H, Sasaki S. The influence of bulges produced by laser surface texturing on the sliding friction and wear behavior. Tribology International. 2013 Apr 1;60:216–223.
[58] Allahyari E, Nivas JJ, Oscurato SL, Salvatore M, Ausanio G, Vecchione A, Fittipaldi R, Maddalena P, Bruzzese R, Amoruso S. Laser surface texturing of copper and variation of the wetting response with the laser pulse fluence. Applied Surface Science. 2019 Mar 15;470:817–824.
[59] Vlădescu SC, Olver AV, Pegg IG, Reddyhoff T. Combined friction and wear reduction in a reciprocating contact through laser surface texturing. Wear. 2016 Jul 15;358:51–61.
[60] Yuan S, Lin N, Wang W, Zhang H, Liu Z, Yu Y, Zeng Q, Wu Y. Correlation between surface textural parameter and tribological behaviour of four metal materials with laser surface texturing (LST). Applied Surface Science. 2022 May 1;583:152410.

3 Analysis of Pre/During/Post Treatment on Laser Surface Textured Materials

Syed Quadir Moinuddin, Suresh Goka, Saad Patel, and Monaf Hodhod

3.1 NOMENCLATURE

AI	Artificial intelligence
AM	Additive manufacturing
ASTM	American Society for Testing and Materials
CBBM	Compliance-based beam method
CFRP	Carbon fiber-reinforced polymer
CVD	Chemical vapor deposition
DED	Direct energy deposition
ENF	End notched flexure
LBT	Laser beam machining
LST	Laser surface texturing
LTHT	Laser texture and heat treatment
MEMS	Micro electromechanical systems
ML	Machine learning
PBF	Powder bed fusion
PVD	Physical vapor deposition
SCHT	Steam condensation heat transfer
SEM	Scanning electron microscope
SMA	Shape memory alloys
WEDM	Wire electric discharge machining

3.2 INTRODUCTION

Laser surface texturing (LST) has become a very effective and adaptable method for changing the surface characteristics of a wide range of materials. The present chapter also highlights laser surface texturing's importance as a potential technique for customizing surface qualities across several materials and industries by examining its features, uses, difficulties, and future possibilities.

DOI: 10.1201/9781003542087-3

Various surface texturing techniques have been developed recently, including wire electric discharge machining, lithography, hot embossing, ion beam etching, micro-milling, electrochemical machining, and laser beam machining (LBT). Several process variables have a major impact on the surface texture quality during material processing. These factors are essential in establishing the material's final textural properties. Regarding LBT, the primary process variables are laser output power, scanning speed, wavelength, beam spot size, repetition rate, and pulse duration.

Laser beam texturing is a method that uses controlled ablation of a laser beam to alter a material's surface roughness and texture. This technology makes it possible to create exact micropatterns with remarkable accuracy and consistency, such as dimples, grooves, and unique forms. With the use of this flexible and adaptive technology, producers may customize surface properties to match particular needs across a range of sectors. Laser texturing is an excellent technique for efficiently and successfully producing required surface qualities because of its high accuracy and repeatability.

Numerous important advantages of laser beam texturing lead to a broad variety of applications in various sectors.

Adherence: Laser texturing is a common option in industrial and automotive applications because it may enhance the adherence of coatings, paints, or adhesives on a material's surface by producing certain textures.

Wettability: The microstructure created by laser texturing has the power to dramatically alter the way liquids stick to the surface, enabling the development of more hydrophobic or hydrophilic characteristics. Water-repellent coatings, biomedical equipment, and applications needing regulated fluid flow can all benefit from this.

Heat conductivity: Surface texture changes can also affect heat conductivity. In electronic technology and heat exchanger applications, the textured patterns may produce channels for better heat retention or dissipation.

Friction reduction: In systems where low friction is desired, such as gears, bearings, or automotive components, laser texturing can be used to generate patterns that decrease friction and improve performance.

3.3 CURRENT START OF ART

Shape memory alloys (SMAs) are smart actuating materials used in robotics, microelectromechanical systems, automobiles, and aerospace due to their ability to recover their deformed shape with thermal stimuli. Nitinol, the most explored SMA, is valued for its shape memory effect, pseudo elasticity, corrosion resistance, biocompatibility, and compatibility with human bone and tissue [1]. However, its poor formability and machinability make conventional manufacturing challenging. Hence, additive manufacturing (AM) processes are cutting-edge manufacturing methods ruling the industries nowadays. Their classification such as the powder bed fusion (PBF) process, where powder is selectively fused by a laser or electron beam, and the direct energy deposition (DED) process melts material using an electric arc and laser. Of course, the laser involves a costly process; it is used for high-precision applications, while arc-based DED is ideal for large-scale parts, repairs, and adding material to existing components.

Thermal management plays a key role in the fabrication of thin wall components with proper deposition and good surface finishing. The cooling method can influence the weld bead shape, which eventually affects surface flatness [2]. Therefore, AM processes have successfully demonstrated the fabrication of complex, thin, and thick steel wall structures, along with the evolution of properties in the build direction [3]. These advancements help overcome challenges related to the poor formability and machinability of various materials. For instance, PBF has recently gained attention for producing intricate nitinol shapes, including the successful fabrication of nitinol gyroid structures [4].

Furthermore, awareness of joining and welding practices within an automated industrial environment is crucial [5]. This includes understanding the challenges, opportunities, and applications of these processes [6]. Additionally, manufacturing industrialists need to be aware of surface texturing techniques to enhance their production capabilities.

There has not been much research done on the combined impacts of heat treatment and laser surface texturing, particularly in terms of improving surface properties in Ti6Al4V. Kashyap and Ramkumar investigated that heat treatment and laser scanning play in improving surface attributes such as oxygen diffusion depth [7]. The overall objective of this study is to improve knowledge of the combined effects of heat treatment and laser surface texturing on Ti6Al4V, with the aim of maximizing surface properties for a range of engineering applications, such as aerospace and biomedical implants.

The process of precisely applying laser energy to a material's surface to generate microscale characteristics is known as laser surface texturing. This method has demonstrated the potential to alter surface attributes including wettability, tribological qualities, and roughness. On Ti6Al4V surfaces, its combined effect with heat treatment is still largely unknown.

It is postulated that significant surface property enhancements may be achieved by post-heating laser-textured Ti6Al4V surfaces due to the interaction between thermal processing and laser-induced microstructures. It is anticipated that the heat treatment procedure will impact the material's mechanical, phase transition, and microstructural development, all of which will work in concert to maximize the benefits of laser surface texturing.

In particular, it is expected that the combined treatment would have a considerable impact on the oxygen diffusion depth, which is critical to titanium alloys' biocompatibility and corrosion resistance. During heat treatment, laser-induced microstructures may act as nucleation sites for phases that are desired, encouraging the development of an improved protective oxide layer.

Thus, the study intended to clarify the underlying mechanisms governing the noted enhancements in surface characteristics using a systematic analysis of laser scanning parameters, heat treatment conditions of use, and surface characterization techniques like scanning electron microscopy, X-ray diffraction, and surface profilometry.

3.4 MATERIAL SURFACE TREATMENTS USING LASER SURFACE TEXTURING PROCESS

By enhancing a material's tribological, hydrophobic, mechanical, and mechanical qualities, the LST procedure can improve a material's performance. Figure 3.1 provides further information about the laser surface texturing treatments in detail.

3.4.1 Pre-Treatment

Preparing the material before the laser texturing method is known as pre-treatment for materials with surface textures created by lasers. Achieving the intended properties in the finished product, improving laser contact with the material surface, and assuring consistent results all depend on proper pre-treatment. An examination of popular pre-treatment techniques and how they affect laser surface texturing is provided now:

Surface cleaning: Cleaning the material's surface to get rid of impurities, oils, or residues is the most fundamental pre-treatment procedure. A spotless surface minimizes undesired artifacts or flaws and guarantees that the laser interacts with the material efficiently. Chemical, solvent-based, and ultrasonic cleaning are examples of common cleaning techniques.

Abrasive blasting: Before laser texturing, the surface may occasionally be roughened using abrasive blasting techniques like shot or sandblasting. Increased surface adherence from this pre-treatment might be advantageous for upcoming coatings or texturing designs. Additionally, it can aid in removing oxidation and surface defects.

Anodizing or chemical etching: These techniques can be used to roughen the surface in preparation for laser texturing or to produce a controlled oxide layer. For magnesium and aluminum alloys in particular, this treatment is helpful because it offers a reliable foundation for additional processing.

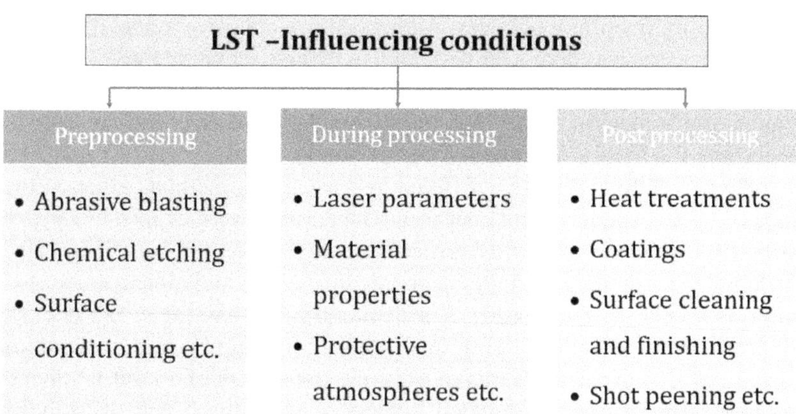

FIGURE 3.1 Factors influencing laser surface texturing.

Heat treatment: Using heat as a pre-treatment can change the microstructure of the material or reduce internal tensions. This can lessen the chance of warping or deformation during the procedure and aid in enhancing the consistency of laser texturing. Heat treatment can also be used to improve the mechanical qualities of the material, such as its hardness.

Application of coatings or films: The texture's final result may vary depending on whether a thin coating or film is applied to the surface before laser texturing. For instance, using a dark coating could enhance laser absorption and provide more noticeable texturing. Using this technique, you may create depth variations or particular patterns.

Protective layers and masking: Masking entails covering parts that shouldn't be laser textured using tapes or protective layers. Precise control over the patterned process and selective texturing are made possible by this pre-treatment. When sophisticated patterns or complex geometries are needed, it is extremely helpful.

Surface conditioning: The material can be ready for laser texturing by using surface conditioning techniques like grinding or machining. This stage guarantees a consistent, level surface and gets rid of any flaws that could get in the way of the laser texturing procedure.

In summary, pre-treatment is essential to the effectiveness of laser surface texturing. Manufacturers may produce consistent results, improve laser contact with the substance, and accomplish the required properties in the final product by carefully choosing and using suitable pre-treatment processes. A strong pre-treatment also contributes to fewer flaws and increased textured surface durability and lays the groundwork for additional post-treatment procedures.

One of the most crucial processes in creating a trustworthy bonded joint is surface preparation for bonding. The study of Parodo and coworkers [8] examined the efficacy of a novel CO_2 pulsed laser pre-treatment for joining carbon fiber reinforced polymer (CFRP). End-notched flexure samples were created using varying laser treatment densities, and the compliance-based beam method was used to determine the corresponding fracture toughness. Additionally, a cohesive model that simulated debonding processes was demonstrated, and an inverse technique was used to determine the cohesive parameters. The obtained outcomes signify a crucial phase toward the creation of a numerical model that may be employed to ascertain the laser texturing concerning the local stress exerted on the bonded joint.

The authors looked at how two different pre-treatment methods affected the performance of CFRP single-lap manner joints when they were exposed to hydrothermal aging [9]. The authors' earlier research demonstrated the beneficial effects of the developed laser texturing on the mechanical function and morphology of CFRP adherents. However, in the context of a demanding environment, the dependability of bonded joints created utilizing this newly found laser therapy was unknown. The purpose of this experiment was to examine the effects of a laser texturing pre-treatment on CFRP single-lap bonded joints that were then aged hydrothermally. This study (9) enables comprehension of the impact that such a pre-treatment has on the joint's dependability when bonded joints are submerged in water and freeze-thaw cycles are applied. The outcomes validated the effectiveness of the suggested method, as bonded joints created with such pre-treatment exhibited a decreased susceptibility to hydrothermal aging.

When examining unaged specimens, it is feasible to see that the laser treatment significantly affected the lap shear strength, resulting in a performance gain of almost 38.7%, as shown in Figure 3.2. This effect was more noticeable when the effects of aging were considered. The lap shear strength of the degreased specimens exposed to hydrothermal aging had a sharp decline of around 240.2 %, whilst the specimens treated with lasers displayed a minor difference in lap shear strength. Perhaps to degreasing treatment, the mechanical interlocking produced by the laser texturing enables a better correction of the residual stress caused by aging circumstances.

Usually the standard, ASTM D5573 was used to examine the specimens' failure surfaces to determine the cause of this variance. The failure surfaces for every pre-treatment and aging environment are displayed in Figure 3.3. One can see that the glue turned green instead of red-purple as a result of the adhesive's deterioration from water diffusion within the adhesive. The authors also noted this occurrence in single-lap bonded joints with composite adherents composed of woven plies in previous publications [10].

Using optical observations, the failure sites of the joints that underwent laser pre-treatment were examined in order to emphasize the impact of mechanical interlocking.

A typical failure surface may be seen on the adherend side, as shown in Figure 3.4.

It's seen to see that the glue stuck to the surface fibers inside the dimple and caused them to rip as the fracture spread. This occurrence has resulted in localized delamination with diameters comparable to or larger than those of the dimples themselves, confirming that the existence of mechanical interlocking and improved adhesive wettability of the surface are associated to the joint's enhanced strength. Based on surface examination, the failure surfaces of the old and non-aged specimens do not significantly differ from one another.

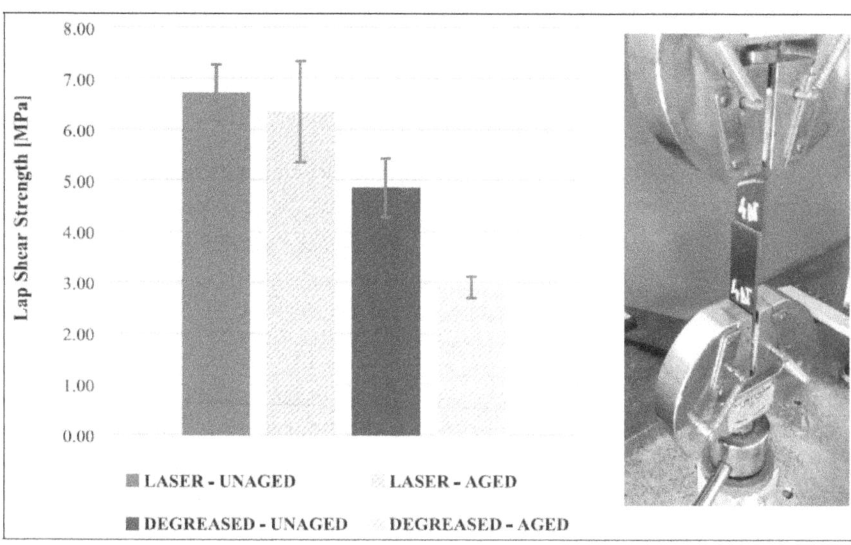

FIGURE 3.2 Lap shear strength analysis [10].

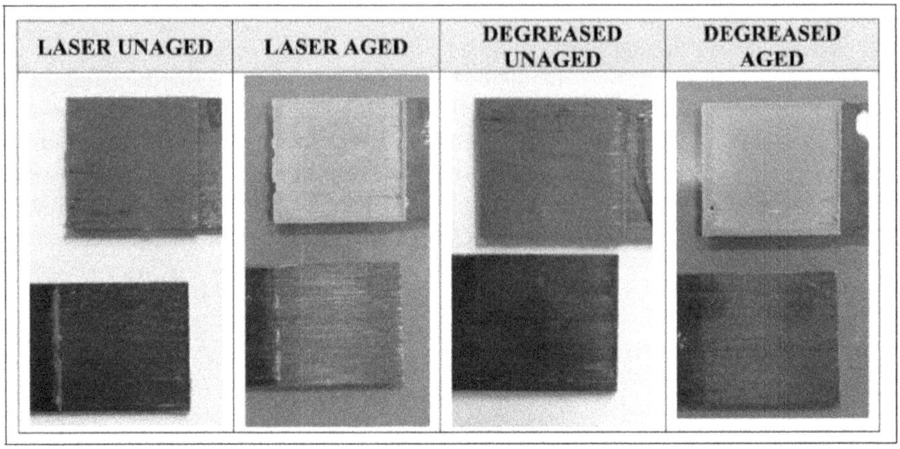

FIGURE 3.3 Failure surfaces of pre-treatment and aging environment [10].

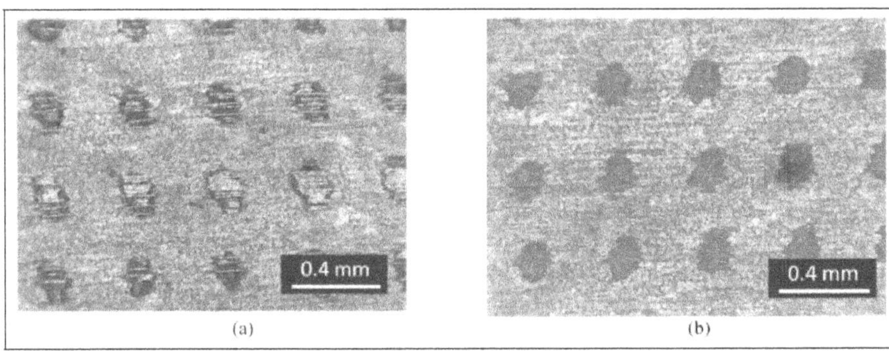

FIGURE 3.4 Microscopic views of laser-textured surfaces (a) adherend side (b) adhesive side [10].

In conclusion, hydrothermal aging provided additional evidence of the efficiency of the chosen laser texturing technique. The authors' future research will examine if laser texturing affects the bonded joints' mechanical resistance even at varying operating temperatures.

3.4.2 During LST

To generate desired patterns, textures, and various other features, laser surface texturing entails changing the surface properties that materials have during the laser treatment process. The examination of the treatment procedure while laser surface texturing centers on variables that impact the result, such as laser settings, material characteristics, and ambient elements. Here's a detailed look at the factors that influence texturing and how various approaches can provide a range of outcomes.

Laser Parameters

Laser power: The power level controls the laser beam's intensity and has an impact on the depth and volume of material removed. Deeper textures are often produced by more power; however, if unchecked, this can cause damage or overheating.

Laser wavelength: The material's absorption rate is influenced by the laser's wavelength. Certain materials respond better to certain wavelengths; longer wavelengths are more effective across wider regions, while shorter wavelengths are more precise.

Frequency and duration of laser pulses: The texturing's accuracy and caliber are affected by the laser pulses' duration and frequency. Longer pulses may produce deeper but less sophisticated patterns, whereas shorter, high-frequency pulses can produce finer textures.

Scan speed: The homogeneity and patterning of the surface are affected by the speed when the laser beam travels across it. Slower rates might result in more uniform texturing but can lead to a build-up of heat.

Material Properties

Material type: The way that various materials respond to laser texturing varies. The specific characteristics of metals, ceramics, and polymers influence how laser energy is absorbed and reflected by them.

Thermal conductivity: By effectively dissipating heat, materials with excellent thermal conductivity lower the possibility of thermal damage while texturing. To prevent overheating, laser settings may need to be carefully adjusted for materials with lower conductivity.

Surface composition: The laser texturing method may be impacted by surface contaminants, oxidation, or preexisting coatings. These elements may result in irregular patterns or necessitate further processing to bring about consistency.

Environmental Factors

Temperature and humidity: Modifications in these factors may have an impact on the material's response as well as the laser's performance. For consistent outcomes, stable circumstances are essential.

Protective atmospheres: When laser texturing materials are prone to corrosion, employing inactive/inert gas atmospheres or vacuum circumstances can help minimize oxidation and other undesired reactions.

Other Conditions

Cooling systems: During texturing, efficient cooling helps stop overheating and preserve the material's integrity. Systems for air or water cooling are frequently utilized to control heat accumulation.

Control systems: Sophisticated control systems make the laser move precisely and consistently, enabling texturing patterns that are reliable.

A complicated interaction of laser settings, material qualities, and ambient circumstances is involved in the handling of laser surface texturing while performing the texturing process. To produce surface textures that are consistent and of good quality, it is essential to comprehend these elements and optimize them accordingly.

3.4.3 Post-Treatment

To improve the material's performance and preserve the intended surface features, post-treatment of laser surface–textured substances is essential. On the material's surface, laser surface texturing produces distinct patterns or roughness; however, these patterns may be enhanced or preserved using a variety of post-treatment methods. The following examines popular post-treatment techniques and how they affect materials with laser surface textures:

1. **Heat treatment:** Laser-textured materials can have their microstructure and hardness changed by heat treatment. The material's mechanical qualities and stress resistance may be enhanced by this procedure. Sometimes it can aid in the microstructure's reorganization or homogenization, which improves durability.
2. **Anodizing and passivation:** An oxide layer of protection is added to aluminum and magnesium alloys by the process of anodizing. It can produce a regulated surface finish and improve resistance to corrosion. By chemically treating the material to produce a protective coating, a process known as passivation, corrosion resistance is frequently increased.
3. **Coating and plating**: Coatings like electroplating, PVD (physical vapor deposition), and CVD (chemical vapor deposition) can be applied to improve the laser-textured surface's resilience to corrosion and wear as well as its visual appeal. Additionally, coatings can offer certain functional qualities like electrical conductivity or hydrophobicity.
4. **Polishing and surface finishing:** To enhance the texture produced by laser texturing, post-treatment may entail polishing or other surface finishing methods. In applications like medical devices or optical components, where smoothness and minimal friction are necessary, this stage might be crucial.
5. **Cleaning and deburring:** Following laser texturing, there could be leftover residues and particles on the surface. For future coatings or treatments to adhere correctly and function as intended, the texture must be clean and clear of undesirable particles. This is ensured by cleaning and deburring.
6. **Functionalization:** It is the process of imparting certain qualities to a surface by adding molecules or functional groups. For instance, the wettability of laser-textured surfaces may be controlled by adding hydrophobic or hydrophilic agents, improving performance in certain applications such as microfluidics.
7. **Mechanical post-treatment:** By creating compressive stresses, mechanical treatments like shot peening can extend the fatigue life of materials with laser textures. High-stress applications such as aerospace or automobile parts benefit greatly from this.

In summary, the final qualities that are wanted and the needs of the laser surface–textured material's application determine which post-treatment is best. The usefulness, endurance, and durability of laser-textured components may be greatly increased by manufacturers via the careful selection and application of post-treatment methods.

3.4.4 Case Studies

Thermal treatments are frequently used to change surfaces to enhance their functional behavior. This can be achieved by generating coatings, introducing foreign materials into alloys, or producing certain microstructures, all of which increase the treated samples' surface integrity. Controlling contours and thickness without compromising the integrity of the material is possible with thermal laser treatments [11]. This study examined differences in the tribological performance of samples that had undergone superficial modification. These differences were attributed to an increase in the hardness and depth of the material that was removed and exposed to oxidation. The authors provide an examination of how the texturing laser process conditions, which are based on changes in the beam's scan speed rate, affect the features and surface characteristics, with a particular emphasis on the tribological behavior of the changed layer.

Vickers hardness values vary as a result of heat treatment and oxidation events that occur in textured regions; these variations tend to grow for larger layer thicknesses with slower laser beam scan speeds. Along with a reduction in the intensity of the texturing process on the alloy's surface, a notable drop in material hardness is seen as the beam scan speed increases. An increase in surface compactness resulting from the geometry generated in the changed material layer may be noticed as a significant source of hardness. Because of the volume difference between the titanium alloy utilized and the oxidized substrate layer that results, unstable conditions in the cooling as well as solidifying process may have an impact on the increased thickness of the altered material layer where oxidation phenomena are present. This is because the process creates mechanical stress in the oxide layer, which makes the outer layer more brittle and increases the likelihood of fractures. In this sense, the created layer's ability to defend can be lost. Microcracks are encouraged to form on the surface and within the textured grooves as a result of the instabilities created during the cooling process by the variations in the volume of the changed layer's components, as seen in Figure 3.5. In contact and frictional settings, the integrity of the alloy's outer layer is weakened by this surface breaking.

The development and appearance of microcracks can be detrimental in tribological wear scenarios when stick slip and contact pressure are present. This drawback has an impact on how much the increases in mechanical qualities like hardness are reduced. In this instance, as Figure 3.6 illustrates, an increase in the friction coefficient and worn material rate may be seen for more severe texturizing operations (where a modified layer of larger thickness is formed).

FIGURE 3.5 Images of textured layer (a, b) with micro cracks and (c) without cracks [11].

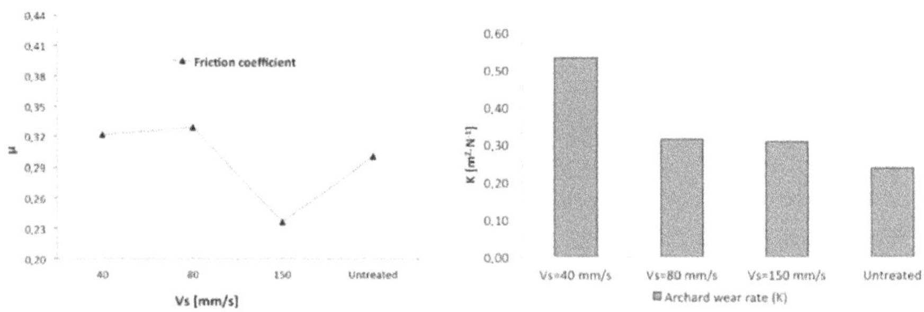

FIGURE 3.6 Plots representing friction co-efficient and Archard coefficient [11].

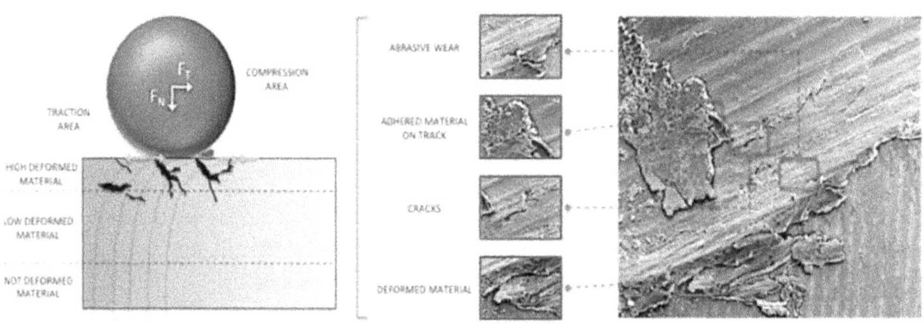

FIGURE 3.7 Localized wear effects over textured layer [11].

In the pin-on-disk test, wear growth is mostly caused by the brittleness of the layer due to microcracks, which results in a trend of debris being released from the changed surface layer across the sliding track.

Because of the increased hardness of the deposited and attached debris on the track compared to the alloy below, the substrate is worn down by abrasive wear. Similar to this, altered layer debris on the sliding track leads to instabilities in the surface-to-surface contact as well as a variation in the friction coefficient values, which in turn causes layer fatigue and cracking in the tribological pair's contact region, as depicted in Figure 3.7. These data point to an increase in the changed alloy's wear rate.

Owing to the material's localized heating in an unprotected environment, the formation of an enriched thin layer of titanium oxides (Ti_xO_y) is encouraged, with titanium dioxide (TiO_2) often predominating in greater concentrations. The UNS R56400 alloy undergoes induced oxidation using laser texturing, resulting in a change of basic qualities and features of the untreated material [12].

Through experimental testing, a direct correlation has been observed between the textured process control parameter, the beam's scanning speed, and the treated layer thickness, which is measured by the laser's material removal depth. When treatments are administered in an environment that is not protective, oxides begin to form in the radiation-exposed regions. The Ti6Al4V alloy's microstructural alterations and oxidation brought on by cooling cause the material's hardness values to rise exponentially. This fact indicates comparable tendencies to the depth of removal material and again creates a direct reliance on the beam's scan speed.

Through the use of SEM microscopy testing, it has been confirmed that the inclusion of oxides into the alloy through texturing a process that involves rapid cooling affects the growth of cracks in the outermost protective layer of the material by increasing its instability. In this regard, it has been shown that there is a critical thickness of the treated layer, below which the treatment's wear strength benefits are reduced. The emergence of surface fissures weakens the textured layer's ability to withstand contact and gliding wear, mostly because of the debris that releases and adheres to the sliding track. Due to this feature, depending on how the beam scans at that speed, the wear rate and friction coefficient increase.

The production of light-trapping "black" silicon via chemical, laser, and hybrid processes is covered by Michael Kovalev et al. [13]. They created a hybrid chemical/laser texturing approach that combines the commonly used laser texturing and chemical etching techniques with laser post-texturing of pyramidal configurations that are generated after chemical etching. The surface morphology was illustrated by a chaotic relief of micro-cones following laser treatment, whereas it took on a chaotic pyramidal relief following chemical treatment. Furthermore, compared to raw flat silicon, laser texturing of silicon wafers with preliminary chemical micro-texturing takes five times less time. In this instance, the average total reflectance in the 250–1100 nm spectral band is 7–10% lower in the chemically/laser-treated samples than in the purely chemically treated samples.

Laser actuation has been proposed as a promising method for the actuation of smart actuators, which are difficult to access in biomedical and other engineering applications [14] due to their non-contact nature. It improves actuator mobility and degrees of freedom by eliminating the need for connectors and wiring. In this process, a specific region of an SMA is irradiated with a laser, generating heat through photothermal interactions, unlike Joule heating, which heats the entire volume of the SMA and does not undergo heat generation. Due to photothermal interactions, the photon energy delivered via laser irradiation is converted into heat, which is transferred to the remaining portions of the irradiated element by conduction. An extensive study on the laser surface texturing of SMAs for improving actuation performance has not yet been reported. Therefore, this case study [15] was designed to investigate the influence of nanosecond laser texturing on the surface properties and laser actuation behavior of nitinol.

Nevertheless, the process has been widespread across various materials based on numerous applications. Table 3.1 shows a list of various materials that have undergone laser surface texturing process from an application point of view.
Similarly, coatings on the surface will be applied using laser surface texturing, which ensures adhesive strength to the base metal surface. The coating can be applied to almost all types of materials and results in improved material properties. Table 3.2 provides a summarized study of the coatings and their results.

TABLE 3.1
Summarized Information of Laser Surface Texturing Materials

Material	Summary	Application	Reference
Black silicon	Laser texturing of bare flat Si wafer. After merely laser texturing, the resultant structure is a chaotic conical relief; after chemical texturing, it becomes a chaotic pyramidal relief.	Broadband light-trapping structures	[13]
Super elastic NiTi	Surface morphology to regulate the proliferation of cells. Depending on the chosen process parameters, the surface topology and Ni/Ti ratio were significantly altered from the original surface after the laser treatments.	Biodevices for biomedical applications	[16]
AA6082	To evaluate the resistance to cavitation erosion. With the best laser texturing treatment, a cavitation erosion volume decrease of around 20% lower was achieved.	Propellers (naval applications)	[17]
Stainless steel	Without using chemicals, the suggested surface treatment approach shows promise for improving the SCHT of SS surfaces. LT combined with green heat treatment (LTHT) is a promising method that may be used to metal surface condensation for water collecting or heat transmission.	Steam condenser, heat exchangers (thermal applications)	[18]
Ti6Al4V	To investigate how better surface wear performance of Ti6Al4V may be achieved via laser-textured surfaces. In the dry sliding wear test, the wear rate of the Ti6Al4V surface with laser texture showed an improvement of 88.31% when compared to the Ti6Al4V as received.	Tribological, biomedical, biological application	[19–20]
Copper	After heat treatment, the laser micro-/nano-textured copper surface increases to 161°, becoming a superhydrophobic surface from a super hydrophilic 4° contact angle.	Heat transfer applications (refrigeration, microelectronics, and batteries)	[21]

TABLE 3.2
Effect of Laser-Textured Coating Types on Various Materials [22]

Base Material	Coating Material	Type of Testing	Result
Zirconia	HAP, 45S5 BG	Fluorescence staining	Laser sintering of HAP and 45S5 BG is a fast coating technique and has no significant difference from conventionally sintering coating.
Ni-based base	Thermal barrier coating	Thermal stability test	The bond coat is replaced by laser texturing and texturing interfaces were not damaged after 1% creep strain on the top coating.
Tungsten and steel substrate	Tungsten powder by plasma spray technique	Shear loading	After coating on the textured surface, shear force was applied to the coating surface. It was found that the best adhesion bond strength obtained on a steel substrate.

3.5 CONCLUSIONS

Overall, laser surface texturing provides an effective means of improving the functionality and adaptability of metals in a variety of applications. It is a very useful technology for increasing adhesion, durability, wear and tear, and corrosion protection because of its ability to precisely and consistently alter surface properties. The potential uses and advantages of laser surface texturing are anticipated to grow as laser technology develops, spurring innovation and enhancing material performance across several sectors.

Laser surface texturing has shown to be an effective method for changing the surface characteristics of a variety of metals, providing noteworthy advantages in many applications. This procedure makes use of laser technology's accuracy to create micro- and nano-scale designs that improve adhesion, friction, corrosion resistance, and other important qualities. LST is a precise and efficient method for modifying material surfaces to enhance properties such as wettability, friction, and corrosion resistance. The integration of artificial intelligence and machine learning with manufacturing industries [23] can significantly improve the process by optimizing parameters, predicting outcomes, and ensuring consistent quality. The current trend may reproduce a great revolution in LST through these treatments for various types of material across all industries and automotive sectors.

The following are some salient conclusions emphasizing the advantages and prospective uses of laser surface texturing for different metals:

- Laser surface texturing provides the biomedical industry with the capability to regulate surface wettability and roughness, which is essential for implants and other medical equipment. Better patient results can result from the technique's ability to decrease bacterial adherence and boost osseointegration.
- Metals' mechanical and tribological qualities, such as their decreased friction and increased wear resistance, can be enhanced by laser surface texturing. This makes it a useful method for parts in industrial, automotive, and aerospace applications, where endurance and durability are essential.
- Creating customized textures on metal surfaces enhances adhesion, making it possible for paints, coatings, and adhesives to attach more strongly. In sectors where dependable adhesion is crucial, like the automobile and construction industries, this property is advantageous.
- Using laser surface texturing gives designers the freedom to create distinctive textures for either utilitarian or ornamental uses. Surface patterns may be tailored to match certain design specifications. This is very helpful for industrial and consumer product design.
- By generating patterns on the surface that offer protection or by making it easier to apply protective coatings later, laser texturing can help increase corrosion resistance. For metals used in severe conditions, such as those found in marine or chemical processing applications, this is especially crucial.
- Laser surface texturing can modify metals' thermal conductivity, which improves heat dissipation in applications like energy systems and electronics. In a variety of sectors, this skill may result in more effective heat management.

3.6 ACKNOWLEDGMENT

This work was supported by the Deanship of Scientific Research, Vice Presidency for Graduate Studies and Scientific Research, King Faisal University, Saudi Arabia. Grant No. KFU261869.

REFERENCES

[1] Nam S, Seo BR, Najibi AJ, McNamara SL, Mooney DJ. Active tissue adhesive activates mechanosensors and prevents muscle atrophy. Nat Mater [Internet]. 2023 Feb 10;22(2):249–259. Available from: https://www.nature.com/articles/s41563-022-01396-x

[2] Jayprakash P, Suresh G, Suryakumar S, Abhay S. Thermal management in manufacture of thin-walled components produced by arc-based additive manufacturing. 10th Int Conf Trends Weld Res. 2016;1–4.

[3] Kulkarni JD, Goka SB, Parchuri PK, Yamamoto H, Ito K, Simhambhatla S. Microstructure evolution along build direction for thin-wall components fabricated with wire-direct energy deposition. Rapid Prototyp J. 2021;27(7):1289–1301.

[4] Jin J, Wu S, Yang L, Zhang C, Li Y, Cai C, et al. Ni–Ti multicell interlacing Gyroid lattice structures with ultra-high hyperelastic response fabricated by laser powder bed fusion. Int J Mach Tools Manuf [Internet]. 2024 Feb;195:104099. Available from: https://linkinghub.elsevier.com/retrieve/pii/S0890695523001074

[5] Goka S, Srirama SN, Divya Jyothi G, Moinuddin SQ, Shaik HS. Welding-Based 3D, 4D, 5D printing. In: Automation in Welding Industry [Internet]. Wiley; 2024. pp. 157–181. Available from: https://onlinelibrary.wiley.com/doi/10.1002/9781394172948.ch10

[6] Goka S, Moinuddin SQ, Cheepu M, Dewangan AK. Welding practices in industry 5.0. In: Automation in Welding Industry [Internet]. Wiley; 2024. pp. 263–279. Available from: https://onlinelibrary.wiley.com/doi/10.1002/9781394172948.ch15

[7] Kashyap V, Ramkumar P. Improved oxygen diffusion and overall surface characteristics using combined laser surface texturing and heat treatment process of Ti6Al4V. Surf Coatings Technol [Internet]. 2022 Jan;429:127976. Available from: https://linkinghub.elsevier.com/retrieve/pii/S0257897221011506

[8] Sorrentino L, Parodo G, Turchetta S. Influence of laser treatment on end notched flexure bonded joints in carbon fiber reinforced polymer: Experimental and numerical results. Materials (Basel) [Internet]. 2022 Jan 25;15(3):910. Available from: https://www.mdpi.com/1996-1944/15/3/910

[9] Parodo G, Polini W, Sorrentino L, Turchetta S, Corrado A. Influence of hydrothermal ageing on CFRP single lap bonded joint pre-treated by laser texturing. Proc Inst Mech Eng Part C J Mech Eng Sci [Internet]. 2024 Apr 21;238(7):2848–2857. Available from: https://journals.sagepub.com/doi/10.1177/09544062231203557

[10] Bellini C, Parodo G, Polini W, Sorrentino L. Experimental investigation of hydrothermal ageing on single lap bonded CFRP joints. Procedia Struct Integr [Internet]. 2018;9:101–107. Available from: https://linkinghub.elsevier.com/retrieve/pii/S2452321618300179

[11] Vazquez-Martinez JM, Botana M, Botana FJ, Del Sol I, Salguero J, Marcos M. Thermal laser oxidation based texturing as finishing treatment for improving surface properties of UNS R56400. Procedia Manuf [Internet]. 2017;13:372–379. Available from: https://doi.org/10.1016/j.promfg.2017.09.022

[12] Lavisse L, Jouvard JM, Imhoff L, Heintz O, Korntheuer J, Langlade C, et al. Pulsed laser growth and characterization of thin films on titanium substrates. Appl Surf Sci [Internet]. 2007 Jul;253(19):8226–8230. Available from: https://linkinghub.elsevier.com/retrieve/pii/S0169433207004060

[13] Kovalev M, Podlesnykh I, Nastulyavichus A, Stsepuro N, Mushkarina I, Platonov P, et al. Efficient broadband light-trapping structures on thin-film silicon fabricated by laser, chemical and hybrid chemical/laser treatments. Materials (Basel) [Internet]. 2023 Mar 15;16(6):2350. Available from: https://www.mdpi.com/1996-1944/16/6/2350

[14] Herath M, Epaarachchi J, Islam M, Zhang F, Leng J, Fang L, et al. Remote actuation of light activated shape memory polymers via D-shaped optical fibres. Smart Mater Struct [Internet]. 2020 Apr 1;29(4):047001. Available from: https://iopscience.iop.org/article/10.1088/1361-665X/ab72ea

[15] Ryu C, Prabu SSM, Palani IA, Nguyen AP, In J Bin. Improving the actuation behavior of nitinol shape memory alloys by nanosecond laser surface texturing. Opt Laser Technol [Internet]. 2024 Sep;176:110957. Available from: https://linkinghub.elsevier.com/retrieve/pii/S0030399224004158

[16] Biffi CA, Fiocchi J, Rancan M, Gambaro S, Cirisano F, Armelao L, et al. Ultrashort laser texturing of superelastic NiTi: Effect of laser power and scanning speed on surface morphology, composition and wettability. Metals (Basel) [Internet]. 2023 Feb 13;13(2):381. Available from: https://www.mdpi.com/2075-4701/13/2/381

[17] Genna S, Leone C, Mingione E, Rubino G. Surface treatments for the improvement of mechanical and cavitation resistance of Al 6082 alloy. Int J Adv Manuf Technol

[Internet]. 2023 Dec 15;129(11–12):5149–5165. Available from: https://link.springer.com/10.1007/s00170-023-12411-z

[18] Lou D, Wang B, Liang E, Lu G, Cheng J, Liao J, et al. Effect of heat treatment on wettability and condensation heat transfer of laser-textured stainless steel surfaces. Phys Status Solidi [Internet]. 2022 Jan 23;219(2). Available from: https://onlinelibrary.wiley.com/doi/10.1002/pssa.202100458

[19] Zul MH, Ishak M, Md. Nasir R, Aiman MH, Quazi MM. Effect of laser-textured surface of Ti6Al4V on frictional wear behavior. Int J Automot Mech Eng [Internet]. 2023 Mar 30;20(1):10165–10174. Available from: https://journal.ump.edu.my/ijame/article/view/8879

[20] Lin N, Li D, Zou J, Xie R, Wang Z, Tang B. Surface texture-based surface treatments on Ti6Al4V titanium alloys for tribological and biological applications: A mini review. Materials (Basel) [Internet]. 2018 Mar 24;11(4):487. Available from: http://www.mdpi.com/1996-1944/11/4/487

[21] Lou D, Mei S, Wang B, Li T, Cao J, Yang Q, et al. Effect of stabilizing heat treatment on condensation heat transfer performance of laser micro-/nano-textured copper surface. J Mater Sci [Internet]. 2021 Feb 11;56(5):3981–3994. Available from: http://link.springer.com/10.1007/s10853-020-05454-0

[22] Kumar V, Verma R, Kango S, Sharma VS. Recent progresses and applications in laser-based surface texturing systems. Mater Today Commun [Internet]. 2021;26(October 2020):101736. Available from: https://doi.org/10.1016/j.mtcomm.2020.101736

[23] Goka S, Narayana GS, Divya Jyothi G, Shaik HS, Moinuddin SQ. AI and ML in welding technologies. In: Automation in Welding Industry [Internet]. Wiley; 2024. pp. 73–90. Available from: https://onlinelibrary.wiley.com/doi/10.1002/9781394172948.ch5

4 Mechanical and Tribological Characteristics of Laser Surface Textured Coating

Lakshmanan Vellaichamy, Durairaj Manikandan, and Durairaj Rajkumar

4.1 INTRODUCTION

In current research work a surface texturing, applying coating and lubricant, was used to improve the tribological performance of the substrates. The surface texturing is highly useful to reducing friction coefficient, enhancing load-carrying capacity, and increasing wear resistance. In this technique applicable for tools and components such as cutting tools, bearings, cylinder liners, and needles were used. An instance of laser texturing technology has been employed to generate pit textures on cylinder liners, resulting in enhanced friction and wear characteristics in comparison to untextured surfaces [1]. Wear-related phenomena are greatly influenced by surface properties, starting from the surface itself. Surface coating technologies have garnered significant attention due to the widespread use of steels and the diverse range of environments in which they are utilized. These technologies offer improved resistance to friction and wear while using a minimal amount of coating material. Surface pretreatments such as grit blasting, shot peening, and surface texturing have proven to be successful in enhancing the tribological characteristics of coatings.

Several research works investigate the efficiency of the pretreatment methods under different operational requirement conditions [2].

The primary techniques employed in this field are laser surface treatment, physical vapour deposition (PVD), chemical vapour deposition (CVD), and anodic oxidation. Magnetron sputtering is a type of physical vapour deposition (PVD) that involves the bombardment of a target material with ions, resulting in the discharge of atoms or molecules. These particulates subsequently settle onto a substrate, resulting in the formation of a thin film. In comparison to other methods, this technique provides numerous benefits compared [3]. Surface texturing can improve the friction and wear characteristics and enhance the ability to bear loads when compared to smooth

DOI: 10.1201/9781003542087-4

surfaces. Textured surfaces have shown an 18% decrease in torque when subjected to rolling contact, as well as a notable reduction in friction and wear rates during dry sliding reciprocating conditions. The automotive industry utilizes two main surface texturing techniques on hypereutectic Al–Si cylinder liner surfaces: mechanical indentation and micro-drilling. Engine friction accounts for approximately 15% of the total fuel energy consumed in an internal combustion engine. The use of texturing can reduce this friction by increasing the thickness of the lubricating film under mixed or boundary lubrication conditions [4]. Laser micro-cladding technology is a cutting-edge technique used to create texture on surfaces. During friction experiments, the lower sample remains stationary while the upper ball moves at a velocity of 10 mm/s under a consistent load of 5 N. The computer algorithm calculates the friction coefficient using the measured friction force and applied load [5]. Surface texturing is a highly effective and commonly employed technique for regulating the tribological properties of a material. Through the alteration of surface interactions, it attains superior lubrication, regulated friction, and improved resistance to wear. Surface texturing was first applied in the 1940s to prevent seizure in diesel engines operating at high speeds. This was achieved through interrupted surface treatment at the piston ring–cylinder liner interface [6]. Applying solid lubricant to surfaces with texture significantly enhances the tribological characteristics of the corresponding components. A MoS2 solid lubricant coating was synthesized through the utilization of laser texturing and hot-pressing methodologies [7]. Solid lubricants play a vital role in addressing tribological requirements by minimizing friction and mechanical interactions between surfaces in motion, thereby ensuring consistently low levels of friction [8]. The laser texturing process enables precise manipulation of texture dimensions, has minimal environmental impact, and provides fast processing times, making it a highly researched method for improving the tribological properties of materials [9].

The main function of surface texturing is to reduce friction and to prevent wear during the cutting process; it also serves as a lubricant and improves the ability to bear load [10]. The surface texture alignment is also very important for chip flow direction, and also to reduce friction and wear. The surface texturing combined with solid lubricant dramatically reduced wear and friction [11]. MoS2 and WS2, which possess comparable hexagonal layer-lattice structures, are utilized as solid lubricants in a variety of forms, including powders, burnished coatings, hot-pressed coatings, and PVD sputtered coatings. In comparison to MoS2, WS2 demonstrates superior thermal stability, withstanding temperatures that are approximately 100°C higher [12].

4.2 INTRODUCTION TO SURFACE TEXTURE

Surface texture differs from surface topography, which plays a vital role in determining the product (what way a product will react with different environmental conditions). It was applied to many engineering areas, such as bumps, grooves, pores, and scratches. It can be visible with the naked eye or through microscopic analysis. It affected various physical properties like friction, wear, heat conduction, and light reflection, and the ability of the surfaces to hold a lubricant or coating. Surface texturing plays major role in optimizing the product performance and durability.

Surface texturing is important in various industries: product assembly, safety, and functionality. In manufacturing, surface texture impacts parts assembly, the overall aesthetic appeal of products, and their operational effectiveness. Furthermore, surface texture plays major role in tire grip, fuel efficiency, and durability of moving components. In aerospace, surface texture plays a major role in optimizing aerodynamics, enhancing fuel efficiency, and ensuring safety. In the biomedical field, surface texture is important to determine the biocompatibility and effectiveness of implants. The surface texture is of universal importance to necessary for precise measurement, analysis, and specific demands of each application [13].

4.2.1 Types of Surface Texture

Research has been conducted to analyse different types of surface pretreatments, such as plasma etching, mechanical polishing, femtosecond laser texturing, and plasma-assisted femtosecond laser texturing. Some research work concentrated on the effect of surface treatment, such as physical and chemical adhesion, between PVD soft and hot coatings, and also various machining processes used to surface texturing processes, such as mechanical processing, abrasive jet machining, laser texturing, and lithography [14]. Another process is the ultrasonic surface rolling process (USRP), a method to improve the tribological characteristics and fatigue durability of mechanical parts by reinforcing the surfaces. This technique provides numerous benefits, such as cost efficiency, straightforward installation, environmental friendliness, and user-friendliness. Furthermore, ultrasonic nanocrystalline surface modification (UNSM) technology improves the tribological properties of materials [15]. Some other methods to create surface texture technologies are laser surface texturing (LST), micro-milling, hot-embossing, mechanical texturing, ion beam etching/milling, photolithography, and electrochemical machining. The LST technique is important because of its high speed, exceptional productivity, great adaptability, environmentally friendly nature, and capacity to produce intricate and precise surface textures. Most of the research work was on frictional and wear characteristics of materials' surfaces that have been treated with laser texturing [16].

4.2.2 Measuring Surface Texture

To measure surface texturing for highly precise operation, primary methods, such as contact methods, involve a stylus moving across the surface to detect variations. The second method, non-contact methods, utilize techniques like optical profilometry and laser scanning to create a surface map without any physical contact (Table 4.1). Each method has certain advantages, which depend on factors, accuracy, and the nature of the surface.

4.2.3 Parameters of Surface Texture

Surface texture is specified using a variety of parameters, each of which provides unique insights into the surface's characteristics. Ra (average roughness) and Rz (average maximal height) are among the most frequently encountered parameters (Table 4.2). Ra is

TABLE 4.1
Comparison of Surface Texture Measurement Methods

Measurement Method	Contact/Non-Contact	Advantages	Disadvantages
Stylus profilometry	Contact	High-precision, direct physical measurement	Can damage delicate surfaces, slow
Optical profilometry	Non-contact	Fast, no risk of damaging the surface	Less effective on reflective surfaces
Laser scanning	Non-contact	High resolution, versatile	Complex set-up, affected by surface material

TABLE 4.2
Parameters of Surface Texture and Their Significance

Parameter	Description	Significance
Ra	Arithmetic average of surface deviations	Indicates general surface roughness, widely used
Rz	Average maximum height of surface deviations	Useful for understanding peak-to-valley changes and impacts on wear resistance

a general indicator of the uniformity of a surface, as it is the arithmetic average of the absolute values of surface height deviations measured from a mean line over a sample length. Conversely, Rz represents the average of the sums of the height of the highest peak and the depth of the deepest valley over a series of sampling lengths, effectively quantifying surface irregularity [17]. These parameters are essential for quantifying surface texture, enabling the comparison and standardization of various applications.

4.2.4 Surface Texture in Industry Applications

The implications of surface texture are extensive, influencing product design, quality control, and functionality. For instance, in the automotive sector, the surface texture of engine components can impact emissions and fuel efficiency. In biomedical applications, the surface texture of implants influences their longevity and efficacy, as well as their body's acceptance of them. Consequently, the comprehension and regulation of surface texture are not merely technical competencies; they are also strategic imperatives for organizations that aspire to market leadership and product excellence, as illustrated in Figure 4.1.

4.2.4.1 Overview

The foundation for investigating the wide range of surface texture approaches, including their measurement, analysis, and enhancement, has been established. The foundational knowledge of surface texture's significance and fundamental measurement

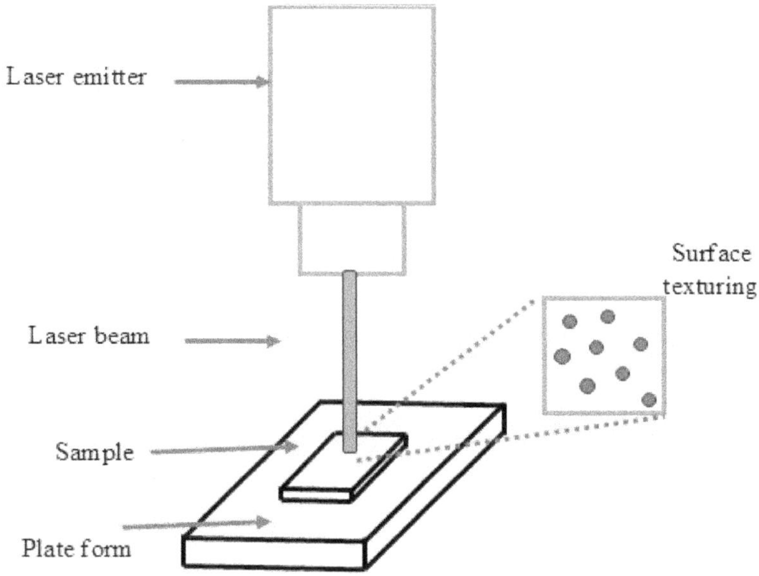

FIGURE 4.1 Basis structure of surface texture.

ideas will be essential for comprehending the advanced procedures and their effects on different industries as we go deeper into specific techniques and applications in the parts that follow.

4.2.4.2 Fundamentals of Surface Texture Techniques

Surface texture is an important feature of material science and engineering, influencing the performance, durability, and usefulness of materials and components across multiple industries. This section delves into the various types of surface textures, measuring methodologies, and critical factors, with the goal of providing a good basis for understanding surface texture approaches

4.2.5 Types of Surface Textures

Surface texture encompasses the variations in the surface's geometry, which can significantly affect a component's mechanical and physical properties, as shown in the Figure 4.2. There are three primary types of surface textures: roughness, waviness, and form errors.

Surface texture is influenced by several key factors, including roughness, waviness, and form errors, each of which contributes to the overall characteristics of a surface. Roughness refers to the fine irregularities that result from the manufacturing process, often due to machining, as shown in Figure 4.3. These irregularities can impact material properties such as friction, wear resistance, and lubrication. Waviness describes the more widely spaced deviations from the nominal shape,

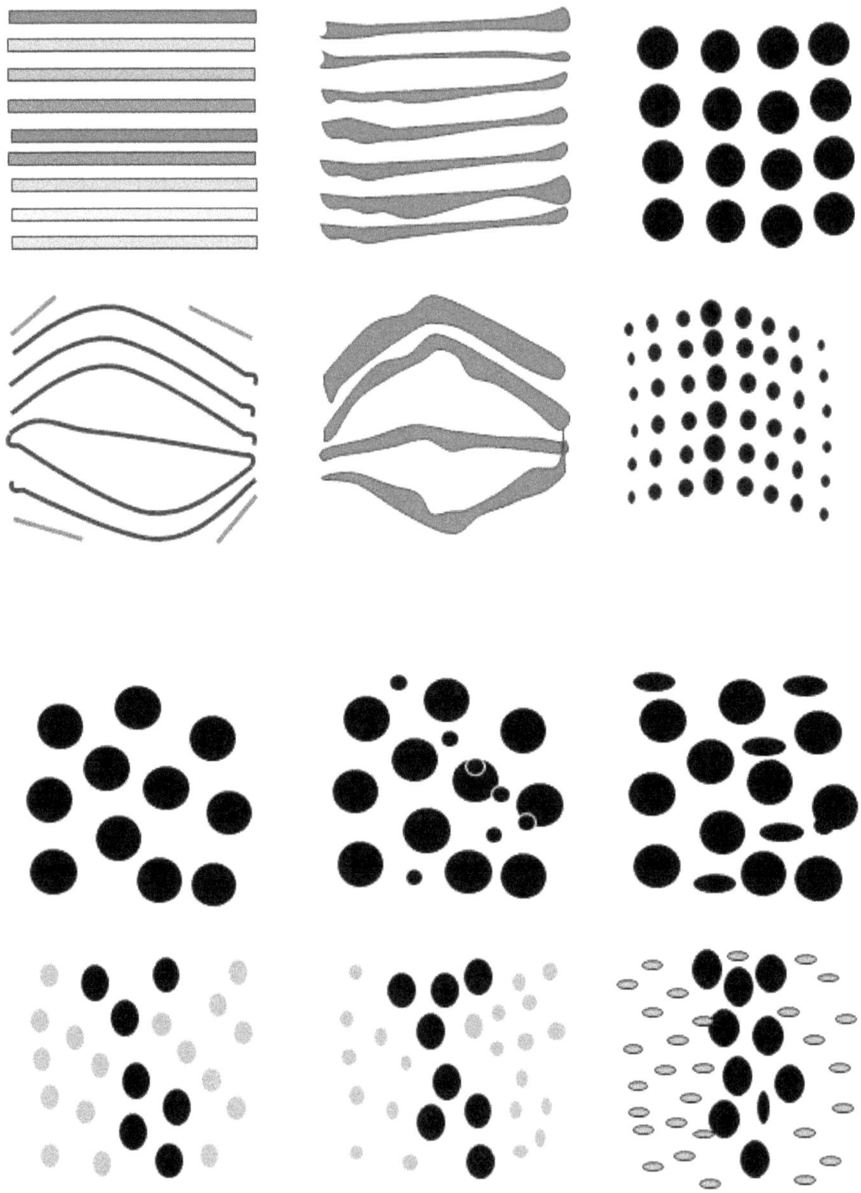

FIGURE 4.2 Types of surface texture.

typically caused by vibrations or deflections during manufacturing. It can affect assembly accuracy and light reflection and contribute to operational noise in moving parts. Form errors represent deviations from the intended geometric shape, including issues like flatness, roundness, and cylindricity errors. These errors are particularly critical in high-precision applications, as they can significantly impact the fit and function of manufactured parts as are given by Table 4.3.

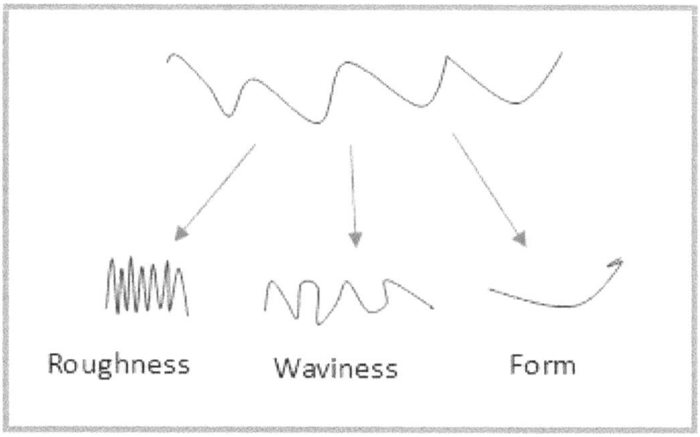

FIGURE 4.3 Surface texture waviness and roughness.

TABLE 4.3
Comparison of Surface Texture Types

Texture Type	Description	Causes	Impacts
Roughness	Fine irregularities from the manufacturing process	Machining, grinding	Friction, wear, and lubrication
Waviness	Wide deviations from the nominal shape	Vibration and deflections in manufacturing	Assembly accuracy and noise
Form errors	Deviations from the intended geometric form	Misalignment, tool wear	Fit and functionality

4.2.6 Preparation of Laser Surface Textures

Laser surface texturing (LST) is considered the most sophisticated technique among existing methods for texturing surfaces [18]. It has a wide range of applications in areas such as tribology, biology, and the wetting properties of materials. as shown in Figure 4.4. LST provides several benefits, such as cost-effectiveness, high productivity, outstanding accuracy, environmental friendliness, and precise control.

These advantages enable the straightforward and efficient creation of diverse texture patterns, such as meshes, chaotic structures, dimples, and grooves. Surface texture fabrication employs three distinct types of ultrafast laser fabrication techniques; laser-induced ablation is a technique that involves the accurate removal of material by harnessing the energy from a laser. The process of material melting through laser interference involves the use of laser beams to selectively heat and melt materials, resulting in the formation of specific patterns. The technique of material shaping and patterning through laser shock-induced surface plastic deformation involves using laser-generated shockwaves to cause plastic deformation on the surface, thereby creating desired shapes and patterns [6].

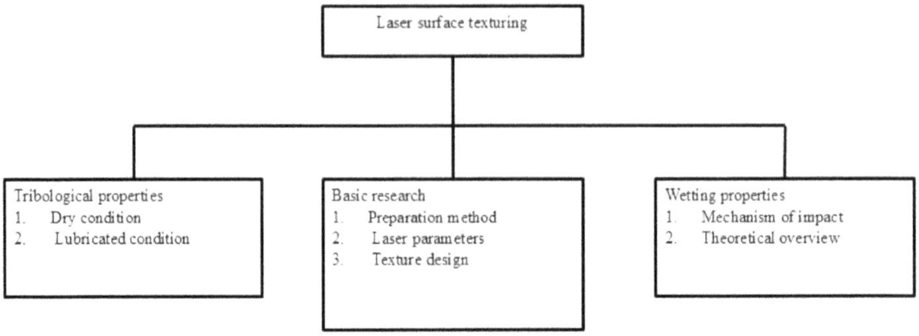

FIGURE 4.4 Method of analysis of laser surface texturing.

4.2.7 Measuring Surface Texture

The measurement of surface texture is crucial for quality control and ensures that components meet their design specifications. Two main categories of measurement techniques exist: contact (stylus) methods and non-contact (optical) methods, as shown in Figure 4.5.

- **Contact (stylus) methods:** In this technique, a stylus mechanically traces the surface, and the vertical movements are recorded to generate a surface profile. It's widely used due to its simplicity and precision, but it can be invasive on delicate surfaces.
- **Non-contact (optical) methods:** These methods involve using light, such as lasers or structured illumination, to measure surface texture. They are advantageous for delicate or soft surfaces and can provide rapid, high-resolution measurements over larger areas, as shown in Figure 4.6.

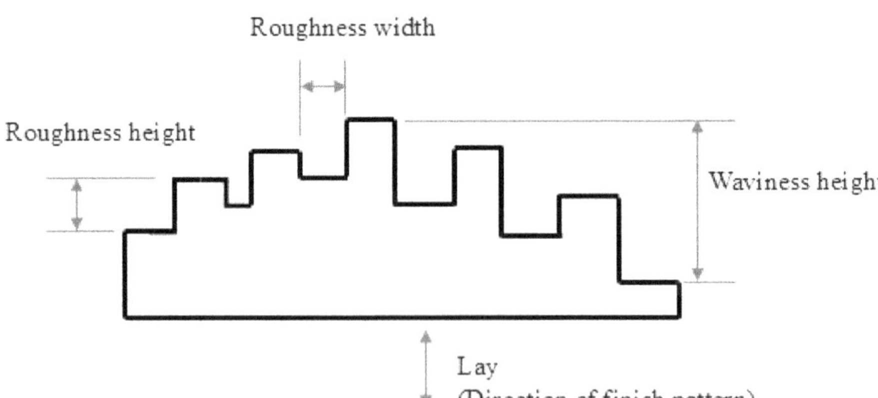

FIGURE 4.5 Types of measurement techniques.

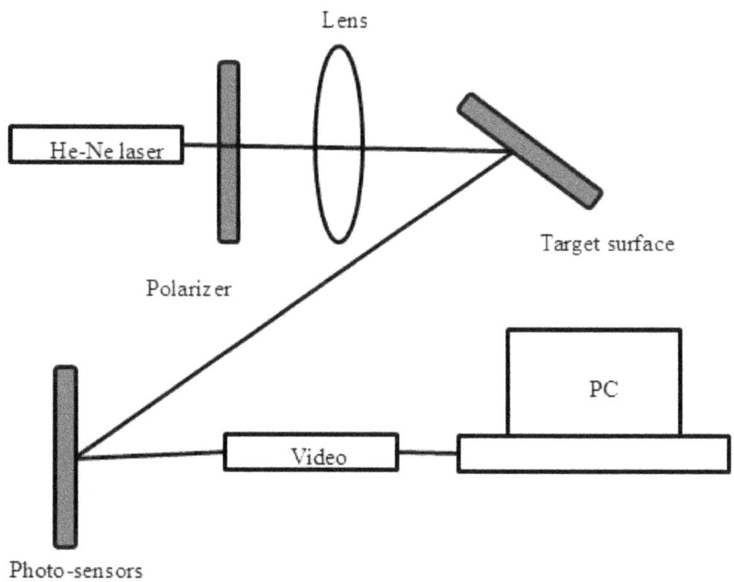

FIGURE 4.6 Non-contact surface-texture-measuring techniques.

TABLE 4.4
Parameters of Surface Texture and Their Significance

Parameter	Description	Significance
Ra (roughness average)	Arithmetic average of surface height deviations	Indicates general surface roughness
Rz (average maximum height)	Average distance from the highest peak to the lowest valley	Captures extreme values, providing a detailed surface texture assessment

4.2.8 Parameters of Surface Texture

Several parameters quantitatively describe surface texture, with roughness average (Ra) and average maximum height (Rz) being among the most common (Table 4.4).

- **Ra (roughness average):** This is the arithmetic average of the absolute values of the surface height deviations from the mean plane. Ra provides a general indication of surface roughness but may not capture extreme values or the distribution of surface irregularities.
- **Rz (average maximum height):** Rz measures the average of the vertical distance from the highest peak to the lowest valley in several sampling lengths. It gives a more comprehensive picture of the surface texture by including extreme values.

Comprehending the foundations of surface texture approaches is crucial for streamlining production procedures, enhancing product quality, and meeting the exacting specifications of diverse applications. Engineers and manufacturers can greatly improve the functionality and dependability of parts and systems by precisely measuring and evaluating surface texture. Taking into account the properties of the material, the manufacturing process, and the operating environment, the measuring method selection and surface texture parameter interpretation should be customized to the particular needs of the application.

4.2.8.1 Techniques for Surface Texture Analysis

In many different industries, surface texture research is essential for assessing and guaranteeing the functionality and quality of materials and components. The principles, benefits, drawbacks, and applications of the three main methods for surface texture analysis—atomic force microscopy (AFM), optical profilometry, and stylus profilometry—are examined in this section.

4.2.8.2 Stylus Profilometry

One of the oldest techniques for determining surface roughness is stylus profilometry. It entails moving a tiny stylus up and down over the surface imperfections while dragging it across a specimen's surface at a steady pace. At a constant speed, the stylus moves up and down over the surface irregularities. These vertical movements are then converted into electrical signals, which are analysed to provide a detailed topographical map of the surface as shown in Figure 4.7. Stylus profilometry parameters are shown in Table 4.5.

Applications

- Automotive and aerospace components for ensuring proper surface finish to reduce friction and wear.
- Manufacturing of medical devices where precise surface textures are required for functional or biological compatibility.

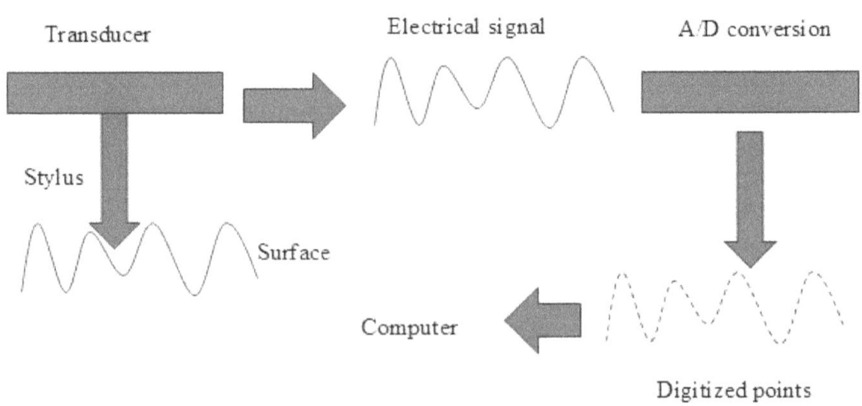

FIGURE 4.7 Stylus profilometry.

TABLE 4.5
Stylus Profilometry Parameters

Parameter	Description
Ra (roughness average)	This is the arithmetic average of the absolute values of the surface height deviations.
Rz (average maximum height)	This measures the average vertical distance between the highest peak and the lowest valley on the surface.
Rq (root mean square roughness)	This is the square root of the mean of the squares of the surface height deviations.

Advantages

- High precision and reliability for a wide range of surface finishes.
- Direct physical contact provides accurate representation of surface irregularities.

Disadvantages

- Contact with nature may damage delicate surfaces.
- Limited by stylus geometry; very fine or deep features may be difficult to measure accurately.

4.2.8.3 Optical Profilometry

Optical profilometry is a non-contact measurement technique that utilizes light to measure surface textures. This method can be divided into several types, including white light interferometry and laser scanning, which use different principles to capture the surface's topography without physical contact, as shown in Figure 4.8.

Applications

- Semiconductor and electronics manufacturing for inspecting surface defects on wafers and PCBs.
- Quality control in precision engineering, where non-contact measurements are preferred to avoid damage.

Advantages

- The non-contact nature prevents damage to the sample.
- Capable of measuring a wide area quickly and providing high-resolution data.

Disadvantages

- Surface properties, such as reflectivity, can affect measurement accuracy.
- More complex and potentially expensive compared to stylus profilometry, as shown in Table 4.6.

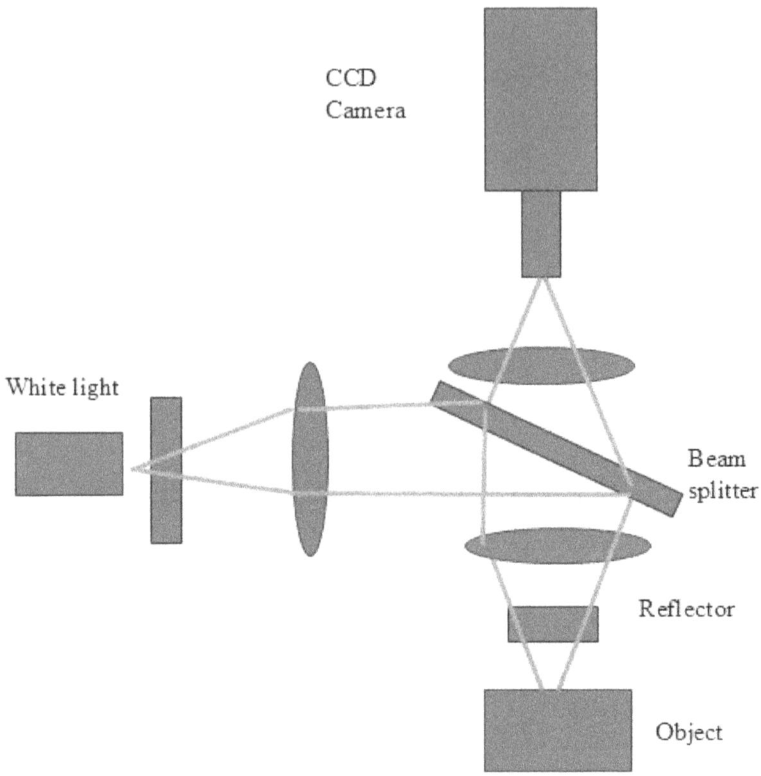

FIGURE 4.8 Optical profilometry.

TABLE 4.6
Comparison of Optical Profilometry Types

Type	Principle	Advantages	Disadvantages
White light interferometry	Uses broadband light to produce an interference pattern	High resolution; suitable for a variety of surfaces	Sensitive to vibrations; complex set-up
Laser scanning	Scans the surface with a laser beam to measure distance	Fast measurements are good for large areas	May have difficulty with very smooth or very rough surfaces

4.2.8.4 Atomic Force Microscopy

AFM represents the pinnacle of surface analysis techniques, providing nanoscale resolution. It operates by scanning a sharp tip mounted on a cantilever, very close to the surface. The tip interacts with the surface atoms, causing the cantilever to deflect. These deflections are measured and used to construct a high-resolution image of the surface at the atomic or molecular level.

Applications

- Nanotechnology and materials science research for studying material properties at the nanoscale.
- Biomedical research for analysing cellular interactions and biomaterials.

Advantages

- Extremely high resolution, capable of visualizing individual atoms.
- Versatile: it can be used in air, vacuum, or under liquids.

Disadvantages

- Slow scanning speed compared to other methods.
- Requires highly specialized equipment and operator expertise, as shown in Table 4.7.

Selecting the appropriate surface texture analysis technique is crucial and depends on various factors, such as the scale of features to be measured, the nature of the sample, the desired resolution, and the specific application requirements. Stylus profilometry offers high precision for a wide range of surfaces but may not be suitable for very delicate or soft materials. Optical profilometry provides rapid, non-contact measurements across large areas, making it ideal for certain industrial applications. AFM, with its nanoscale resolution, is unparalleled for research and development in nanotechnology and materials science.

Incorporating these techniques into the material and component evaluation processes ensures that products meet their specified performance criteria, contributing to the advancement of technology and the improvement of quality in various fields. Each technique has its own unique advantages and limitations.

4.2.8.5 Surface Texture Improvement Methods

Surface texture modification and control are essential for improving component performance, longevity, and functionality in a variety of sectors. This section examines various techniques for enhancing surface texture, including as chemical, mechanical, and laser texturing techniques, outlining their uses, benefits, and drawbacks.

TABLE 4.7
Key Parameters in Atomic Force Microscopy

Parameter	Description
Amplitude	Measures the height difference between peaks and valleys.
Phase shift	Indicates material composition or mechanical properties at the surface.
Lateral resolution	Determines the smallest feature that can be resolved.

4.2.9 Mechanical Treatments

According to Table 4.8, mechanical treatments entail physically changing a material's surface to enhance its texture. Depending on the required surface finish and the properties of the material, common techniques like polishing, grinding, and blasting are appropriate for various applications.

In order to give a material the texture and quality desired, surface finishing techniques like polishing, grinding, and blasting are essential. Polishing creates a flawless, mirror-like finish by removing material from a surface with abrasive particles. This technique is frequently used to improve surface accuracy and attractiveness in metals, optics, and jewellery creation. In contrast, grinding is a crucial production method for the accurate shape and finishing of components because it uses abrasive wheels to remove material from a work piece. Depending on the blast medium and particular parameters used, blasting is the process of projecting abrasive particles against a surface under high pressure. This technique can be used for cleaning, roughening, or finishing. In various industrial settings, each of these procedures is essential to the final goods' functioning and look.

4.2.10 Chemical Methods

Chemical techniques for improving surface texture entail etching or altering a material's surface using chemicals, as shown in Table 4.9. These techniques are especially helpful for materials that are challenging to mechanically manufacture or when a particular micro-texture is needed.

Etching is the technique of using chemical processes to remove material from a surface. This method is frequently used in metalworking, semiconductor fabrication, and microfluidic device manufacturing. In contrast, electrochemical machining (ECM) removes material from the workpiece by using an electrical current in

TABLE 4.8
Mechanical Treatment Methods

Method	Description	Applications
Polishing	Uses fine abrasives to achieve a smooth finish.	Metal parts, optics, and jewellery.
Grinding	Employs abrasive wheels for material removal.	Precision manufacturing and shaping.
Blasting	Projects abrasive particles under pressure.	Surface cleaning, texturing, and preparation.

TABLE 4.9
Chemical Improvement Methods

Method	Description	Applications
Etching	Uses chemicals to remove material.	Semiconductor manufacturing, microfluidics.
ECM	Utilizes electrical current to remove material.	Hard materials have complex geometries.

a conductive fluid. ECM provides exact control over the surface texture, making it especially well suited for complicated forms and hard materials. Both procedures offer special abilities for moulding and purifying materials, making them indispensable in a variety of sectors.

4.2.11 Laser Texturing

The laser beam mainly alters the substrate surface to get a textured pattern for attaining the specific function. The different patterns are used for different applications. In order to reduce wear, oil retention, enhance load and prevent seizure, cross-hatch groove, triangular array, hexagonal array and chevron grooves are used in engine components. The spiral mixed radial groove is used in automobile brakes and its clutches. The herringbone groove uses seals. IN718 and TC4 are used in turbine blades. The parallel grooves are formed in the blade for reducing fretting fatigue, coating adhesion, and improving cooling. In defence firearms, a narrow groove is used for reducing erosion and its fouling. A primary merit of laser texturing lies in its remarkable precision and reproducibility, enabling the fabrication of intricate surface geometries over a wide variety of substrates [13,19]. Nevertheless, the technique also presents certain challenges; it demands advanced instrumentation, expert handling, and stringent control of operational parameters to prevent thermal or structural deterioration of the workpiece. Table 4.10 illustrates laser texturing parameters.

Enhancing surface morphology is crucial for improving component efficiency and extending operational life in various industrial sectors. Conventional mechanical processes provide reliable means of surface modification and can be applied to many engineering materials. Chemical processing routes serve as effective alternatives when substrates are difficult to machine or when sub-micron level modifications are required. Among the available approaches, laser texturing distinguishes itself by enabling finely tuned and function-oriented surfaces designed for next-generation technologies. Each approach possesses distinct benefits, operating limits, and areas of application; thus, the selection of a specific technique should correspond to the intended function and desired surface attributes. Through the appropriate choice of method and accurate adjustment of process parameters, substantial improvements in surface characteristics can be achieved, thereby contributing to superior product durability and operational performance

TABLE 4.10
Laser Texturing Parameters

Parameter	Description	Impact
Pattern	The geometric arrangement of the textured surface.	Influences material interactions like adhesion and lubrication.
Depth	The vertical distance from the peak of the texture to its base [20].	Affects surface area and mechanical interlocking.
Distribution	The spatial arrangement of textured features.	Determines the uniformity of surface properties.

4.2.11.1 Applications of Surface Texture

The deliberate modification and assessment of surface texture are fundamental to improving functionality, reliability, and service life of components used across diverse sectors. In biomedical implants, to enhance bacterial adhesion, healing response, and nutrient flow, micro pits made on TC4 hip joint, stent, and dental screws. In the energy and power system, with help of textured fuel cell, turbine blade, and heat exchanger, heat transfer and condensation efficiency are improved. Also, the texturing is used to enhance service life of fuel cells. In optics and photonic devices, to enhance light trapping, decrease reflection loss, and improve solar efficiency, the textured surface is made on the solar cell, photodetector, and LEDs. In lab-on-chip, the textured surface is applied in DNA analysis and diagnosis to improve the liquid movement.

4.2.12 Friction and Wear Test

Tribological assessments were carried out under ambient conditions using a ball-on-disc tribometer (model UMT-2, Bruker Co., Ltd., USA) operating in a unidirectional rotary sliding configuration. In experimental studies related to biomedical materials, friction and wear tests are conducted under conditions that resemble human body functions as closely as possible. The magnitude of applied load depends mainly on the nature of the component under study. Tests on small parts such as dental screws or miniature joint pins are usually performed under loads of about 5 to 20 newtons, while large joint pairs such as hip or knee replacements are examined under much higher loads, sometimes exceeding 1000 newtons. This variation is necessary to represent the actual contact pressure experienced during routine movement and load transfer inside the body. The sliding motion between the test specimen and the counter face is maintained at relatively low speeds so that it matches the slow movement of natural joints. Laboratory-scale tribometers often work within a range of 5 to 150 millimetres per second, whereas in full joint simulators the motion corresponds to a walking rhythm of roughly 1 cycle per second. To understand long-term behavior, the number of cycles is increased from a few thousand to several hundred thousand depending on the purpose of the study. In certain endurance tests, the sliding motion is continued until a cumulative distance equivalent to years of service is achieved. Since the interior of the human body is a moist environment, the test always includes a simulated body fluid. Researchers often use mixtures such as phosphate-buffered saline, Ringer's solution, or diluted bovine serum because they represent the chemical and lubricating nature of synovial fluid. The entire set-up is kept at a constant temperature near 37°C, and the solution is monitored to maintain a physiological pH. This ensures that any corrosion or chemical interaction taking place during the test will reflect what happens inside the body rather than in open air. The mating material used against the test specimen, commonly called the counter face, is selected according to the implant application. Examples include titanium alloy sliding against zirconia, cobalt–chromium alloy sliding against ultra-high molecular weight polyethylene, or stainless steel sliding against ceramic surfaces. These combinations simulate the real contact conditions of orthopaedic or dental implants. The friction coefficient is continuously monitored throughout the test, and the recorded values usually remain between 0.05 and 0.25 when lubrication is adequate. Any sudden increase is taken as an indication of coating failure or wear film breakdown. After the wear run, the surfaces are examined under microscopes to determine the amount and type of material removal. The wear rate is calculated by measuring the lost volume and dividing it by the applied load and total sliding distance. Microscopic analysis helps in identifying whether the wear is adhesive, abrasive, or corrosive in nature. Most laboratories follow the testing guidelines provided by standards such as ASTM G99, ASTM F732, or ISO 14242, which describe suitable specimen geometry, motion type, and data interpretation methods. The live thermography display presented the spatial temperature distribution resulting from sliding friction, while the software automatically captured the peak interface temperature during each test sequence. The wear was calculated through formula [14].

4.3 MANUFACTURING INDUSTRY

Texturing or surface finish in manufactured components is essential to get a desired function. The small variations in surface are called surface texturing, which includes the topographical features, flaw, waviness and roughness. The micro-crack, hardness, residual stress are also included in the surface. In friction, wear, and tribology, increased surface roughness generally leads to higher friction due to enhanced asperity interaction and abrasive mechanisms, resulting in greater energy dissipation; however, the overall behavior also depends on material properties, contact conditions, and lubrication. But a fine surface generates reduced friction and increases the life part. In the impact of fatigue, stress concentration and crack initiation, the micro-notch gives a stress concentration. The crack forms more easily on a flawed surface. Hence, a better surface finish is needed in specific components. Texturing on sealing and leakage on gasketed joints prevents leakages. In the effect of lubrication, the texturing helps to degrade lubrication. In the view of thermal performance, the controlled surface finish gives an enhanced heat transfer. The surface irregularities affect the tight tolerances in dimensional control. Moreover, on the impact of market and economic footprint, coating industry/surface treatment/finishing are significant impacts. Table 4.11 shows the impact of surface texturing with examples.

4.3.1 Biomedical Applications

Within biomedical engineering, surface texture is a decisive factor determining the functional success of implants and medical devices. It dictates the biological response of surrounding tissues and their interaction with implant surfaces, thereby influencing integration and overall in vivo performance [19]. The texture of an implant surface governs osseointegration, the critical process by which bone tissue forms a direct

TABLE 4.11
Impact of Surface Texture in Manufacturing

Application	Surface Texture Impact	Example
Friction and wear	Reduces friction and enhances durability.	Automotive engine components.
Adhesion	Improves bond strength and reliability.	Coating applications in aerospace.

mechanical and biological bond with an artificial load-bearing implant. This phenomenon is vital for ensuring the longevity and stability of orthopaedic and dental implants. Similarly, in tissue engineering, surface microtopography governs cellular adhesion, proliferation, and differentiation. Through the deliberate modification of scaffold texture, cellular activity and tissue regeneration can be directed to yield superior healing responses and enhanced biological outcomes, as illustrated in Table 4.12.

In aerospace, the roughness is increased due to the functional effect of texturing with drag and flow loss. In the functional effect of heat transfer/thermal performance, the rough surface reduces the effective contact area due to the nonuniform thermal field. In the fretting, wear abrasion, and tribological interactions, the splined surface is influenced by the sliding contact. In addition, rough surfaces wear fast, create debris, and decrease the mating surfaces. The Ra average roughness, Rq mean square roughness, Rz, Rt, Rsk, Rku, and waviness parameters are important parameters to study the surface based on the standard of ISO 25178. The designer needs a special skill to perform the texturing works. Based on the texturing limits, the parameters are selected in turbine blades. The residual stress with a smooth surface gives a big benefit to increase the life of components. To improve the retention of lubricant, decrease friction, and manage film formation, the texturing on automobile parts is an important step. In aerospace, surface texture is crucial because small deviations cause drag penalties, increased fuel consumption, aerodynamic instability, heat transfer inefficiencies, noise, and lower fatigue life. In automotive (powertrain, engines, bearings, transmissions), surface texture impacts friction, wear, lubrication regime, seal performance, component durability, and NVH (noise/vibration/harshness). The significant parameters are more than just Ra—engineers also use Rz, skewness, PSD, areal texture measures, waviness, and surface integrity metrics (residual stress, subsurface damage). The art is in optimizing (rather than minimizing) texture: selecting finishing processes, residual stress treatments, coatings, and metrology to hit performance goals at acceptable cost. Table 4.13 shows applications within the aerospace and automotive industries.

TABLE 4.12
Biomedical Applications of Surface Texture

Application	Surface texture impact	Example
Implant integration	Enhances bone in growth and stability	Orthopaedics and dental implants
Tissue engineering	Influences cell behavior and tissue formation	Scaffolds for tissue regeneration

TABLE 4.13
Aerospace and Automotive Applications

Application	Surface texture impact	Example
Aerodynamic efficiency	Reduces drag and improves fuel efficiency.	Aircraft wings, automotive exteriors.
Heat transfer	Enhances cooling and thermal management.	Engine parts, heat exchangers.

4.3.2 Renewable Energy

In solar applications, texturing is used for enhancing the light absorption, antireflective. In wind turbines, texturing is used for improving the anti-fouling and flow behavior. The texturing is performed on PV cells to reduce reflection and hydrophilic surfaces. For improving the drag and delay flow separation in wind turbine blades, the surface modification is made on the turbine. The scaling, durability, aging, environment robustness, avoiding laser damage, and integrating manufacturing with coatings are the main risks. Laser-based surface modification is steadily gaining importance in renewable energy sectors for improving the efficiency and durability of components. In solar applications, controlled laser patterning on absorber or cover surfaces helps increase the capture of sunlight and minimizes dust accumulation, which reduces output. For wind turbine blades, fine surface structuring by laser treatment has been shown to delay ice formation and lower aerodynamic drag, thereby supporting smoother operation. The same principle is also being applied in heat exchangers, thermal plates, and energy-storage modules to improve bonding, heat transfer, and corrosion resistance. What makes this technique particularly attractive is its precision and cleanliness. Since the process does not depend on acids or solvents, it avoids chemical waste and environmental hazards. In a single operation, it can produce micro- and nano-scale patterns that enhance light absorption, wettability, and mechanical stability of the surface layer. Yet, industrial implementation faces notable difficulties. Treating large components such as solar panels or wind blades demands high-speed scanning systems and robust process control to maintain uniformity. The treated surfaces must also remain stable against sunlight, moisture, dust, and temperature cycling. Additionally, the initial investment and processing cost continue to limit wider use in mass production. For successful large-scale adoption, manufacturers are exploring hybrid techniques that combine laser structuring with thin protective coatings. Long-term field evaluations, optimization of laser parameters, and realistic cost–benefit analysis are essential before this technology can become a standard part of renewable energy manufacturing. These innovations in texturing significantly enhance the durability and productivity of renewable energy systems, as presented in Table 4.14, thereby promoting sustainable energy utilization and performance consistency.

4.3.3 Consumer Products

In consumer-oriented industries, surface texture dictates both tactile and aesthetic qualities, influencing how users perceive and interact with products. From electronic

TABLE 4.14
Renewable Energy Applications

Application	Surface texture impact	Example
Solar panels	Increases light absorption and efficiency.	Photovoltaic cells.
Wind turbines	Improves aerodynamic properties and enhances efficiency.	Turbine blade design.

devices to household items and textiles, the tactile feedback derived from well-engineered surface finishes enhances grip comfort and user confidence. In addition to ergonomics, surface topography shapes the product's visual and functional appeal. Specific texturing strategies are employed to produce desirable optical effects and tactile sensations, harmonizing aesthetics with performance. The synergy between design, texture, and function determines product success, customer satisfaction, and long-term usability. Collectively, surface texture is a decisive factor uniting aesthetics, safety, and operational efficiency across consumer product categories [21,22].

The industrial significance of surface texturing is extensive, extending its influence from high-performance aerospace components to consumer devices. The consistent optimization of surface morphology continues to enhance efficiency, durability, and sustainability across sectors, as summarized in Table 4.15.

TABLE 4.15
Consumer Product Applications

Application	Surface texture impact	Example
Touch sensation and grip	Enhances usability and comfort.	Smartphone cases, tool handles.
Aesthetics and function	Influences on consumer preference and usability.	Textured fabrics, home appliances.

4.4 CASE STUDY ON THE ENHANCEMENT OF AUTOMOTIVE COMPONENT LONGEVITY THROUGH LASER SURFACE TEXTURING

The automotive industry persistently explores innovative technologies to elevate the reliability and operational life of its components. Among the notable advancements is the utilization of laser texturing, a precise surface engineering technique aimed at enhancing tribological performance. This case study examines its successful application in improving the lifespan of piston rings—critical elements in maintaining engine efficiency and sealing performance. Piston rings operate under extreme thermal and mechanical stresses, making them susceptible to accelerated wear. In order to overcome the recurring problem of frictional wear in engine components, a car manufacturer introduced a new approach by applying laser surface texturing on piston rings. This modification aimed to create tiny grooves on the surface so that lubricant could spread more evenly during engine operation.

Process Description

The process involved directing a concentrated laser beam over the ring surface to remove small amounts of material in a planned pattern. As a result, a series of narrow, shallow grooves were formed. These micro-grooves acted like small pockets where engine oil could collect and stay in place. During movement, the trapped lubricant helped separate the sliding surfaces, which in turn reduced friction, controlled surface temperature, and prevented early surface damage.

Observed Performance

When the textured rings were tested under normal operating conditions, the results were clear. The wear on the laser-treated rings was reduced by nearly one-fourth compared with untreated parts. Because the lubricating film lasted longer, the components required less frequent servicing. The smoother motion also meant less energy was lost to friction, which gave a slight but measurable improvement in fuel efficiency.

Practical Significance

The success of this effort encouraged further investigation into how the same principle could be applied to other moving parts in the vehicle, such as crankshafts, bearings, and gear components. Each of these relies heavily on stable lubrication and surface integrity. By improving both factors, laser texturing offers a dependable way to extend the service life of critical parts and reduce mechanical losses. Over time, such improvements can contribute to more reliable engines, longer maintenance intervals, and better overall system performance within the automotive sector.

4.5 THE IMPORTANCE OF CONTINUOUS INNOVATION IN SURFACE TEXTURE TECHNIQUES

The continuous innovation in texturing is a basic step because it is an integration and platform for sharing the innovation and new strangles. Such integration not only accelerates data interpretation but also enhances predictive accuracy, reducing experimental uncertainty. These developments pave the way for next-generation design frameworks in which surface characterization, modelling, and improvement occur concurrently and intelligently [23,24].

4.5.1 Future Prospects

The outlook for surface-texture technologies remains highly encouraging. The industrial movement towards sustainability and resource efficiency will stimulate the emergence of environmentally responsible texturing methods and the increased utilization of renewable or recyclable materials. In parallel, the extension of texture-engineering principles into nanotechnology and biotechnology promises breakthroughs in biomedical devices, renewable-energy systems, and micro-mechanical actuators. Continued progress in metrological instruments and analytical algorithms will deepen our understanding of the interrelationship between surface geometry and functional performance. As a result, future materials and components will be designed with unprecedented precision, offering greater wear resistance, improved energy efficiency, and superior adaptability to specific operating environments [20].

In essence, sustained research and interdisciplinary collaboration will ensure that surface-texture innovation continues to advance the frontiers of science and technology, enabling the creation of materials and systems that are not only more efficient but also more sustainable and responsive to global challenges.

4.6 CONCLUSION

Laser-based patterning of surfaces is steadily carving out a central place in modern engineering practices, with applications spreading into numerous sectors. What distinguishes it from older surface treatment methods is its ability to sculpt the microscopic contours of a material precisely, while leaving the underlying substrate unchanged. By adjusting laser parameters—such as pulse timing, beam intensity, scanning paths, and wavelength—engineers can embed surface geometries tailored to a variety of functions, from friction control to optical absorption to corrosion resistance.

Within the transportation sector, this method has already demonstrated tangible gains. Texturing elements like piston rings, camshafts, and gear faces produce microscopic grooves or dimples that trap lubricants and limit direct metal contact. As a result, wear rates drop and fuel efficiency climbs—offering a compelling argument for adoption in automotive powertrains. In aviation, textured surfaces on compressor vanes, cooling channels, and structural parts help manage heat flow, reduce stress concentrations, and optimize airflow around components, contributing to both durability and performance under demanding conditions.

When it comes to renewable energy technology, laser surface patterning is also proving its worth. In solar systems, carefully arranged micro- and nano-features can reduce surface reflectivity and enhance photon capture by scattering light inward. On wind turbine blades, tailored textures discourage ice formation and improve how air flows across the blade's surface—delivering gains in reliability and energy output. Elsewhere, in thermal collectors, heat exchangers, or energy storage modules, surface structuring improves heat transfer interfaces, adhesion of coatings, and resistance to environmental degradation.

In medical and life science fields, laser texturing enables implants to mimic natural tissue surfaces, promoting better integration into bone or soft tissue and reducing infection risk. And in electronics or battery systems, this technique enhances interfaces—boosting contact area for conductive paths, improving mechanical anchoring of coatings, and helping thermal dissipation in tight spaces.

That said, the technique is not without its hurdles. Implementing patterning over large surfaces—such as full solar panels or full-length turbine blades—with uniform precision remains challenging. The up-front cost of ultra-fast lasers, beam control systems, and precise motion set-ups is high. Ensuring that the textured surfaces retain their functional performance in real-world conditions—exposure to moisture, UV, temperature swings, abrasion—is demanding. And achieving consistent results across curved, irregular, or complex geometries further complicates adoption.

To make industrial use practical, a few strategies are emerging. Integrating laser texturing with complementary thin coatings or treatments can offer hybrid performance, alleviating durability constraints. The optimization of scanning patterns, beam shaping, and motion control is helping accelerate processing speeds. Field testing over realistic lifetimes is essential to confirm that performance gains persist. Detailed cost–benefit studies—factoring in lower maintenance, longer lifetimes, and energy savings—are needed to justify investment.

REFERENCES

[1] Youqiang Xing, Mingyu Zhu, Ze Wu, Zhao Li, Shuowei Bai, Kedong Zhang, Lei Liu, High-temperature tribological properties of Si3N4/TiC ceramic with bionic surface textures and DLC coatings, Tribology International 186 (2023) 108648.

[2] Ying Meng, Jianxin Deng, Dongliang Ge, Jiaxing Wu, Wei Sun, Ran Wang, Surface textures fabricated by laser and ultrasonic rolling for improving tribological properties of TiAlSiN coatings, Tribology International 164 (2021) 107248.

[3] Ankit Kumar, Rahul S. Mulik, Improving tribological behavior of titanium nitride (TiN) hard coatings via zirconium (Zr) or vanadium (V) doping, Tribology International 189 (2023) 108997.

[4] Ajith Kurian Baby, PK Rajendrakumar, Deepak Lawrence K, Comparative study on the tribological performance and wear behaviour of hypereutectic Al–Si cylinder liner surfaces textured using micro-drilling and mechanical indentation-based approaches, Wear 540–541 (2024) 205240.

[5] Yichen Bao, Jianxin Deng, Shenghan Cao, Kexin Ma, Zhihui Zhang, Yang Lu, Laser micro-cladding in situ forming textured surface to improve the tribological performance, Wear 550–551 (2024) 205422.

[6] Peixuan Lei, Peilei Zhang, Shijie Song, Zhenyu Liu, Hua Yan, Tianzhu Sun, Qinghua Lu, Yu Chen, Victor Gromov, Haichuan Shi, Research status of laser surface texturing on tribological and wetting properties of materials, Optik—International Journal for Light and Electron Optics 298 (2024) 171581.

[7] Tianchang Hua, Yongsheng Zhang, Litian Hu, Tribological investigation of MoS2 coatings deposited on the laser textured surface, Wear 278–279 (2012) 77–82.

[8] G. Mariani, Selection and application of solid lubricants as friction modifiers, in: R. Leslie, Rudnick (Eds.), Lubricant Additives: Chemistry and Applications, CRC Press, 2003, pp. 171–201.

[9] Leslie R. Rudnick, "Lubricant Additives—Chemistry and Applications", CRC Press, 2017.

[10] Youqiang Xing, Jianxin Denga, Xingsheng Wang, Rong Meng, Effect of laser surface textures combined with multi-solid lubricant coatings on the tribological properties of Al2O3/TiC ceramic, Wear 342–343, 15 November (2015) 1–12.

[11] T.C. Hu, Y.S. Zhang, L.T. Hu, Tribological investigation of MoS2 coatings deposited on the laser textured surface, Wear 278–279 (2012) 77–82.

[12] F. Gustavsson, F. Svahn, U. Bexell, S. Jacobson, Nanoparticle based and sputtered WS2 low-friction coatings-differences and similarities with respect to friction mechanisms and tribofilm formation, Surface and Coatings Technology 232 (2013) 616–626.

[13] Rashi Tyagi, Ranvijay Kumar, Nishant Ranjan, "Tribological Aspects of Additive Manufacturing", CRC Press, 2024.

[14] Jiawei He, Wenyuan Zhou, Shuqun Chen, Anran Wu, Yuqing Zhou, Yanhui Chen, Jinshu Wang, Hongyi Li, Tribological properties of MoS2 nano-flowers supported by porous alumina aperture array, Tribology International 161 (2021) 107093. https://doi.org/10.1016/j.triboint.2021.107093

[15] Amanov, Auezhan, Young-Sik Pyun, A comprehensive review of nanostructured materials by ultrasonic nanocrystal surface modification technique, The Journal of Engineering 13 (2015) 144–149. https://doi.org/10.1049/joe.2015.0067

[16] Ravi K. Dwivedi, Premanand S. Chauhan, Avadesh K. Sharma, Madhavi Singh, Anupma Agarwal, "Biomedical Implants—Materials, Design, and Manufacturing", CRC Press, 2024.

[17] Carolina Fernandez-Gonzalez, Bopeng Zhang, Antonio Dominguez-Ramos, Raquel Ibafez, Angel Irabien, Yongsheng Chen, Enhancing fouling resistance of polyethylene anion exchange membranes using carbon nanotubes and iron oxide nanoparticles, Desalination 411 (2017) 19–27. https://doi.org/10.1016/j.desal.2017.02.007

[18] Na Guo, Xiaomin Mao, Tao Liu, Xinrui Hui, Zhangwei Guo, Bowen Tan, Kaiyuan Shao, XiaoFeng Li, Zhenshun Zeng, Corrosion mechanism of copper in seawater containing the bacterial pyomelanin with redox activity, Corrosion Science 204 (2022) 110407. https://doi.org/10.1016/j.corsci.2022.110407

[19] Rayan Coutinho, Loan D. Marinescu, Methodology to compare 3-d and 2-d parameters for the optimization of hard turned surfaces, Machining Science and Technology 9 (3) (2005) 383–409. https://doi.org/10.1080/10910340500196330

[20] Shuai Zhang, Houfu Dai, Atomic-scale simulations of constant-load scratching of gallium nitride under monolayer graphene lubrication, Diamond and Related Materials 142 (2024) 110731. https://doi.org/10.1016/j.diamond.2023.110731

[21] N Jeyaprakash, CH Yang, KR Ramkumar, Microstructural, mechanical and wear behaviour of Inconel-718 produced through laser-powder bed-fused additive manufacturing, Materials Science and Technology 37 (3) (2021) 326–337. https://doi.org/10.1080/02670836.2021.1893457

[22] S Nirnjalkumar, P Gopal, Design and development of an automobile radiator test rig and development of heat transfer correlations for nanofluid solutions, Journal of Chemical and Pharmaceutical Sciences 8 (6) (2015) 297–299. (ISSN: 0974-2115).

[23] N Jeyaprakash, IA Alnaser, Y Cheng, SS Karuppasamy, Effect of heat treatment on microstructure and mechanical characteristics of Laser Powder Bed Fusion (LPBF) produced CuCrZr alloy, Tribology International197 (2024) 109826. https://doi.org/10.1016/j.triboint.2024.109826

[24] B Kumaragurubaran, P Gopal, T Senthil Kumar, M Prasanna Mugunthan and NH Mohamed Ibrahim, Optimization of turning parameters of EN-9 steel using design of experiments concepts, International Journal of Mechanical Engineering and Robotics Research 2 (3) (2013) 182–190.

5 Surface Heat Transfer Analysis Using Laser Surface Texturing

G. Prabu, Jeyaprakash Natarajan, and Che-Hua Yang

5.1 INTRODUCTION

Power plants and other heat-transfer-based industries depend on the efficiency of the boiler, condenser, and other supporting components [1,2]. However, the performance of the components is subject to many factors. The transfer of thermal energy from the heat source to the medium is done through the surface of the contact material. It reveals that the surface plays an important role in controlling the heat transfer. The improvement of heat transfer is done by different techniques [3]. Surface modification is an effective technique to improve the efficiency of heat transfer under difficult conditions [4,5]. The different patterns scales such as micro, nano, and a combination of micro and nanostructure are used to modify the surface [6–8]. The long-term performance of heat transfer material with modified patterned surfaces is rarely reported. The appropriate size of micro- and nano textures on the surface of the heat transfer system is important for ensuring the formation of a continuous and efficient system. This texture plays a key role in enhancing the overall performance of the heat transfer system. Laser has the capability to create micro- and nano-scale structures on the heat transfer surface. The texture is produced through the process of machining by using laser source.

The surfaces that have been modified through laser texturing have been designed to have specific properties such as large surface area, high surface roughness, and ability to spread the heating medium such as liquid, superior wettability, and good capillarity. These unique properties allow the surfaces to improve the process's efficiency significantly. The intense laser beam produces a significant amount of thermal energy. It causes the metal to evaporate through the interaction of phonons and electrons. It leads to producing the symmetric micro and nanostructure. The processes of laser texturing involve melting and evaporating of metal. These processes can be regulated by adjusting the parameters of the laser [9, 10]. The process of laser texturing is becoming increasingly significant. Large number of researchers are working on developing various approaches in this area. The detailed study of the existing literature on the heat transfer of laser texture surfaces is important and relevant to industrial needs.

DOI: 10.1201/9781003542087-5

5.2 NEED FOR LASER TEXTURE TREATMENT IN HEAT TRANSFER SYSTEM FOR BOILING ENHANCEMENT

The process of boiling is intricate and involves various factors such as the properties of the fluid, contact surface, and the atmospheric conditions. It is characterized by the transition of a fluid into vapor [11]. The energy needed for bubble nucleation in the heat transfer system is crucial in determining the characteristics of the nucleation process. The laser energy controls the texture formation on the specimen surface [12]. Schematic representations of linear groove patterns, square patterns, and scanning electron microscopy images of linear grooves are shown in Figure 5.1. When a fluid is heated, homogeneous nucleation starts within the fluid. Moreover, heterogeneous nucleation initiates at the interface between the heat transfer contact surface and the fluid medium [13]. A solid surface from the heat transfer system is placed in a large body of stagnant fluid medium during the heat transfer process. On the other hand, flow boiling involves moving a volume of liquid over the heated surface, either an open channel or a closed channel [14]. When exposed to an intense heat flux, a vapor blanket forms, enveloping the entire surface. This effectively hinders the transfer of heat to the surrounding liquid and results in the accumulation of thermal energy on the heat transfer surface. The abrupt increase in temperature not only is undesirable but also poses a significant and hazardous risk.

The critical value of heat flux also known as the burnout heat flux: refers to the maximum heat flux at which the heat transfer system should operate. Operating

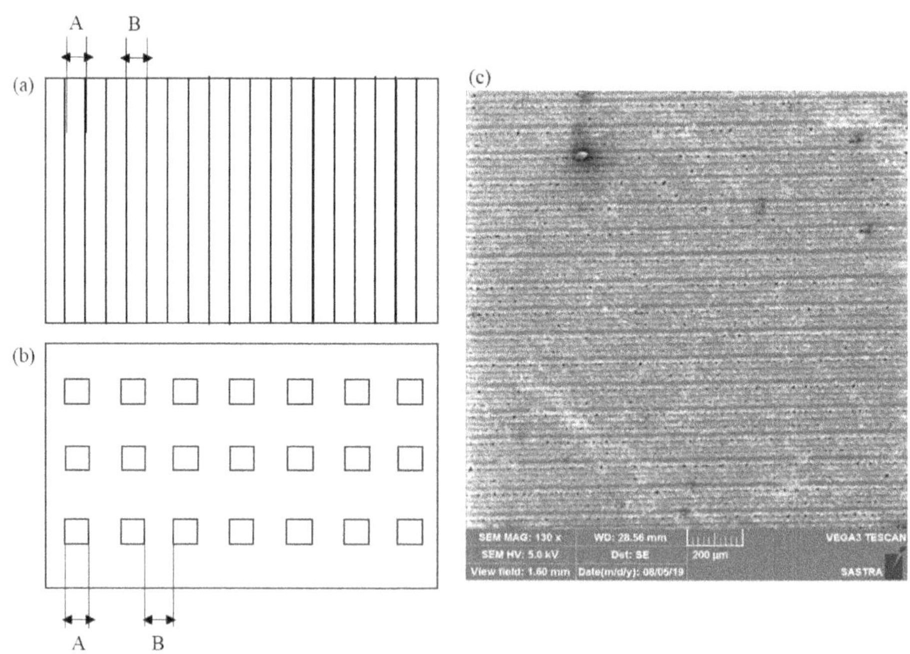

FIGURE 5.1 Schematic representation of laser texture (a) linear grooves pattern, (b) square patterns, and (c) scanning electron microscopy image of linear grooves.

beyond this limit can lead to adverse effects. The critical value of heat flux is an essential parameter that plays a significant role in the heat transfer process [15,16]. A significant amount of work has been dedicated to preventing the build-up of congestive heat transfer system failure and thereby guaranteeing the safe operation of the systems [17–19]. In the past, numerous correlations and models have been suggested to describe the behavior of fluids and surfaces in diverse conditions [20–22]. In order to create a cavity of a specific size on a heated wall, it is necessary to supply at least the minimum amount of superheat to the wall [23–25]. Once the process of nucleation begins, a formation of cavities on the heat transfer surface can be triggered. The inclusion of nanoparticles or other substances in the standard fluids can improve the heat transfer characteristics of the fluids, as mentioned in references [26–28]. Enhancing the heat transfer performance is possible not only with water, but also with other types of fluids [29]. Enhancing heat transfer system performance through surface modification is a passive technique that can be implemented across various scales [30–32].

The change of the heat transfer surface induces different surface-related properties like wettability, surface area, nucleation site density, roughness, capillarity, and permeability [33,34]. The process of altering the surface of a material has an impact on the characteristics and behavior of the fluid medium. These changes, in turn, have an effect on the efficiency of the boiling process. This relationship between surface modification and boiling performance has been studied and documented in previous research [35,36]. One of the key factors that significantly affects the characteristics of heat transfer is wettability. The ability of a surface to be wetted depends on its physical features, chemical properties, and the molecules interacting at the boundary between the solid and liquid [37,38]. Increasing the heat transfer during the process of pool boiling is essential for numerous industrial and technological applications. The performance of the application in the boiling process should not be affected by any constraints of the improvement method. The suggested method for improvement should not impact the system's performance and safety.

It requires a simultaneous enhancement of heat transfer coefficient and critical heat flux. The heat transfer phenomenon, in addition to immersion cooling, is utilized across various industries. Its applications span heat exchangers, space-cooling systems, nuclear power generators, petrol-chemical processing, and cryogenic, power conversion, and desalination industries [39]. Hence, an effective method is required to modify the heat transfer surface to enhance the evaporation property. Laser material processing is one of the reliable methods that can modify the surface topography at the micron and nano levels [40].

5.3 CHARACTERIZATION OF LASER-TEXTURED SURFACES

The laser is used to create different textures such as nano-sized periodic surface textures and micron-sized textures. The confocal microscopy is used to measure the roughness of the laser-textured specimen and the untextured specimen. Different structures such as rectangular, circular, square, trapezoidal, and waviness are used to study the heat transfer behavior. In addition, the texture is made on three different specimens to confirm the repeatability of the heat transfer output. The roughness of the texture surface is higher than the untextured surface [41]. In addition, the surface area of the heat transfer surface increases. Moreover, the production of the vapor bubbles increases when the roughness increases. As a result, the nucleation formation density increases, leading to a higher heat transfer rate. The texture depth and its roughness increase with the increase of laser fluence. The laser parameter affects the fluence setting during the texturing process, which lead to influence on the formation of high-depth texture. The unit cells prepared through the laser processing are in the range of 50 μm × 50 μm to 200 μm × 200 μm. Moreover, the pitch is varied from 5 μm to 30 μm in the x and y directions. These laser texture processes overcome the drawback of the other process such as dimensional accuracy, repeatability, and processing time. The textures are in the form of micropillar, micro dimples, micro wavy paths, nano-sized wavy paths, and so forth.

The concave and convex textures are also used to modify the wettability of pain surfaces. Hence, the micropillar, micro dimples, and other textures are made with the concave and/or surfaces to improve its performances. The texture pattern becomes a three-dimensional micropattern after increasing the depth of texture during the laser texturing process. In addition, the wettability, roughness, and other parameters are not likely maintained if the laser parameter is changed. The increasing or decreasing depth of the texture also affects the heat transfer property of the surface. Hence, each parameter affects the wettability, roughness, heat transfer, contact angle, and so forth. The checkerboard patterns such as square textured patterns are used to increase the depth of the textures. However, the roughness of the textured surface is maintained constant. In addition, the number of square patterns is increased to retain the same roughness value. Nanoforest structure is used to form the checkerboard pattern due to higher heat transfer and cooling rate. The increase in the heat transfer rate of checkerboard pattern is nearly 10% higher than that of the rectangular texture patterns. The percentage of heat transfer changes when the contact medium is changed due to a significant change in wettability of the heat transfer surface. The major concept for increase in the heat transfer on the texture surface is the presence of edges.

The increase in the number of edges in the texture region directly increases the heat transfer rate. This phenomenon is known as the edge effect to improve the heat transfer rate in the texture region. The inter-structural region in the laser texture surface is the main source of control for vapor formation during the heat transfer process [42]. The heat transfer rate increases when the inter-structural region increases [43]. It acts as a channel and allows the vapor to escape from the heat transfer region. As a result, the vapor is released from the heat transfer region and enhances the boiling rate. Hence, the presence of texture edges and the vapor-escaping region are used to enhance the heat transfer rate. The depth of the laser checkerboard pattern and high roughness explain the improvement of heat transfer rate over the polished specimen surface. Moreover, increasing the roughness of the specimen is carried out through the laser inscribing method such as the formation of nanotexture. The nanotexture improves the roughness of the specimen higher than that of the laser checkerboard patterns. The nano-sized square pillars are considered in the laser inscribing patterns. Each nano-sized square pillar acts as an effective edge to improve the heat transfer process.

5.4 SURFACE MORPHOLOGY OF THE HEAT-TRANSFERRED LASER TEXTURE REGION

The micro- and nanometer-sized regimes form between the two linear grooves, waviness patterns, and trapezoidal patterns. The texture is made on the Ti-6Al-4V specimen and its chemical composition is shown in Table 5.1. The heat transfer rate of the textured surface in the nanofluid is superior than that of the ethanol. The percentage increase in the heat transfer rate is above 50% and below 60% compared to the polished surface. The surface morphology of the laser-textured heat transfer surface is analyzed by using scanning electron microscopy. The surface morphology of different patterns are evaluated. The confocal microscopy is used to study the 3D topography of the laser-textured surface (Figure 5.2).

TABLE 5.1
Chemical Composition of Ti-6Al-4V

Elements	Ti	Al	V	Fe	Others
Wt (%)	Bal.	5.92	3.86	0.40	< 0.3

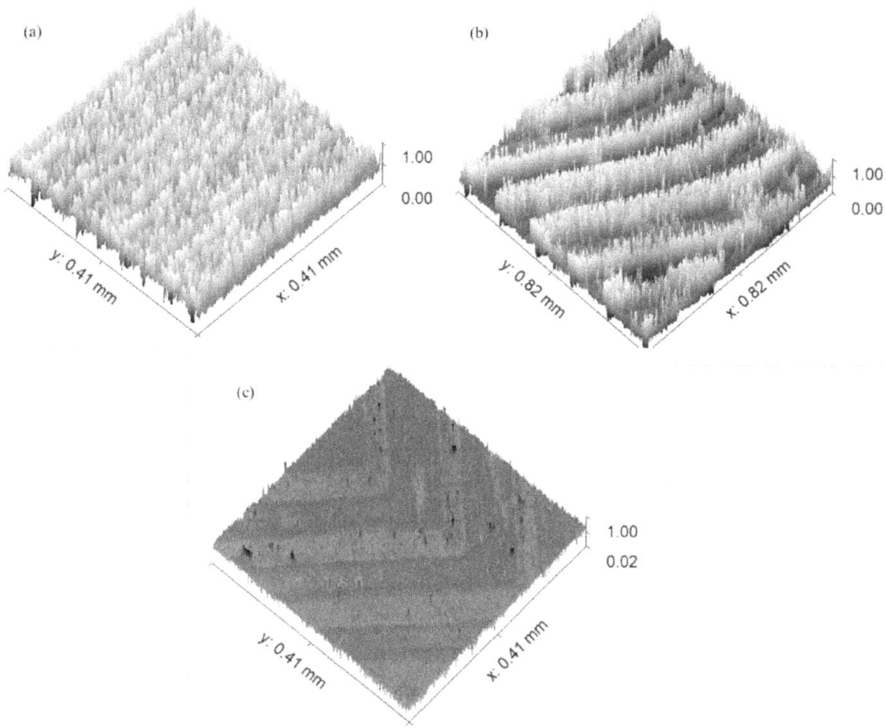

FIGURE 5.2 3D topography of laser texture (a) linear grooves pattern, (b) waviness patterns, and (c) trapezoidal patterns.

The overall surface area of the waviness texture is higher than the other two textures. However, the width of the waviness texture is lower than the width of the trapezoidal texture. Moreover, the high-depth square pillar texture has a longer microscale edge than the lower-depth square pillar texture. The heat transfer depends on vapor bubble formation in the heating region. In addition, the resistance to the formation of the vapor bubble in the textured surface is less when compared to the plain untextured surface [44]. As a result, the efficiency of the formation of vapor bubbles in the textured surface is larger than that of the plain untextured surface. The micron-sized layer is evaporated at the heat transfer nucleation region.

5.5 DROPLET FORMATION ON THE LASER TEXTURE SURFACE

Droplet evaporation is one of the natural processes in the day-to-day life. The bioscience involved in the evaporation of droplets is one such interesting subject [45,46]. The temperature, pressure, roughness wettability, surface chemistry, surface homogeneous property, heating rate, the chemical composition of a droplet, and so forth involve evaporating a droplet from the heat transfer surface [47–49]. The homogeneous surface induces heat to the droplet through the elongation of the contact line between the surface and the droplet [50–52]. In addition, heat transfer is an essential process in the boiling power plant, chemical industry, and nuclear reactors and in power electronics, fuel cells, computer chips, and spacecraft electronics. A thermal energy of 10^3 to 10^5 W/cm^2 must be removed from the heat flux devices to maintain its operation [53,54]. Existing heat exchangers are unable to remove the required heat, even at maximum efficiency [55–57]. To overcome this drawback, the two-phase heat exchanger is used. It is a safe and stable heat exchanger to produce reliable cooling effect in the heat flux system. However, the heat transfer efficiency is affected by three factors such as superheating capability of the heat transfer surface, heat transfer from surface to liquid medium, and maximum heat flux value.

The maximum heat flux controls the transformation of nucleate heat transfer to film boiling-based heat transfer [58–60]. The two-phase system slows down the transfer of heat, which causes the temperature on the surface of the heat exchanger to rise [61–63]. The improvement in the heat exchange is focused on the modification of the surface at micrometers and nanometers [64]. Wettability, roughness, and porosity of the surface control the droplet formation on the heat transfer surface. It stabilizes the boiling conditions and improves the boiling efficiency. The bubbles in the laser texture region experience the hydrophobic surfaces. The nucleation frequency and the departure bubble diameter are used for analyzing the behavior of departing bubbles. The bubble diameter at the laser-treated surface is smaller than that of the untreated surface when the laser treatment is performed at 37 kW/m^2 energy. Moreover, the nucleation frequency of the bubble is higher than that of the untreated surface. The bubbles form and then depart from the laser texture surface repeatedly due to the cyclic process of bubble nucleation. This cyclic process of bubble formation is greater than the untextured surface. The trend of bubble formation on the textured surface increases with the increase of heat flux. The heat flux against the contact angle and the roughness is shown in Figure 5.3. The performance of the laser texture surface is measured by calculating the removed heat flux during the heat

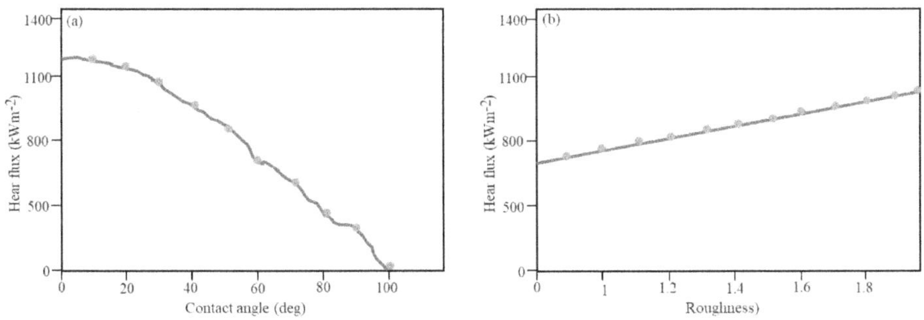

FIGURE 5.3 Graph between the heat flux and (a) contact angle and (b) roughness.

transfer process [65–67]. Superior nucleation rate and high nucleation region density of bubbles are the important reasons for the improvement of performance of laser texture surfaces. The untextured surface exhibits low nucleation temperature due to decreasing in the bubble nucleation frequency is a reason for its low performance.

5.6 ENHANCEMENT OF EVAPORATION DURING HEAT TRANSFER PROCESS

The evaporation of the liquid is one of the natural phenomena which occurs in day-to-day life [68–70]. Droplet cooling, biosciences and inkjet printing are some of the examples of droplet evaporation [71–73]. Moreover, the chemical composition of the droplet, temperature, pressure, roughness of the contact surface, and wettability between the droplet and the contact surface affect the droplet evaporation. In some of the cases, the wettability changes with respect to the location of the contact region. The heterogeneous surface texture affects the evaporation of liquid due to changes in the heat transfer from the surface to the contact liquid region [74]. The laser texture surface produces a heterogeneous surface, which increases the surface area than the plane surface. It leads to an increase in the heat transfer from the surface to the contact medium. The hydrophilic laser texture surface exhibits a higher heat transfer rate than the hydrophobic laser texture surface due to the self-propelled contact material. This type of heat transfer occurs at the heterogeneous laser texture wetting patterns [75]. It improves the evaporation of the contact medium through the various types of heat transfer. The formation of nanostructures on the surface through the laser texturing process is superior to the other methods such as oxidation, chemical, ion implantation, sol–gel, and carburization [76–78].

Ultra-short pulse laser is used to fabricate this type of nanotexture on the surface to enhance the heat transfer. The laser pulse produces the periodic structure on the surface region and increases the roughness of the surface. The production of periodic structures through a laser system has been proved in different studies [79–81]. As a result, the wettability of the surface significantly increased on the laser-irradiated surface. The formation of superhydrophobic on the SS material is controlled by the laser texture surface [82]. The adhesion of the contact medium is reduced at the laser texture surface when compared to the untextured surface. It significantly changes the contact angle between the laser texture surface and the contact medium. Ta et al. [83] analyzed the heat transfer property of laser-textured SS material and found that the laser-textured specimen exhibited higher heat transfer efficiency than that of the untextured specimen. Zaitsev et al. [84] evaluated the heat transfer property of the laser-textured Al material. The results revealed that the heat transfer depended on the wetting and roughness properties of the material. These properties depended on the surface morphology of the tested specimen. The laser-textured process changes the surface property of the material.

The heat transfer property varies with respect to the morphology such as homogeneous and heterogeneous surfaces. The laser texturing process is an easy way to modify the surface property into a homogeneous or heterogeneous surface. The heterogeneous laser texture can be induced on homogeneous or heterogeneous surfaces. Similarly, a homogeneous laser texture can be formed on homogeneous or heterogeneous surfaces. As a result, the contact angle of the laser texture surface differed from the untextured surface. The Gaussian distribution of laser energy removes the material from the plain surface and produces the periodic texture surface. The intensity of the laser beam is controlled to obtain the required texture surface. The laser pulse energy is varied in terms of J/cm^2 to increase or decrease the deposition of thermal energy on the testing surface [85,86]. The laser pulse width is changed in the unit of picosecond at a constant wavelength. The scanning speed in mm/s and repetition rate in kHz are also the process parameters that affect the texturing process. The laser irradiation occurs throughout the homogeneous surface to produce a homogeneous laser texture surface. However, the heterogeneous laser texture is formed when the laser irradiation occurs at a particular location alone. The wettability and roughness of the laser-textured and untextured specimens are used to evaluate the heat transfer behavior of the material.

The presence of the ripple structure on the laser texture surface increases the roughness of the specimen. As a result, the roughness is modified from a single-digit micrometer value to a double-digit micrometer value in the laser texture region. The laser texture surface possessed more micron-sized structures such as dimples, square patterns, rectangular patterns, trapezoidal, and waviness when compared to the untextured surface. The comparison of heat transfer during the evaporation of water droplets from the texture and un-textured surfaces is shown in Figure 5.4. These surface characteristics control the heat transfer of the textured surface. The higher laser fluences lead to a large amount of irradiation on the specimen surface [87]. It increased the surface irregularity, which results in the increase of hydrophobic properties of laser-textured surface. The heat transfer of the texture surface is dependent on the wettability of the surface, which is measured with the use of contact angle measurement. The contact angle is increased from the double-digit to the triple-digit value on the textured surface. The contact angle is measured at the atmosphere temperature of 25°C and the humidity of 55% to obtain a reliable value. The deionized water is one of the effective methods to measure the contact angle using a sessile droplet method. The textured surface is used to measure the contact angle of the droplet. The quantity of liquid is large to test the behavior of laser texture surface wettability.

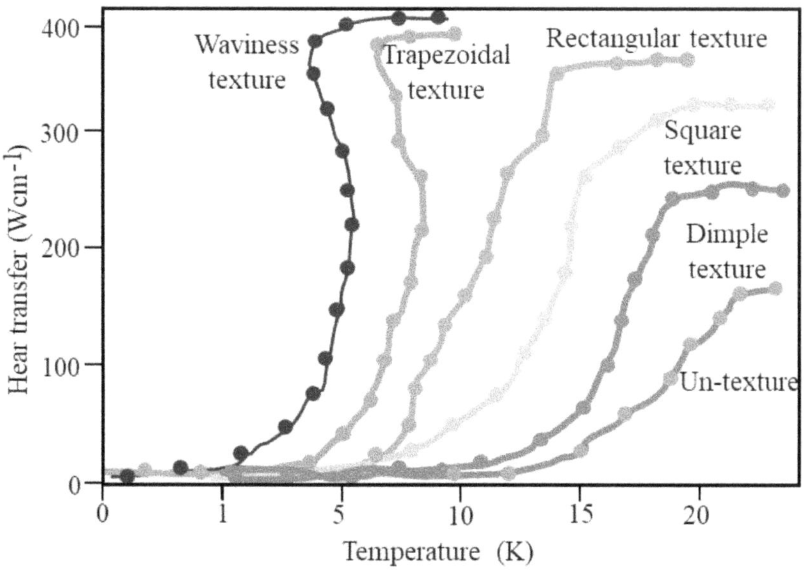

FIGURE 5.4 The comparison of heat transfer of different textures against the surface temperature.

In this analysis, the contact area is constant as the matting region between the evaporating medium and the laser texture surface remains the same. Moreover, the analysis of the heat transfer at the laser texture surface includes the different temperatures. The heat flux varies when the temperature at the laser texture surface is changed. In this analysis, the measuring medium absorbs heat to reach equilibrium between the laser texture surface and the measuring liquid medium. This stability of the measuring liquid medium on the laser texture surface is known as a quasi-steady state. The measuring medium changes from the droplet shape to the thin film shape during contact with the laser texture surface. Based on this analysis, the heat transfer rate and the corresponding evaporation time on the textured surface can be found.

5.7 SURFACE WETTABILITY

The sample undergoes processing using a variable laser scan line separation technique, resulting in the formation of uniform surface texture. Surface wettability measurements are taken on the textured surface, which is used in heat transfer experiments and treated at optimized laser parameters. The contact angle of an untextured surface is hydrophilic when utilizing the water droplet. Following the laser texture processing, the textured surfaces show superhydrophobic characteristics and display non-uniform wettability owing to their heterogeneous microstructure. The wettability of the textured samples changes over time as they are left in the open air at room temperature. The wetting property of all the surfaces decreases as time passes, which is consistent with findings from other researchers [88–90]. An elevated concentration of carbon is formed on the textured surface of the stainless steel due to the treatment with a femtosecond laser. The current amount of carbon is insufficient to completely cover the laser structure, resulting in a highly hydrophilic surface. This is caused by a combination of iron oxides and the surface roughness. Over time, the process of carbon dioxide breaking down into carbon occurs, causing the accumulation of carbon on the textured surface. The combination of this dual-scale roughness results in the creation of a superhydrophobic surface.

Following the laser treatment, the superhydrophobicity took approximately 30 days to develop on this steel surface. The contact angle of the liquid on the surface is enhanced by 20° to 35° based on the specific surface texture. Interestingly, the contact angle stabilizes after around 20 days from the laser texture processing. There are no noticeable distinctions between surfaces with texture alone and surfaces that are utilized in heat transfer experiments. The study demonstrates that laser texturing holds significant promise for enhancing heat transfer applications by ensuring the durability of the resulting texture surfaces.

5.8 NUCLEATION SITES CONTROL: LASER TEXTURING

Controlling the formation of active nucleation sites poses a significant challenge in the field of heat transfer systems. The identification of active nucleation sites and their correlation with the laser-textured microstructure is of utmost importance. Specific image-processing techniques are utilized to pinpoint the locations of active nucleation sites. The high-speed thermography recording is used to examine the texture surfaces for evaluating the active nucleation sites. The heat flux is varied from 50 to 1400 KW/m^2 during the thermography test. The results of heat flux and the volume ratio are shown in Figure 5.5. Corresponding laser processing parameters are listed in Table 5.2. When each bubble forms, the surface cools down quickly because of the intense heat when the liquid turns into vapor. The surface cools down quickly because of the intense heat generated during vaporization, which results in the formation of bubbles. The cold spot is easily found in the thermograph and measures the dimension of each bubbles with the support of pixel resolution.

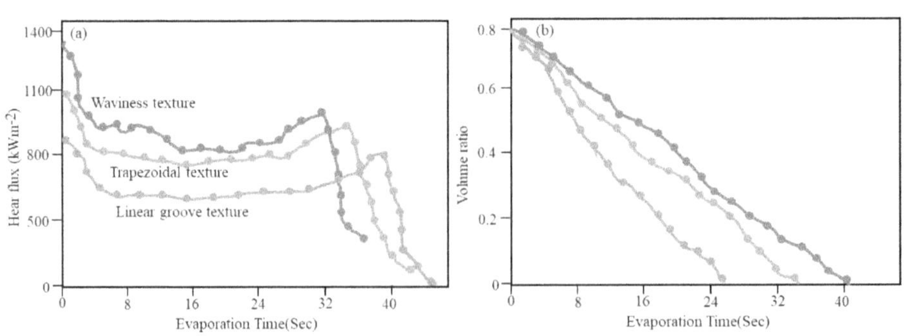

FIGURE 5.5 A comparison of (a) heat transfer of different textures against the evaporation time and (b) volume ratio against the evaporation time.

TABLE 5.2
Laser Processing Parameters

Samples	Linear Grooves Pattern	Waviness Patterns	Trapezoidal Patterns
Power (W)	3	3	3
Scanning speed (mm/s)	4	6	4
Number of passes	7	5	6

The density of the nucleation region is determined by dividing the heated region with the bubble counts. The maximum size of the bubble is found in the grayscale image, which is of a circular shape. Large diameter bubbles form at high temperatures due to shallow microstructure, which leads to a low frequency of nucleation. It results in the reduction of the total number of nucleation site formations. However, some of the laser texture surfaces exhibit smaller bubble diameters than the shallow microstructure laser texture surface. The size of the texture plays an important role in controlling the nucleation density and contact angle between the bubble and the content heat surface [91]. The high nucleation density and low contact angle lead to high wettability and improve the heat transfer from the textured surface to the medium. Hence, nucleation density also affects the heat transfer rate of texture surface along with the contact angle. The nucleation of bubbles randomly spread at the top surface of the heat transfer region. Generally, the bubble diameter of 1 to 1.6 mm forms at the heat transfer rate of 300 kW/m^2 due to the textured surface. Moreover, the periodic distribution of nucleation density occurs when the laser texture is smaller in size and matches with the spatial distribution. These heat transfer regions are possessed with microcavities that have different scales. The variable wettability with the micro- and nano-sized cavities increases the formation of active nucleation regions on the heat transfer area. Moreover, the microcavities in the laser texture region increase the nucleate boiling by 5 K when compared to the untextured heat transfer surface. The temperature mapping graph is used to determine the location of nucleation sites as IR thermography has limited resolutions [92]. In addition, the density of nucleation sites is higher within the laser-textured microcavities than in the untreated regions.

5.9 INFLUENCE OF LASER TEXTURE REGION ON THE HEAT TRANSFER BEHAVIOR

The micron-level structure and nano-level shape on the micron structure, such as 100 to 200 nm burrs, significantly affect the heat transfer behavior of the specimen. The nanostructure is fabricated by using picosecond or femtosecond lasers. In addition, the nano-sized particles can be deposited on the micro-sized laser texture to fabricate the nanostructure on the micro-texture surface. Moreover, the oxidation of the laser texture surface leads to a reduction in the contact between the evaporating medium and the texture surface. It reduces the heat transfer efficiency from the textured surface to the surrounding medium. The laser-treated surface reveals that the rate of heat transfer increases at the given heat flux range when compared to the untextured surface. The major reason for the increase of heat transfer in the textured surface is the earlier nucleation of the boiling medium than the untextured surface. The stability of the microstructure of the texture region and surface property preservation such as roughness, capillarity, wettability, and so forth are the reasons for the repeatable achievement of the same result from the textured surface. The improvement of the heat transfer in the textured surface is 30% to 40% higher than the untextured surface. At low-rate heat flux conditions, the efficiency of heat transfer increases more than the high flux conditions on the laser-textured surface. The higher density of the heat flux reduces the heat transfer efficiency of the texture region.

5.10 MECHANISM BEHIND THE HEAT TRANSFER AT THE LASER TEXTURE REGION

The buoyancy force acts between the vapor bubbles and the laser texture surface during the heat transfer process. As a result, the microlayer of the fluid forms on the laser texture surfaces. It converts from the liquid state to vapor state as the heat energy exceeds the latent heat of the liquid. Hence, phase transformation occurs on the laser texture surface and forms vapor bubbles [93,94]. The schematic representation of the vapor bubbles formation at the textured surface is shown in Figure 5.6.

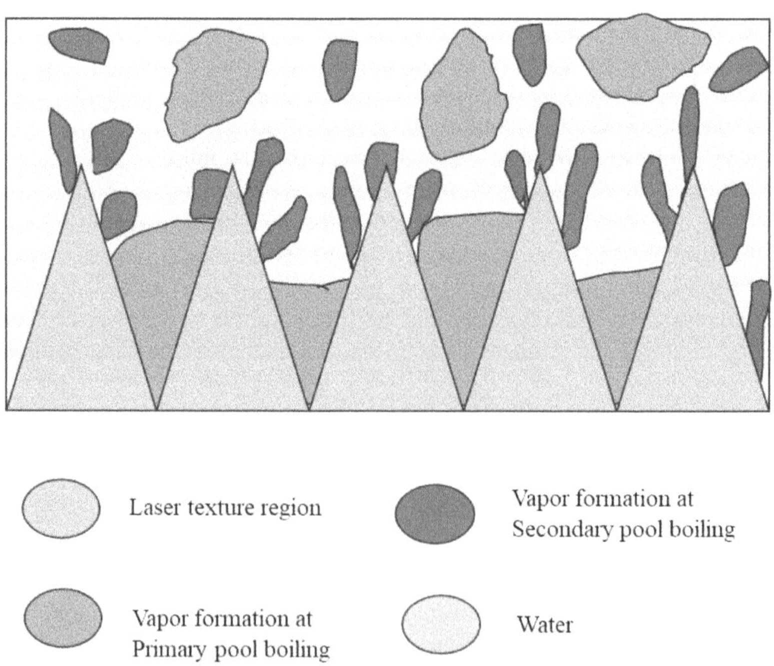

FIGURE 5.6 Schematic representation of the mechanism involved in the heat transfer on the textured surface.

Surface tension starts to form between the vapor bubbles and the laser texture surface. It keeps on rising as the heat energy in the bubble's increases. The vapor bubbles start to leave from the laser texture surface when the moving force exceeds the surface tension force. The heat energy is continuously adding into the heat region. As a result, the moving force increases and the vapor bubbles leaves from the laser texture surface. The difference in the vapor process at the laser texture surface and the untextured surface is the contact of the vapor bubbles. The vapor bubbles touch the edge region of the laser texture surface. However, on the untextured surface, the vapor bubbles touch the whole region. Hence, detachment of vapor bubbles from the laser texture surface is easier than from the untextured surface. The liquid region receives heat from the heat transfer surface from conduction as well as convection in the laser texture region.

The untextured surface receives heat through the conduction process only. The vapor bubbles vibrate for a shorter period of time in the laser texture region than the untextured surface. The thermal expansion in the laser texture region is more than in the untextured surface. The square pillar edges easily transfer the thermal energy from the textured surface to the liquid region. The pushing force is induced at the edges and nearer to the edges, which act toward the bubble in the laser texture region. Hence, the pushing force occurs from the side of the square pillar texture region and the bottom of the pillar. The vapor bubbles leave the square pillar region and improve the heat transfer process.

5.11 CONCLUSION

The influence of laser texture surface on the heat transfer process was studied and the following conclusions are derived.

- The ability of a laser texture surface to be wetted depends on its physical features, chemical properties, and the molecules interacting at the boundary between the solid and liquid
- The heat transfer rate increases when the inter-structural region increases. Wettability, roughness, and porosity of the lase texture surface control the droplet formation on the heat transfer surface.
- The laser texture surface produces a heterogeneous surface, which increases the surface area than the plane surface.
- The buoyancy force is acting between the vapor bubbles and the laser texture surface during the heat transfer process.
- The microlayer of the fluid forms on the laser texture surfaces, which leads to converting from the liquid state to vapor state as the heat energy exceeds the latent heat of the liquid.
- The square pillar edges easily transfer the thermal energy from the textured surface to the liquid region. The force is induced at the edges and nearer to the edges, which act toward the bubble in the laser texture region.
- The high nucleation density and low contact angle lead to high wettability and improve the heat transfer from the textured surface to the medium.

5.12 ACKNOWLEDGMENT

We are grateful for the support from the Ministry of Science and Technology, Taiwan, for providing financial support for this study.

REFERENCES

[1] M. McCarthy, K. Gerasopoulos, S.C. Maroo, A.J. Hart, Materials, fabrication, and manufacturing of micro/nanostructured surfaces for phase-change heat transfer enhancement, Nanoscale Microscale Thermophys. Eng. 18 (2014) 288–310.

[2] H.J. Cho, D.J. Preston, Y. Zhu, E.N. Wang, Nanoengineered materials for liquid–vapour phase-change heat transfer, Nat. Rev. Mater. 2 (16092) (2016) 1–17.

[3] C.R. Kharangate, I. Mudawar, Review of computational studies on boiling and condensation, Int. J. Heat Mass Transf. 108 (2017) 1164–1196.

[4] S.A. Khan, F. Tahir, A. Ali, B. Baloch, Review of micro-nanoscale surface coatings application for sustaining dropwise condensation, Coatings. 9 (2019) 117.

[5] Y.-W. Lu, S.G. Kandlikar, Nanoscale surface modification techniques for pool boiling enhancement: A critical review and future, directions, Heat Transf. Eng. 32 (2011) 827–842.

[6] S. Mori, Y. Utaka, Critical heat flux enhancement by surface modification in a saturated pool boiling: A review, Int. J. Heat Mass Transf. 108 (2017) 2534–2557.

[7] C. S. Sujith Kumar, G. Udaya Kumar, Mario Mata Arenales, Chin-Chi Hsu, S. Suresh, Ping-Hei Chen, Elucidating the mechanisms behind the boiling heat transfer enhancement using nano-structured surface coatings, App. Thermal Engg. 137 (2018) 868–891.

[8] F. Peter Prakash, N. Jeyaprakash, Muthukannan Duraiselvam, G. Prabu, C.-H. Yang, Droplet spreading and wettability of Laser textured C-263 based Nickel superalloy, Surf Coat Techn. 397 Jun (2020) 1260552.

[9] A. Malinauskas, A. Žukauskas, S. Hasegawa, Y. Hayasaki, V. Mizeikis, R. Buividas, S. Juodkazis. Ultrafast laser processing of materials: from science to industry, Light Sci Appl. 5 (8) (2016 Aug) e16133.

[10] G. Prabu, Muthukannan Duraiselvam, Manickam Matheswaran, P. Amrish, Tamilmani Jayabalan, M. Varatharajulu, Laser surface texturing for enhancing microbial fuel cell-based electricity generation from wastewater, Part A: Journal of Power and Energy. 236 (5) (2022) 937–948.

[11] M. Dadhich, O.S. Prajapati, A brief review on factors affecting flow and pool boiling. Renew Sustain Energy Rev. 112 (2019 Sep 1) 607–625.

[12] N. Jeyaprakash, G. Prabu, Che-Hua Yang, Manikanta Prasad Banda, E. Mohan, Structural and nanohardness of laser melted CM247LC nickel-based superalloy, Transactions of the Indian Institute of Metals 76 (2) (2022) 287–295.

[13] Y.A. Cengel, A. Ghajar, Heat and mass transfer (a practical approach, SI version), McGraw-Hill Education, 2011.

[14] S.G. Kandlikar, M. Shoji, V.K. Dhir, Handbook of phase change: Boiling and condensation, 1st ed., CRC Press, 1999.

[15] M.S. Kamel, F. Lezsovits, A.M. Hussein, O. Mahian, S. Wongwises, Latest developments in boiling critical heat flux using nanofluids: A concise review, Int. Commun. Heat Mass Transf. 98 (2018) 59–66.

[16] G. Prabu, Che-Hua Yang, Ibrahim A. Alnaser, N. Jeyaprakash, Nanowear characterization of LPBF formed CuCrZr alloy, Tribology International, 194 (2024) 109430

[17] G. Liang, I. Mudawar, Pool boiling critical heat flux (CHF)—Part 1: Review of mechanisms, models, and correlations, Int. J. Heat Mass Transf. 117 (2018) 1352–1367.

[18] G. Udaya Kumar, S. Suresh, M.R. Thansekhar, Deepkumar Halpati, Role of inter-nanowire distance in metal nanowires on pool boiling heat transfer characteristics, J. Colloids and Interface Science 532 (2018) 218–230.
[19] G. Udaya Kumar, S. Suresh, M. R. Thansekhar and Dinesh Babu, Effect of diameter of metal nanowires on pool boiling heat transfer with FC-72, App. Surface Science. 423 (2017) 509–520.
[20] R. Kamatchi, S. Venkatachalapathy, Parametric study of pool boiling heat transfer with nanofluids for the enhancement of critical heat flux: A review, Int. J. Therm. Sci. 87 (2015) 228–240.
[21] H. Hu, J.A. Weibel, S.V. Garimella, A coupled wicking and evaporation model for prediction of pool boiling critical heat flux on structured surfaces, Int. J. Heat Mass Transf. 136 (2019) 373–382.
[22] Y. Hu, K. Huang, J. Huang, A review of boiling heat transfer and heat pipes behaviour with self-rewetting fluids, Int. J. Heat Mass Transf. 121 (2018) 107–118.
[23] G. Prabu, N. Jeyaprakash, Che-Hua Yang, S. Sivasankaran, Synergetic effect of piled-up material around the nanoindentation cavity on additively manufactured Cu-Cr-Zr alloy, Material Today Communication. 38 (2024) 107691.
[24] A. Jaikumar, S.G. Kandlikar, Pool boiling inversion through bubble induced macroconvection, Appl. Phys. Lett. 110 (2017) 094107.
[25] N. Jeyaprakash, G. Prabu, C.H. Yang, The influence of different phases on the microstructure and wear of inconel-718 surface alloyed with AlCuNiFeCr hard particles using plasma transferred arc. J Mater Eng Perform. 31 (12) (2022 Dec) 9921–9934.
[26] X. Li, I. Cole, J. Tu, A review of nucleate boiling on nanoengineered surfaces: The nanostructures, phenomena and mechanisms, Int. J. Heat Mass Transf. 141 (2016) 20–33.
[27] S. Ahangar Zonouzi, H. Aminfar, M. Mohammadpourfard, A review on effects of magnetic fields and electric fields on boiling heat transfer and CHF, Appl. Therm. Eng. 151 (2019) 11–25.
[28] G. Liang, I. Mudawar, Review of pool boiling enhancement with additives and nanofluids, Int. J. Heat Mass Transf. 124 (2018) 423–453.
[29] S. Tsang, Z. Wu, C. Lin, C. Sun, On the evaporative spray cooling with a selfrewetting fluid: Chasing the heat, Appl. Therm. Eng. 132 (2018) 196–208.
[30] G. Prabu, N. Jeyaprakash, Che-Hua Yang, Ibrahim A. Alnaser, Investigation on deformation of nano-twins of LPBF produced Cu alloy through Triboindenter, Tribology International. 191 (2024) 109117.
[31] C.G. Jothi Prakash, R. Prasanth, Enhanced boiling heat transfer by nano structured surfaces and nanofluids, Renew. Sustain. Energy Rev. 82 (2018) 4028–4043.
[32] G. Udaya Kumar, Khushboo Soni, Suresh Sivan, Kaushik Ghosh, M. R. Thansekhar, P. Dinesh Babu, Modified surfaces using seamless graphene/carbon nanotubes based nanostructures for enhancing pool boiling heat transfer, Exp.Thermal and Fluid Science. 96 (2019) 493–506.
[33] I.L. Pioro, W. Rohsenow, S.S. Doerffer, Nucleate pool-boiling heat transfer. I: Review of parametric effects of boiling surface, Int. J. Heat Mass Transf. 47 (2004) 5033–5044.
[34] N. Jeyaprakash, G. Prabu, Che-Hua Yang, Inclusion of nano silicon particle in SS316L through LPBF and its responses on corrosion behaviour, Ceramics International. 49 (16) (2023) 26450–26468.
[35] R.L. Mohanty, M.K. Das, A critical review on bubble dynamics parameters influencing boiling heat transfer, Renew. Sustain. Energy Rev. 78 (2017) 466–494.
[36] G. Liang, I. Mudawar, Review of pool boiling enhancement by surface modification, Int. J. Heat Mass Transf. 128 (2019) 892–933.

[37] M.N. MacGregor-Ramiasa, K. Vasilev, Questions and answers on the wettability of nano-engineered surfaces, Adv. Mater. Interfaces. 4 (2017) 1–24.
[38] N. Jeyaprakash, G. Prabu, Che-Hua Yang (2021), Formation of different phases and their influences on the mechanical and tribological properties of surface alloyed FeCoCrNiAl particles using PTA technique, Intermetallics. 142 (2022) 107457.
[39] T.S. Emery, A. Jaikumar, P. Raghupathi, I. Joshi, S.G. Kandlikar, Dual enhancement in HTC and CHF for external tubular pool boiling: A mechanistic perspective and future directions, Int. J. Heat Mass Transf. 122 (2018) 1053–1073.
[40] G. Prabu, N. Jeyaprakash, Che-Hua Yang, Ibrahim A. Alnaser, Evaluation of friction and wear depth during the cyclic loading on partly melted LPBF particles of LPBF Cu alloy, Materials Characterization. 217 (2024) 114386.
[41] M.M. Rahman, M. McCarthy, Boiling enhancement on nanostructured surfaces with engineered variations in wettability and thermal conductivity, Heat Transfer Eng. 38 (14–15) (2017) 1285–1295.
[42] A. Mourgues, V. Hourtané, T. Muller, M. Caron-Charles, Boiling behaviors and critical heat flux on a horizontal and vertical plate in saturated pool boiling with and without ZnO nanofluid, Int. J. Heat Mass Transf. 57 (2) (2013) 595–607.
[43] G. Prabu, Che-Hua Yang, Ibrahim A. Alnaser, N. Jeyaprakash, Study on the influence of 1μm and 15μm pitch values on the apparent nanomechanical properties of Laser Powder Bed Fusion produced CuCrZr alloy, Surf Rev Lett. 7 (2024 Oct 7) 2550045.
[44] S. Malik, A.K. Nayak, Buoyancy driven heat transfer in nanofluids due to wall mounted heat source, Alexandria Eng. J. 55 (2) (2016) 797–810.
[45] S.Y. Misyura, Evaporation of a sessile water drop and a drop of aqueous salt solution, Sci. Rep. 7 (14759) (2017).
[46] N. Jeyaprakash, G. Prabu, Che-Hua Yang, Microstructure and tribological behaviour of FeCoCrNiAlCuMn particles alloyed on Inconel-718 substrate using plasma transferred arc technique, Intermetallics. 139 (2021) 107360.
[47] S. Dash, S.V. Garimella, Droplet evaporation on heated hydrophobic and superhydrophobic surfaces, Phys. Rev. E 89 (4) (2014) 042402.
[48] M.A. Kadhim, N. Kapur, J.L. Summers, H. Thompson, Experimental and theoretical investigation of droplet evaporation on heated hydrophilic and hydrophobic surfaces, Langmuir 35 (19) (2019) 6256–6266.
[49] K. Sefiane, L. Tadrist, Experimental investigation of the de-pinning phenomenon on rough surfaces of volatile drops, Int. Commun. Heat Mass Transf. 33 (4) (2006) 482–490.
[50] M. He, D. Liao, H. Qiu, Multicomponent droplet evaporation on chemical micropatterned surfaces, Sci. Rep. 7 (2017) 41897.
[51] G. Prabu, N. Jeyaprakash, Che-Hua Yang, N. Radhika, Analysis on the transformation of melt wear to self-healing crack of wire arc additive manufactured Al 5356 alloy, Tribology International. 188 (2023) 108802.
[52] T.S. Lin, Y.H. Zerig, R.Y. Tsay, S.Y. Lin, Roughness-induced strong pinning for drops evaporating from polymeric surfaces, J. Taiwan Inst. Chem. Eng. 62 (2016) 54–59.
[53] I. Mudawar, Recent advances in high-flux, two-phase thermal management, J. Therm. Sci. Eng. Appl. 5 (2) (2013) 021012, https://doi.org/10.1115/1.4023599.
[54] M.A. Ebadian, C.X. Lin, A review of high-heat-flux heat removal technologies, J. Heat Transfer. 133 (11) (2011) 110801, https://doi.org/10.1115/1.4004340.
[55] S. Fan, F. Duan, A review of two-phase submerged boiling in thermal management of electronic cooling, Int. J. Heat Mass Transf. 150 (2020) 119324, https://doi.org/10.1016/j.ijheatmasstransfer.2020.119324.
[56] B.P. Benam, A.K. Sadaghiani, V. Ya˘gcı, M. Parlak, K. Sefiane, A. Kos¸ar, Review on high heat flux flow boiling of refrigerants and water for electronics cooling, Int. J.Heat Mass Transf. 180 (2021) 121787, https://doi.org/10.1016/j.ijheatmasstransfer.2021.121787.

[57] C. Roe, X. Feng, G. White, R. Li, H. Wang, X. Rui, C. Li, F. Zhang, V. Null, M. Parkes, Y. Patel, Y. Wang, H. Wang, M. Ouyang, G. Offer, B. Wu, Immersion cooling forlithium-ion batteries: A review, J. Power Sources. 525 (2022) 231094, https://doi.org/10.1016/j.jpowsour.2022.231094.

[58] G. Prabu, Muthukannan Duraiselvam (2020), Tribological studies on AlCrFeCuCoNi high entropy alloy surface coated on Ti-6Al-4V using plasma transferred arc technique, Archives of Metallurgy and Materials. 67 (2) (2022) 409–420.

[59] S. Mori, Y. Utaka, Critical heat flux enhancement by surface modification in a saturated pool boiling: A review, Int. J. Heat Mass Transf. 108 (2017) 2534–2557, https://doi.org/10.1016/j.ijheatmasstransfer.2017.01.090.

[60] A.V. Dedov, A review of modern methods for enhancing nucleate boiling heat transfer, Therm. Eng. 66 (2019) 881–915, https://doi.org/10.1134/S0040601519120012.

[61] G. Liang, I. Mudawar, Review of pool boiling enhancement by surface modification, Int. J. Heat Mass Transf. 128 (2019) 892–933, https://doi.org/10.1016/j.ijheatmasstransfer.2018.09.026.

[62] O.A. Volodin, N.I. Pecherkin, A.N. Pavlenko, Heat transfer enhancement at boiling and evaporation of liquids on modified surfaces: A review, High Temp. 59 (2021) 405–432, https://doi.org/10.1134/S0018151X21020140.

[63] U. Sajjad, A. Sadeghianjahromi, H.M. Ali, C.C. Wang, Enhanced pool boiling of dielectric and highly wetting liquids: A review on surface engineering, Appl. Therm. Eng. 195 (2021) 117074, https://doi.org/10.1016/j.applthermaleng.2021.117074.

[64] X. Yuan, Y. Du, J. Su, Approaches and potentials for pool boiling enhancement with superhigh heat flux on responsive smart surfaces: A critical review, Renew. Sustain. Energy Rev. 156 (2022), 111974, https://doi.org/10.1016/j.rser.2021.111974.

[65] M.Z. Podowski, Toward mechanistic modeling of boiling heat transfer, Nucl. Eng. Technol. 44 (8) (2012) 889–896, https://doi.org/10.5516/NET.02.2012.720.

[66] R.J. Benjamin, A.R. Balakrishnan, Nucleate pool boiling heat transfer of pure liquids at low to moderate heat fluxes, Int. J. Heat Mass Transf. 39 (12) (1996) 2495–2504, https://doi.org/10.1016/0017-9310(95)00320-7.

[67] M. Kim, S.J. Kim, A mechanistic model for nucleate pool boiling including the effect of bubble coalescence on area fractions, Int. J. Heat Mass Transf. 163 (2020) 120453, https://doi.org/10.1016/j.ijheatmasstransfer.2020.120453.

[68] E. Alberts, C. Warner, E. Barnes, K. Pilkiewicz, E. Perkins, A. Poda, Genetically tunable M13 phage films utilizing evaporating droplets, Colloids Surf. B-Biointerfaces. 161 (2018) 210–218.

[69] T.V. Kasyap, D.L. Koch, M.M. Wu, Bacterial collective motion near the contact line of an evaporating sessile drop, Phys. Fluids. 26 (11) (2014) 111703.

[70] Y.J.P. Carreon, J. Gonzalez-Gutierrez, M.I. Perez-Camacho, H. Mercado-Uribe, Patterns produced by dried droplets of protein binary mixtures suspended in water, Colloids Surf. B-Biointerfaces 161 (2018) 103–110.

[71] Y.J.P. Carreon, M. Rios-Ramirez, R.E. Moctezuma, J. Gonzalez-Gutierrez, Texture analysis of protein deposits produced by droplet evaporation, Sci. Rep. 8 (2018) 9580.

[72] H. Yoo, C. Kim, Experimental studies on formation, spreading and drying of inkjet drop of colloidal suspensions, Colloids Surf. A-Physicochem. Eng. Asp. 468 (2015) 234–245.

[73] S. Semenov, V.M. Starov, M.G. Velarde, R.G. Rubio, Droplets evaporation: Problems and solutions, Eur. Phys. J. Spec. Top. 197 (1) (2011) 265–278.

[74] M. He, D. Liao, H. Qiu, Multicomponent droplet evaporation on chemical micropatterned surfaces, Sci. Rep. 7 (2017) 41897.

[75] C.C. Hsu, Y.A. Lee, C.H. Wu, C.S.S. Kumar, Self-propelled sessile droplets on a superheated and heterogeneous wetting surface, Colloids Surf. A: Physicochem. Eng. Asp. 612 (126074) (2021).

[76] L. Ge, J. Yin, D. Yan, W. Hong, T. Jiao, Construction of nanocrystalline cellulosebased composite fiber films with excellent porosity performances via an electrospinning strategy, ACS Omega 6 (7) (2021) 4958–4967.

[77] E. Andrews, J.A. Lopez-Ruiz, J.D. Egbert, K. Koh, U. Sanyal, M. Song, D. Li, A. J. Karkamkar, M.A. Derewinski, J. Holladay, O.Y. Guti´errez, J.D. Holladay, Performance of base and noble metals for electrocatalytic hydrogenation of bio-oil derived oxygenated compounds, ACS Sustain. Chem. Eng. 8 (11) (2020) 4521–4536.

[78] Y. He, R. Wang, T. Jiao, X. Yan, M. Wang, L. Zhang, Z. Bai, Q. Zhang, Q. Peng, Facile preparation of self-assembled layered double hydroxide-based composite dye films as new chemical gas sensors, ACS Sustain. Chem. Eng. 7 (12) (2019) 10888–10899.

[79] P. Gregorcic, M. Sedlacek, B. Podgornik, J. Reif, Formation of laser-induced periodic surface structures (LIPSS) on tool steel by multiple picosecond laser pulses of different polarizations, Appl. Surf. Sci. 387 (2016) 698–706.

[80] P.M. Fauchet, A.E. Siegman, Surface ripples on silicon and gallium arsenide under picosecond laser illumination, Appl. Phys. Lett. 40 (9) (1982) 824–826.

[81] T.T.D. Huynh, A. Petit, N. Semmar, Picosecond laser induced periodic surface structure on copper thin films, Appl. Surf. Sci. 302 (2014) 109–113.

[82] K. Sun, H. Yang, W. Xue, A. He, D. Zhu, W. Liu, Y. Cao, Anti-biofouling superhydrophobic surface fabricated by picosecond laser texturing of stainless steel, Appl. Surf. Sci. 436 (2018) 263–267.

[83] A. Dunn, T.J. Wasley, J. Li, R.W. Kay, J. Stringer, P.J. Smith, J.D. Shephard, Laser textured superhydrophobic surfaces and their applications for homogeneous spot deposition, Appl. Surf. Sci. 365 (2016) 153–159.

[84] D.V. Zaitsev, K.A. Batishchevab, G.V. Kuznetsov, E.G. Orlova, Prediction of water droplet behavior on aluminum alloy surfaces modified by nanosecond laser pulses, Surf. Coat. Technol. 399 (2020) 126206.

[85] G. Prabu, Muthukannan Duraiselvam, N. Jeyaprakash, C.-H. Yang, Microstructural evolution and wear behavior of AlCoCrCuFeNi high entropy alloy on Ti-6Al-4V through laser surface alloying, Metals and Materials International. 27 (2021) 2328–2340.

[86] Periyakaruppan Murugesan, V. Satheeshkumar, N. Jeyaprakash, G. Prabu, Che-Hua Yang, Nano-level mechanical and tribological behavior of additively manufactured AlSi10Mg plates, Archives of Civil and Mechanical Engineering. 23 (2023) 62.

[87] H.C. Cheng, Z.X. Jiang, T.L. Chang, P.H. Chen, Roughness and wettability properties of plain and silica-coated copper surfaces textured with picosecond laser, Appl. Surf. Sci. 514 (2020) 145918.

[88] A.M. Kietzig, S.G. Hatzikiriakos, P. Englezos, Patterned superhydrophobicmetallic surfaces, Langmuir. 25 (2009) 4821–4827, http://dx.doi.org/10.1021/la8037582.

[89] V.D. Ta, A. Dunn, T.J. Wasley, J. Li, R.W. Kay, J. Stringer, P.J. Smith, E. Esenturk, C. Connaughton, J.D. Shephard, Laser textured superhydrophobic surfaces and their applications for homogeneous spot deposition, Appl. Surf. Sci. 365(2016) 153–159, http://dx.doi.org/10.1016/j.apsusc.2016.01.019.

[90] D.V. Ta, A. Dunn, T.J. Wasley, R.W. Kay, J. Stringer, P.J. Smith, C. Connaughton, J.D. Shephard, Nanosecond laser textured superhydrophobic metallic surfaces and their chemical sensing applications, Appl. Surf. Sci. 357 (2015) 248–254, http://dx.doi.org/10.1016/j.apsusc.2015.09.027.

[91] M.M. Rahman, E. ¨Olc¸erog˘lu, M. McCarthy, Role of wickability on the critical heat flux of structured superhydrophilic surfaces, Langmuir. 30 (2014) 11225–11234, http://dx.doi.org/10.1021/la5030923.

[92] I. Sedmak, I. Urbanˇciˇc, R. Podlipec, J. ˇStrancar, M. Mortier, I. Golobiˇc, Submicronthermal imaging of a nucleate boiling process using fluorescence microscopy, Energy 109 (2016) 436–445, http://dx.doi.org/10.1016/j.energy.2016.04.121.
[93] J. Bi, K. Vafai, D.M. Christopher, Heat transfer characteristics and CHF prediction in nanofluid boiling, Int. J. Heat Mass Transfer. 80 (Supplement C) (2015) 256–265.
[94] J.G. Collier, J.R. Thome, Convective boiling and condensation, 3rd ed., Clarendon Press, 1994.

6 Laser Surface Texturing on Tool Surfaces for Conventional/Unconventional Machining

Jayanth N., T. Sathies, and Kumaran M.

6.1 INTRODUCTION

Laser surface texturing (LST) is a method of enhancing surface properties of materials by changing their texture and surface finish. The laser beam generates micropatterns on the workpiece's surface using laser ablation, which removes layers of material with high accuracy and repeatability. Dimples, grooves, and free forms are the usual patterns made using this process. LST can be used to improve surface characteristics such as adherence, wettability, electrical conductivity, thermal conductivity, and wear resistance.

6.2 APPLICATIONS OF LST IN CONVENTIONAL MACHINING

6.2.1 Turning

The texturing was made on the rake face of the cutting tool to lower friction while doing machining. The micro scale dimples array was made on the HSS tool surface via LST. Then it was used in dry machining. The tools with textures decrease the cutting forces at both elevated and slow speeds; however, it was better at elevated speeds [1]. The textured tools can also be used to reduce machining zone temperature and tool wear. Also better surface finish can be obtained when compared to unstructured tools because of the development of lubricating film and it leads to tool life enhancement [2]. The nano-scale textures were produced using femtosecond laser technology on the rake face nearer to the WC/Co-based TiAlN deposited tool's cutting edge. Also, the coating of WS_2 solid lubricant improved the cutting performance of both textured and untextured tools considerably when compared to the uncoated tool. The cutting forces, temperature, and friction coefficient obtained at tool-chip contact area were small for WS_2-coated nano-scale textured tool. This is because of the drop in the tool-chip contact length (l_f) and improvement in the shear deformation (Figure 6.1a). The adhesive strength of WS_2 film improved by

DOI: 10.1201/9781003542087-6

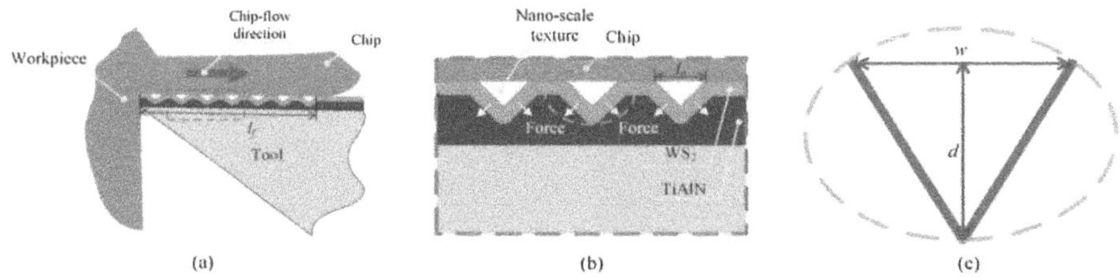

FIGURE 6.1 Schematic representation of the textured cutting tool: (a) tool-chip contact and chip-flow direction on the textured rake face; (b) nano-scale textured WS_2/TiAlN coating structure and its friction-reducing mechanism; (c) geometry of a triangular surface texture characterized by width *(w)* and depth *(d)*.

spreading of the film and contact pressure (Figure 6.1b). The computation of groove interface length was feasible for LST from a cross-sectional view and to determine the CNT-W tool interface area, as given in Figure 6.1c [3].

Micro-textures were made on the rake face of HSS cutting tool using electric discharge machining (EDM) and it was observed that the cutting tool's textures decrease the cutting force and avoids the formation of buildup edge on the cutting tool when performing dry machining on the aluminum alloy [4]. The machinability of cutting processes was improved by using micro- or nano-scale textured tools made using femtosecond laser. Also, the tool surface's friction and cutting forces needed got reduced [5]. Micro dimples made on the tool's rake face improved the crater wear resistance while machining medium carbon steels by acting as a micro-storage and a micro-trap for cutting fluid and wear debris respectively. Also, this tool's performance was better under heavy lubrication when compared to micro-grooved tools [6]. Two types of patterns, namely areal and linear textures, were made on ISO K20 CNMA 120408 tungsten carbide cutting insert's rake face through EDM processes (Figure 6.2). Perpendicular textured tool exhibited better wear resistance and lubrication owing to less adhesion of work material on tool surface. Also, it reduced specific cutting energy and cutting force while cutting Al-Cu/TiB_2 composite and prevents the development of buildup edge [7].

The machining performance was influenced by micro holes textured tool and its design parameters, namely, hole diameter, depth, and pitch distance, while cutting Al metal matrix composites This micro-hole texture along with solid lubricant can act as a chip breaker and provides better cooling and lubrication. Also, the coated textured tool reduced the flank wear, surface roughness, and power consumption [8]. Textured tools with perpendicular grooves (PP) performed better and increased the tool life when compared to the parallel grooves (PR). Also, it enhanced the coating wear rate. The micro-abrasion tests performed at slurry concentration 0.5 g/cm³ revealed the different wear mechanisms (Figure 6.3) [9]. Micro-grooved texture was made on the surface of WC-10Ni_3Al and WC-8Co cermets. It was found that the wear endurance of textured WC-10Ni_3Al is higher than WC-8Co when in contact with Ti6Al4V [10].

6.2.2 Milling

Micro-circles and micro-channels were made on the tool surface's rake face by using LST and a montmorillonite (MMT) nanoclay-additive-based lubricant was used as a

FIGURE 6.2 SEM images of (a) non-textured, (b) areal, (c) linear texture with perpendicular to chip flow direction, and (d) linear texture with parallel to chip flow direction.

cutting fluid. The milling parameters and MMT nanoclay concentration used in this experiment are given in Table 6.1. The textured channels along with the lubrication improved the wear resistance and reduced the friction when compared to micro-holes while milling AISI 1018 steel [11].

The laser texturing was performed on cemented carbide coated with AlCrN/AlTiSiN, which enhanced the quality of the coating by enhancing its phase structure. The best meso-geometry factors of the ball-end milling carbide tool got using genetic algorithm were blunt round edge radius of 20.15 µm and micro-texture distance of 119.463 µm from cutting edge. These factors lead to lesser tool wear, milling force, and better surface finish of the workpiece [12].

The cutting characteristics of cemented carbide YG6X milling tool without coating have been improved by the mixture of surface textures and nanofluids. The textured tools with nanofluids lubrication exhibited lesser cutting force and better surface finish than textured and untextured tools with conventional cutting fluid while milling Ti6Al4V workpiece. Also, the tool wear rate and material adhesion got reduced [13]. The micro-texturing on ball end milling cutter decreased the surface roughness, tool wear, and cutting force required and cutting temperature while milling titanium alloy [14]. Nano/micro-textures were made on the cemented carbide cutting tool's rake face by femto-second laser. The face-milling trials on the aluminum alloys revealed that the chips' adhesion to the tool surface increased for dry machining and decreased for

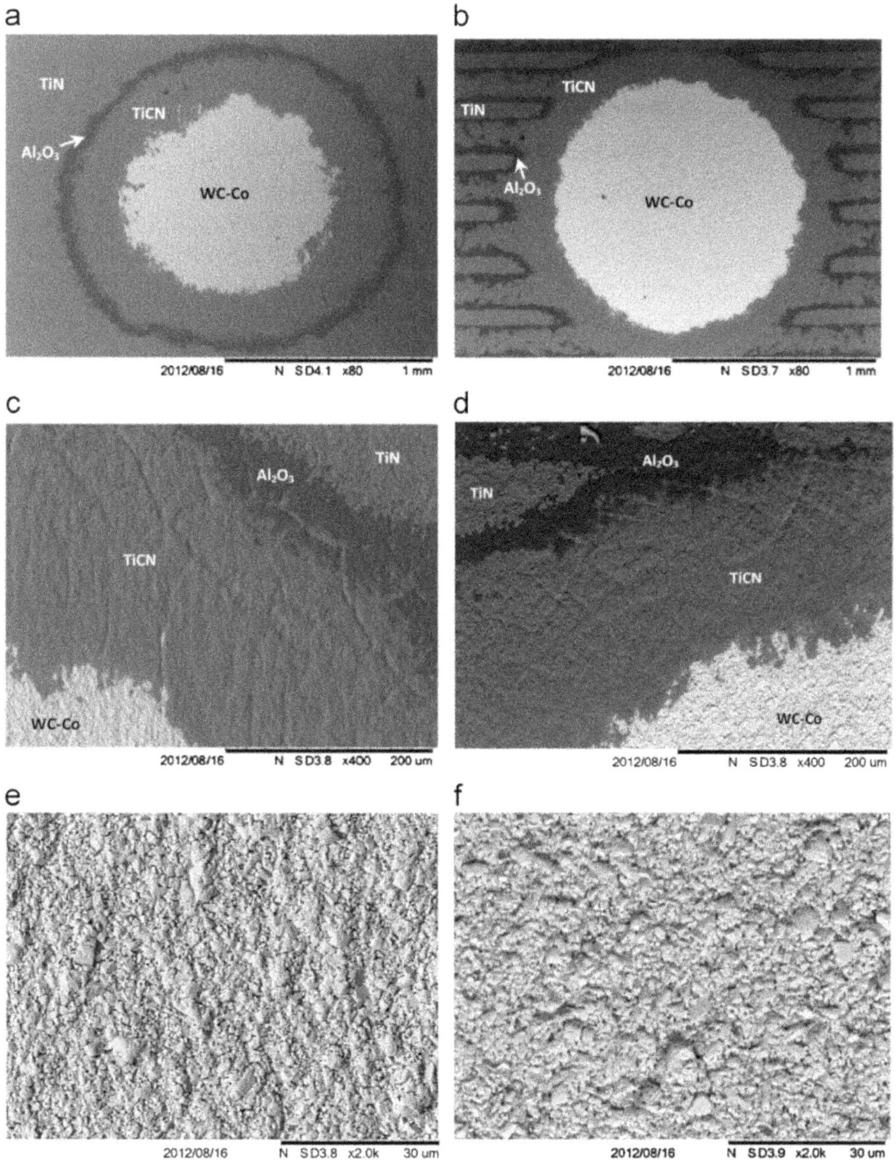

FIGURE 6.3 (a) Wear crater in untextured rake surface; (b) PP texture on rake surface; (c) coating region, untextured; (d) coating region, PP texture; (e) substrate, untextured, and (f) substrate, PP texture. Images by SEM, secondary electrons (SE) and backscattered electrons (BSE).

wet machining. The anti-adhesive property and tool surface lubrication enhanced by using banded nano/micro-textured milling tool [15].

The nano/micro-textured tools were made and milling trials were done on aluminum alloy A5052. It was observed that the nano/micro-textured tools with diamond-like carbon (DLC) coating increased the anti-adhesive property and tool surface

TABLE 6.1
Milling Parameters and MMT Clay Nanoparticle Concentration

Milling Parameter	Level
Cutting speed (rpm)	1165
Depth (in)	0.025
Feed (in/min)	7.5
MMT clay NP concentration (wt.%)	0.13

lubrication. The maximum anti-adhesive behavior was obtained for the tools with nano/micro-grooves parallel to the major cutting edge [16]. The micro-stripe grooves made parallel to the major cutting edge and TiAlN coating on the cemented carbide tool reduced tool wear resistance and its surface lubrication while performing face milling on carbon steel S53C via vertical milling machine [17].

Two types of surface texturing were performed on the double-edge milling carbide tool's rake face, such as linear grooves parallel and perpendicular to the cutting edge respectively. The milling performance of these tools was compared with convention milling tool while performing dry machining of unidirectional laminated carbon fiber reinforced plastics (CFRP) having many orientations (0°, 45°, 90°, and 135°). It was found that the textured tool having linear grooves parallel to the main cutting edge performed better than the conventional milling tool and surface textured tool having grooves perpendicular to the main cutting edge. Also, it decreased the length of burr and the tearing phenomenon of milled surface by changing the bending deformation of fiber chips. The surface texture also acts as a reservoir for carbon fiber fragment and it prevented the fiber chip slippage on the tool's rake face [18].

The ball-end mill cutting tool having multiple distribution density due to the cutting edge action was fabricated and it was used to machine titanium alloy. It was found that the main parameters influencing cutting performance and surface roughness were blunt-round edge radius and the stickiness region's micro texture diameter and it was the major tool-chip interface. The micro textures have the capability of storing chips, which in turn reduces tool-chip contact area and enhance the heat dissipation. The significant factor affecting the impact strength, resistance, and tool's chip removal was blunt-round edge radius [19]. The surface texturing of both the rake face and the flank face of the milling tool was performed and the milling experiments revealed that the surface texturing on both the faces improved the surface finish and machining performance of the tool [20].

6.2.3 Drilling

The circular dimple micro textures were made on the flute and the drilling tool's margin side as given in Figure 6.4a and 6.4b. The average diameter and depth of

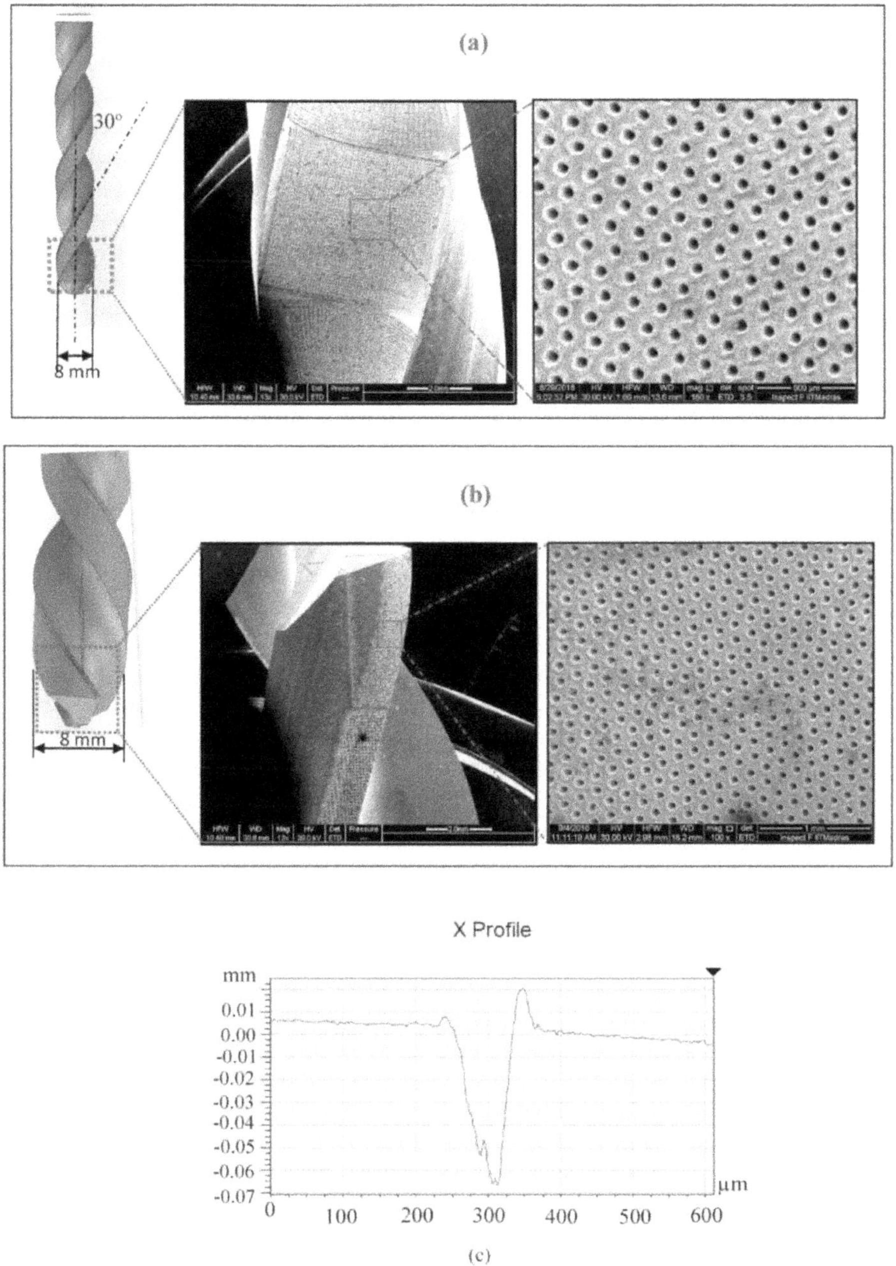

FIGURE 6.4 (a) Micro dimples at the flute side. (b) Micro dimples at the margin side. (c) Depth profile of micro texture measured using 3D optical profilometer.

micro dimple obtained using 3D optical profilometer were 90 μm and 60 μm respectively (Figure 6.4c). The micro texturing reduced sliding friction among the tool and the workpiece. The friction tests revealed that better performance is obtained for micro dimples when compared to untextured and micro grooved surfaces. Also, a decrease in thrust force and torque is obtained for margin and flute textured carbide drill tools. The micro-textured tools reduce the surface defects of the drilled holes [21]. The surface texturing on high speed steel drilling tool was performed and TiN coating was applied using physical vapor deposition. The drilling experiments were performed on low-alloy carbon steel and stainless steel. It was found that the laser surface treated tool with TiN coating has the highest tool life when compared to the conventional tool [22].

The drilling of Ti-6Al-4V workpiece was done by the flute and margin textured tool and it was observed that the textured surface's wear resistance enhanced and the drilling experiments revealed that the torque and thrust force requirement for margin textured tool was less than the untextured and flute textured tools. The main reason for improvement in the tribological properties of drilling tool was decrease in length of contact and wear debris entrapment [23]. The drill bits with rectangular surface textures were made using laser and drilling experiments were performed on titanium workpiece; it was found that the tool life improved due to the texturing and also it decreases the adhesion of chip to the tool surface and prevents the formation of build-up edge [24].

The micro- and nano-scale texturing on the drill bit of 0.5 mm were performed using laser and the textured drill bit was used for machining aluminum. The experimental results revealed that the micro- and nano-scale texturing on the drill flute reduced the thrust and its fluctuation range. The best matching performance was obtained for the drilling tool at high feed rate and while the texture is perpendicular to the chip flow direction. The influence of texture on thrust changes with change in the shape of the texture; it does not change when the depth of the texture varies [25]. The groove on the flank face was generated parallel to the cutting edge using laser melting. The modification of the carbide drill bit's flank face improved the tool life while machining Inconel 718. The tool life of the drill bit improved by 50% and coolant supply also enhanced. Hence, the machined surface quality also improved and it was determined by measuring roundness deviation and average surface roughness [26].

Micro pattern texturing was done on tungsten carbide drill tool using laser micromachining for accommodating solid lubricant while drilling aluminum alloy. The drilling experiments were performed using solid lubricant filled laser surface textured tool and it was found that the formation of build-up edge was reduced with easy removal of chip, which leads to the improvement of accuracy at both the entry and the exit of the hole. More precise holes were obtained for textured tools filled with graphite and mixture of graphite and molybdenum; also, better surface finish and form of the hole were obtained for those tools (Figure 6.5). Graphite-coated drill bits have higher tool life because of slight reactive behavior and easier adhesion to the tool's textured surface [27].

The laser surface texturing on the high-speed steel drilling tool was performed and TiN coating was applied on it via physical vapor deposition. The drilling trials

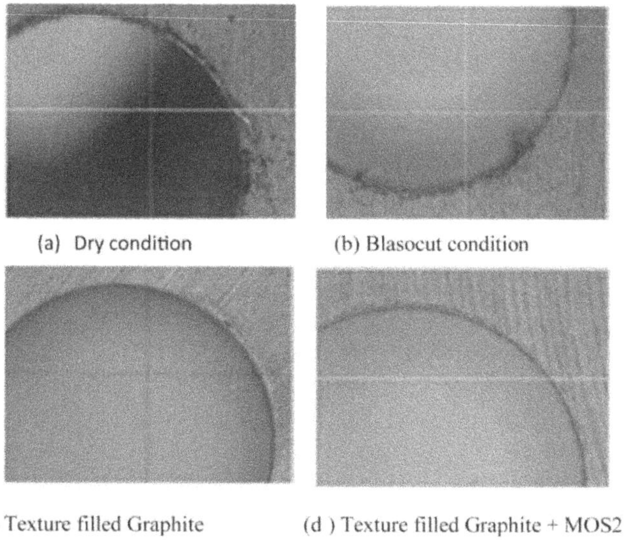

FIGURE 6.5 2D image of hole geometry.

were performed on ABNT 304 steel and its performance was compared with untextured TiN coated tool. It was observed that better adhesion of TiN was obtained for laser surface textured tool; it also improved tool life. The textured tool exhibited lesser tool wear and also the increase in feed force and torsion movement was less with rise in the quantity of holes drilled by them [28].

The friction behavior of different kinds of and sized micro-structures developed on the cemented carbide drilling tool's surface using laser was investigated and their effect on resin adhesion analyzed while drilling printed circuit boards. It was found that the micro-scale texturing using laser improved the machining performance of the drilling tool. The microstructure type and size are the most influencing parameters, which affect coefficient of friction; It was altered by changing the kind, spacing, and depth. Also, the surface texturing improved the chip flow by resin lubrication. The micro-structures with pit and vertical groove on the drilling tool surface can decrease axial drilling force and torque, which reduces tool wear and resin adhesion [29].

The micro textures, namely, star shaped with 6-sharp apex, rectangular slots parallel, and perpendicular to drill axis, were made by laser on the drilling tool. The strength and wear resistance was improved by doing deep cryogenic treatment of the textured tool before filling the solid lubricant. The drilling experiments were done to examine the drilled hole's dimensions and surface integrity. The accuracy of hole was improved for drilling tool with star-shaped texture when compared to other textured and non-textured tools. A considerable enhancement in surface quality and hardness was obtained and also cylindricity error; hole's burr height was less for this condition. It was also found that 96% hole accuracy was obtained for lesser feed rate and higher cutting speed [30].

6.2.4 Grinding

Bionic textures such as shark skin rib-groove and an insect wing vein fractal structure were made on the surface of zirconia ceramics and it exhibited fewer grinding scratches, pits, brittle fracture, and better surface finish than the untextured zirconia ceramics. The insect-wing-vein-textured zirconia showed the highest surface finish and lowest subsurface damage. But the excellent surface morphology was achieved for the bionic shark skin structure. Also, these textures enhanced the cooling conditions and lubrication among the tool and the workpiece surface. Hence, the bionic texturing improved the grinding characteristics and zirconia ceramic's surface integrity. The lubrication mechanisms of various bionic texture are illustrated in the Figure 6.6 [31].

The scaling cross-helix textures were made on resin-bonded cubic boron nitride (CBN) grinding wheels using laser and its influence on grinding performance while machining 390 Böhler steel was studied. It was found that the grinding forces

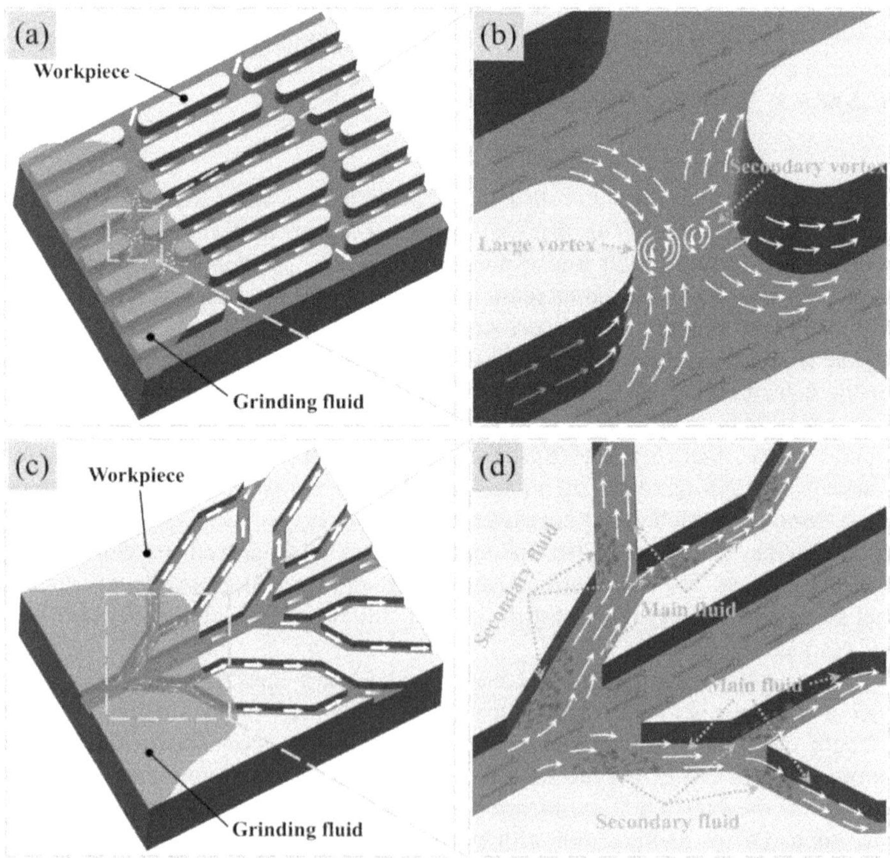

FIGURE 6.6 Grinding fluid flow characteristics of textured grinding wheel surfaces: (a) parallel-groove texture; (b) vortex formation in the parallel-groove texture; (c) branched groove texture; and (d) fluid splitting and redistribution mechanism in the branched groove texture.

decreased by 26% and 24% for grinding wheels with wide grooves (165 μm) and narrow grooves (35 μm), respectively, when compared to conventional grinding wheel. The average maximum temperature of the workpiece was decreased by 20% and 18% for grinding wheels with wide grooves and narrow grooves, respectively, when compared to conventional grinding wheel. The better surface finish of workpiece was obtained while grinding with wheel having narrow grooves than conventional grinding wheel and wheel with wide grooves. Also, the narrow grooved grinding wheel exhibited higher wear resistance [32].

The grinding behavior of CBN grinding wheels while machining laser surface textured Ti-6Al-4V was investigated. It was found that different types of textured patterns have minor effect on surface roughness. But they have significant effect on grinding force, temperature, and wear. The textured Ti-6Al-4V workpieces have better cutting fluid storage capability when compared to non-textured workpiece. This also reduced the mechanical properties of the workpieces, which leads to ease of material removal using CBN grinding wheel. The better grinding performance was obtained for cross-helix pattern and the grinding force decreased from 45% to 56%. Also, the grinding temperature decreased 41% to 52% [33].

The laser ablation was used to make micro- and macro-grooves on the silicon nitride ceramics and diamond grinding wheel, respectively. The cutting fluid flow in the working area and debris-capturing capability of grinding wheel was improved due to the presence of macroscopic grooves in the diamond wheel. The laser surface modification of the silicon nitride decreased the hardness of the material, normal grinding force, and the ratio of the grinding force, which in turn reduced the sample subsurface damage. Also, this condition improved the surface quality by preventing the scratching of debris against the workpiece. Hence, the combination macro and micro grooves improved the grinding wheel life and decreased the wear. The 3D factors of two distinctive structured patterns of diamond wheel surface and the probable thermal deterioration are given in Figure 6.7 [34].

The multiple-pass laser ablation process was used to generate 3D textures on abrasive tools and their dimensional accuracy and structural aspects were investigated in order to accomplish required 3D texture on resin-bonded diamond abrasive tool. This method was used to generate micro structures with considerable geometric precision having 2D sinusoidal shape and 3D staggered hemisphere. The maximum depth and width remain constant in multiple-pass laser ablation, which were related to the resin bond agent's ablation fluence threshold and laser spot size, respectively. The three important material characteristics observed were resin loosening, thermal-expansion-induced diamond breakage, and diamond graphitization with increase in the number of laser passes. The grinding experiments performed using textured wheels reduced the grinding temperature by 28% and a small increase in the grinding force when compared to the conventional grinding wheels [35].

The micro textures were made on the coarse-grained grinding wheel via nanosecond pulsed laser. This method produced micro-groove arrays on the diamond grains; the investigation on the effect of laser properties on surface characteristics of micro textures was carried out and the evaluation of the textured wheel's grinding performance was done. It was observed that the focal point must be in the middle position of wheel surface run-out range so as to confirm the entire grinding wheel surface to

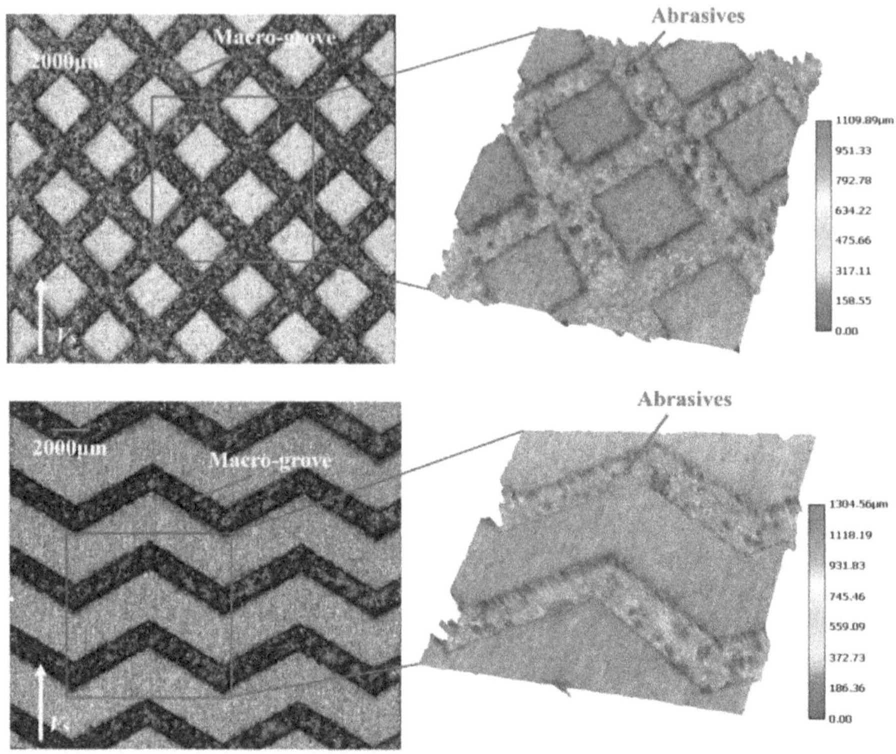

FIGURE 6.7 Optical microscope images of laser-structured grinding wheels and corresponding 3D topography measurements.

be micro textured. The redeposited material on the micro-grooves got reduced with increase in the laser scanning passes; it leads to the formation of sharp micro edges of diamond grains. The grinding experiments revealed that the micro textured wheel decreased the subsurface cracks of the optical glass workpiece [36].

The laser micro structures were developed on the coarse-grained diamond grinding wheel through different processing routes: (1) laser profiling and laser sharpening of the grinding wheel; (2) laser profiling, laser micro-structuring, and sharpening of the grinding wheel; (3) laser profiling, laser sharpening, and micro-structuring of the grinding wheel. The influence of different processing routes on the topographical features specimen of the grooves and the groove angle's influence on the surface finish of the ground was studied. Also, the surface finish of the ground specimens by micro-structured (V-shaped and W-shaped patterns) and non-structured diamond grinding wheels and their grinding ratio were compared. The grinding wheel that has undergone sharpening and micro-structuring illustrated better integral groove structure when compared to the grinding wheels where the micro structuring was performed before sharpening. The average surface roughness values obtained for grinding wheel with V-shaped pattern and W-shaped pattern micro structures were 17–31% and 32–41% less than the non-textured grinding wheel. The geometric factors of the grooves and micro-structure pattern have a significant effect on the grinding performance. The better cooling and chip removal rate was obtained for patterns with W-shape when compared to pattern with V-shape. The wear resistance of the V-shaped and W-shaped patterns on the grinding wheels reduced by 26% and 23%, respectively, in comparison with the non-textured grinding wheel [37].

The macro patterns were developed on the resin-bonded diamond grinding wheel's surface by using pulsed laser. The different patterns were made on the same diamond grinding wheel surface and their effect on the grinding wheel's machining performance was studied. It was observed that kind of pattern had considerable effect on the grinding performance. The grinding force was reduced by 2.5% to 24.5% in comparison with the non-structured grinding wheel. This leads to the better lubrication or cooling effect and less wear among the tool and the sample. The grinding force ratio obtained for laser macro-structured grinding and conventional grinding were 3.45:6.5 and 3.63:6.02, respectively. The grain graphitization occurred due to laser ablation; it was observed using 3D ULDF microscope and there is no considerable rise in the grinding wheel wear was obtained for the structured grinding wheel [38].

The macro-structure patterns were fabricated on the resin bonded diamond grinding wheel's surface using laser. Five types of patterns were made on the grinding wheel surface by accurately controlling laser parameters. The grinding trials revealed that textured grinding wheels can reduce the mechanical and thermal load in the interaction area. The grinding force and temperature also reduced up to 28.9% and 32.0%, respectively. The grinding burn and tensile stresses were reduced by enhanced micro cutting methods and flow rates of coolant through the interaction area. The subsurface cracks were also less for the laser macro structured grinding condition [39]. Three types of textures were fabricated by the laser-textured diamond grinding wheel. The on-machine conditioning such as truing, dressing, and profiling or texturing was possible due the integration of nanosecond pulse laser beam. The surface structuring by laser-conditioned grinding wheel had the ability to make microstructures on tough and fragile samples. Riblets, rectangular grooves, and pillar array were effectively made with varying widths, heights, and periods. The wear failures observed on the structured diamond grinding wheels were macro and micro wear while generating surface structures on ceramics. The grinding ratio obtained was 6.3 to 6.5, and a wear of 40 μm was observed for the laser-shaped texture for the assessed grinding wheels [40].

6.3 APPLICATIONS OF LST IN UNCONVENTIONAL MACHINING

Laser surface texturing was used to create different complex 2D and 3D shapes such as grooves, half spheres, and hexagonal pyramids on electrochemical machining (ECM) tools or cathodes made up of stainless steel. These patterns were reproduced on the surface of the Ti-6Al-4V workpiece via ECM. The process parameters used for the pattern reproduction are given in Table 6.2. It was observed that the ECM reproduced two- and three-dimensional patterns with acceptable geometric features when compared to the tool. But there were some deviations in the geometrical parameters due to the sharp edges present in the patterns. This will cause anisotropic

TABLE 6.2 PECM Process Parameters for Pattern Reproduction

Geometric Characteristic	Level
Process voltage U [V]	18.70
Feed rate vf [mm/min]	0.11
Pulse on-time ton [ms]	1.00
Mech. frequency fmech. [Hz]	50.00
Electr. frequency felectr. [Hz]	50.00
Electrolyte pressure p [kPa]	100.00
Electrolyte temperature T [°C]	22.00 ± 1.50
Electrolyte conductivity σ [mS/cm]	110.00 ± 5.00
pH value [-]	8.50 ± 0.20

electric field and also affect smooth electrolyte fusing condition; it leads to different material rates at those places. This can be rectified by decreasing melting in LST and removing sharp edges from the pattern. Also, the utilization of pico/femtosecond lasers having minimal thermal effects might be considered [41].

The picosecond laser was used to make concave or convex patterns on pulse electrochemical machining (PECM) cathodes; shape reproduction was done on PECM anodes. The groove and half-sphere patterns were replicated on the surface of high-strength material Inconel 718 by integrating LST and by a picoseconds laser and PECM. This method was proved by measuring geometric parameters and analyzing the surface integrity after both LST and PECM surface treatments. The patterns produced on the PECM cathodes using LST have higher accuracy when compared to the patterns reproduced by the PECM on the anodes. The deviations of geometric features were due to the edge rounding and widening effects. This was due to the sharp positions, like hill top on the groove patterns, which significantly increased the local material removal. But the variations were less in the case of half-sphere patterns; this is due to the round patterns without sharp edges, which can be produced more accurately using PECM process when compared to the groove patterns with sharp edges [42, 43].

The microgroove and 3D microstructure were made on titanium alloys using hybrid laser and electrochemical machining (LECM). The material removal mechanism and machining behavior were analyzed while combining laser processing and ECM. The effect of parameters such as voltage, laser power, and feed rate on factors, namely, aspect ratio, accuracy, efficiency, and surface finish, was investigated. The electrochemical corrosion rate became faster due to the laser irradiation and also it improved the ability of passive film removal. There was a considerable improvement in the machining performance such as the aspect ratio, material removal rate, and side gap: they increased by 394.4%, 140% and 172.8% respectively. The titanium alloys were activated by laser-induced temperature rise, which in turn improves the efficiency of electrochemical corrosion by increase in the local electrolyte temperature. The laser power and voltage of 4 W and 18 V in the LECM processing leads to the increase in width and depth by 11.18% and 441.79%, respectively, when compared to normal ECM with 18V voltage. The better LECM accuracy and surface finish were obtained for lower laser power of 3 W. The slot formation occurred in the central area while using direct laser beam with 6 W power. The increase in the laser power to 9 W leads to deep high aspect ratio slot formed through jet-assisted laser beam combined reduced surrounding electrochemical dissolution [44].

Ultrasonic vibration-assisted laser surface processing, which makes use of vertical ultrasonic vibrations to the substrates made up of Ti-6Al-4V alloy during the irradiation with a CO_2 laser was done for the fabrication of laser melted and textured surfaces having probable applications in biomedical implants. The laser processing led to extremely uniform recurring undulating grooved surfaces, and it was notably more evident in the specimens made with greater ultrasonic power outputs. The x-ray diffraction study shows that the phase evolution which proved that the laser processing induced the alteration of globular α to acicular α and martensitic α' and also increased amounts of retained β phases. These phases were also found in the microscopic analysis. The surface texture made by laser processing leads to increase in surface wettability with increase in ultrasonic power output. The textured surfaces showed obvious reduction in friction coefficients while performing sliding wear test with simulated body fluid because of lubricant entrapment inside the textured grooves. The texturing also leads to a major decline in surface contact area during the wear process, which significantly decreased the overall wear rates caused by abrasive wear [45].

The microstructure grooves were made on surfaces of aluminum alloy using electrochemically assisted laser processing. SEM, energy dispersive spectroscopy (EDS), and white light interferometer were used for analyzing textured surfaces. The surface wettability was evaluated by measuring contact angles. Also, the tribological behavior of surface textured samples was investigated by various methods under dry and oil-lubricated conditions. Better surface finish and oil film adherence were obtained for textured made by electrochemically assisted laser processing. Also, the lower coefficient of friction and reduced wear was obtained for oil-lubricated condition when compared to dry machining under same frictional conditions at lower loads. This was because of the existence of electrochemical conditions, which further enhances the accuracy of laser processing. The material removal rate and surface finish were improved by the electrochemical reaction, which further enhanced the wear resistance of the textured surface [46].

6.4 CONCLUSION

LST is a new method which can enhance surface quality and texture, especially improving tribological, biological, or adhesive properties of materials. As LST can make a microscale texture on surface of material, which can be used in different applications, it is regarded as a best method in all industries. But the efficiency of this method is severely affected by process parameters: wavelength of laser, pulse duration, laser power, and scanning speed. Besides, the types of textures made on the surface of material contribute to the improvement in various properties of the materials. In this chapter, the applications of LST in conventional machining processes like turning, milling, drilling, and grinding have been discussed. The LST process can improve the performance and life of the tools and also increases the surface finish and accuracy of the workpiece by improving the lubrication and reducing the cutting force requirement and ease of chip removal. Also, it can be used for machining hard to machine alloys. LST can be used in unconventional machining processes to improve the material removal rate and reproduce the features on the tool on the workpiece with high accuracy. In future, properties of different materials will be studied to obtain the best possible characteristic of surface texture. The major challenge in LST is to create surface texture with smaller sizes in several microns or submicron. After optimization of process parameters, texture and morphology can be the important parameters to enhance the performance of tool and workpiece materials for numerous applications in different fields.

REFERENCES

[1] Sasi, R., Subbu, S. K., & Palani, I. A. (2017). Performance of laser surface textured high speed steel cutting tool in machining of Al7075-T6 aerospace alloy. Surface and Coatings Technology, 313, 337–346.

[2] Sivaiah, P., Revantha Kumar, M., Bala Subramanyam, S., & Prasad, K. L. V. (2021). A comparative study on different textured and untextured tools performance in turning process. Materials and Manufacturing Processes, 36(8), 926–935.

[3] Zhang, K., Deng, J., Ding, Z., Guo, X., & Sun, L. (2017). Improving dry machining performance of TiAlN hard-coated tools through combined technology of femtosecond laser-textures and WS2 soft-coatings. Journal of Manufacturing Processes, 30, 492–501.

[4] Singh, B., Sasi, R., Kanmani Subbu, S., & Muralidharan, B. (2019). Electric discharge texturing of HSS cutting tool and its performance in dry machining of aerospace alloy. Journal of the Brazilian Society of Mechanical Sciences and Engineering, 41, 1–11.

[5] Kawasegi, N., Sugimori, H., Morimoto, H., Morita, N., & Hori, I. (2009). Development of cutting tools with microscale and nanoscale textures to improve frictional behavior. Precision Engineering, 33(3), 248–254.

[6] Sugihara, T., & Enomoto, T. (2017). Performance of cutting tools with dimple textured surfaces: A comparative study of different texture patterns. Precision Engineering, 49, 52–60.

[7] Arulkirubakaran, D., Senthilkumar, V., Chilamwar, V. L., & Senthil, P. (2019). Performance of surface textured tools during machining of Al-Cu/TiB2 composite. Measurement, 137, 636–646.

[8] Devaraj, S., Malkapuram, R., & Singaravel, B. (2021). Performance analysis of micro textured cutting insert design parameters on machining of Al-MMC in turning process. International Journal of Lightweight Materials and Manufacture, 4(2), 210–217.

[9] Da Silva, W. M., Suarez, M. P., Machado, A. R., & Costa, H. L. (2013). Effect of laser surface modification on the micro-abrasive wear resistance of coated cemented carbide tools. Wear, 302(1–2), 1230–1240.

[10] Liang, L., Yuan, J., Li, X., Yang, F., & Jiang, L. (2018). Wear behavior of the micro-grooved texture on WC-Ni3Al cermet prepared by laser surface texturing. International Journal of Refractory Metals and Hard Materials, 72, 211–222.

[11] Peña-Parás, L., Maldonado-Cortés, D., Rodríguez-Villalobos, M., Romero-Cantú, A. G., & Montemayor, O. E. (2020). Enhancing tool life, and reducing power consumption and surface roughness in milling processes by nanolubricants and laser surface texturing. Journal of Cleaner Production, 253, 119836.

[12] Tong, X., & Wang, S. (2024). Effect of laser texturing process on properties of composite coated cemented carbide ball-end milling cutter milling titanium alloy. Proceedings of the Institution of Mechanical Engineers, Part B: Journal of Engineering Manufacture, 09544054241229496.

[13] Zhou, C., Guo, X., Zhang, K., Cheng, L., & Wu, Y. (2019). The coupling effect of micro-groove textures and nanofluids on cutting performance of uncoated cemented carbide tools in milling Ti-6Al-4V. Journal of Materials Processing Technology, 271, 36–45.
[14] Li, Q., Yang, S., Zhang, Y., Zhou, Y., & Cui, J. (2018). Evaluation of the machinability of titanium alloy using a micro-textured ball end milling cutter. The International Journal of Advanced Manufacturing Technology, 98, 2083–2092.
[15] Sugihara, T., & Enomoto, T. (2009). Development of a cutting tool with a nano/micro-textured surface: Improvement of anti-adhesive effect by considering the texture patterns. Precision Engineering, 33(4), 425–429.
[16] Enomoto, T., & Sugihara, T. (2010). Improving anti-adhesive properties of cutting tool surfaces by nano-/micro-textures. CIRP Annals, 59(1), 597–600.
[17] Enomoto, T., Sugihara, T., Yukinaga, S., Hirose, K., & Satake, U. (2012). Highly wear-resistant cutting tools with textured surfaces in steel cutting. CIRP Annals, 61(1), 571–574.
[18] Chen, Y., Guo, X., Zhang, K., Guo, D., Zhou, C., & Gai, L. (2019). Study on the surface quality of CFRP machined by micro-textured milling tools. Journal of Manufacturing Processes, 37, 114–123.
[19] Yang, S., Guo, Y., Xiao, Z., & Wang, D. (2024). Study on cutting performance of micro-textured ball-end milling cutter with multiple distribution density under the action of cutting edge. Journal of Mechanical Science and Technology, 38(4), 1753–1764.
[20] Yang, S., Xing, S., Yu, Y., Han, P., Guo, C., & Liu, L. (2023). Study on the milling performance of ball-end milling cutter under the combined action of micro-texture of rake and flank face. Industrial Lubrication and Tribology, 75(5), 510–525.
[21] Niketh, S., & Samuel, G. L. (2018). Drilling performance of micro textured tools under dry, wet and MQL condition. Journal of Manufacturing Processes, 32, 254–268.
[22] De Lima, M. S. F., Ladario, F. P., Neves, D., & Diniz, A. E. (2011). Machining performance of laser surface micro-textured drilling tools. International Journal of Surface Science and Engineering, 5(2–3), 98–115.
[23] Niketh, S., & Samuel, G. L. (2017). Surface texturing for tribology enhancement and its application on drill tool for the sustainable machining of titanium alloy. Journal of Cleaner Production, 167, 253–270.
[24] Ling, T. D., Liu, P., Xiong, S., Grzina, D., Cao, J., Wang, Q. J., . . . & Talwar, R. (2013). Surface texturing of drill bits for adhesion reduction and tool life enhancement. Tribology Letters, 52, 113–122.
[25] Kawasegi, N. (2009). Development of cutting tools utilizing surface functions induced by micro/nanometer-scale textures-turning experiments of aluminum alloy using DLC coated tools. Journal of the Japan Society for Precision Engineering, 75(10), 1245.
[26] Beer, N., Özkaya, E., & Biermann, D. (2014). Drilling of Inconel 718 with geometry-modified twist drills. Procedia Cirp, 24, 49–55.
[27] Dheeraj, N., Sanjay, S., Bhargav, K. K., & Jagadesh, T. (2020). Investigations into solid lubricant filled textured tools on hole geometry and surface integrity during drilling of aluminium alloy. Materials Today: Proceedings, 26, 991–997.
[28] Neves, D., Diniz, A. E., & de Lima, M. S. F. (2006). Efficiency of the laser texturing on the adhesion of the coated twist drills. Journal of Materials Processing Technology, 179(1–3), 139–145.
[29] Zheng, L., Sun, Y., Luo, C., Zhao, Y., Xu, X., Wei, X., . . . & Wang, C. (2025). Enhancing the tribological properties of cemented carbide drills for printed circuit boards through surface micro-structures. Wear, 564, 205667.
[30] Mishra, P. K., & Kalidas, P. (2021). Hole geometry and surface integrity assessment in drilling of inconel 718 using laser texture filled solid lubricant tools. Industrial Lubrication and Tribology, 73(10), 1267–1274.

[31] Zhang, X., He, T., Wen, D., Li, T., Chen, X., Li, C., . . . & Yang, Z. (2024). Improving zirconia ceramics grinding surface integrity through innovative laser bionic surface texturing. Ceramics International, 50(18), 32081–32097.
[32] Ćwikła, M., Dziedzic, R., Kołodziej, M., & Reiner, J. (2025). Impact of cross-helix texture scaling on grinding performance using resin-bonded CBN laser-textured wheels. The International Journal of Advanced Manufacturing Technology, 1–16.
[33] Zhang, X., Jiang, J., Li, S., & Wen, D. (2019). Laser textured Ti-6Al-4V surfaces and grinding performance evaluation using CBN grinding wheels. Optics & Laser Technology, 109, 389–400.
[34] Zhang, X., Wen, D., Shi, Z., Li, S., Kang, Z., Jiang, J., & Zhang, Z. (2020). Grinding performance improvement of laser micro-structured silicon nitride ceramics by laser macro-structured diamond wheels. Ceramics International, 46(1), 795–802.
[35] Li, H. N., Zhao, Y. J., Cao, S., Chen, H., Wu, C., Qi, H., . . . & Liu, G. (2021). Controllable generation of 3D textured abrasive tools via multiple-pass laser ablation. Journal of Materials Processing Technology, 295, 117149.
[36] Guo, B., Zhao, Q., & Yu, X. (2014). Surface micro-structuring of coarse-grained diamond wheels by nanosecond pulsed laser for improving grinding performance. International Journal of Precision Engineering and Manufacturing, 15, 2025–2030.
[37] Deng, H., & He, J. (2017). A study of the grinding performance of laser micro-structured coarse-grained diamond grinding wheels. The International Journal of Advanced Manufacturing Technology, 93, 1989–1997.
[38] Zhang, X. H., Kang, Z. X., Li, S., Wu, Q. P., & Zhang, Z. C. (2018). Experimental investigations on the impact of different laser macro-structured diamond grinding wheels on alumina ceramic. The International Journal of Advanced Manufacturing Technology, 96, 1959–1969.
[39] Zhang, X., Zhang, Z., Deng, Z., Li, S., Wu, Q., & Kang, Z. (2019). Precision grinding of silicon nitride ceramic with laser macro-structured diamond wheels. Optics & Laser Technology, 109, 418–428.
[40] Geng, Z., Tong, Z., Huang, G., Zhong, W., Cui, C., Xu, X., & Jiang, X. (2022). Micro-grooving of brittle materials using textured diamond grinding wheels shaped by an integrated nanosecond laser system. The International Journal of Advanced Manufacturing Technology, 119(7), 5389–5399.
[41] Fang, S., Ernst, A., Llanes, L., & Bähre, D. (2020). Laser surface texturing of PECM tools and the validation. Procedia CIRP, 95, 891–896.
[42] Fang, S., Frank, A., Schäfer, M., Guo, Y. B., & Bähre, D. (2025). Reproduction of three-dimensional microstructures on Inconel 718 utilizing laser-textured PECM cathodes. Journal of Manufacturing Science and Engineering, 147(3), 031001.
[43] Fang, S., Frank, A., Schäfer, M., & Bähre, D. (2022). Implementation of a picosecond laser for micromachining the cathode of pulse electrochemical machining and a case study. Journal of Micro and Nano-Manufacturing, 10(4).
[44] Yang, Y., Wang, Y., Sun, C., Wu, Q., Yan, J., Liu, Y., . . . & Zhang, W. (2024). Processing of titanium alloys with improved efficiency and accuracy by laser and electrochemical machining. The International Journal of Advanced Manufacturing Technology, 130(7), 4013–4025.
[45] Biswas, S., Alavi, S. H., Sedai, B., Blum, F. D., & Harimkar, S. P. (2019). Effect of ultrasonic vibration-assisted laser surface melting and texturing of Ti-6Al-4V ELI alloy on surface properties. Journal of Materials Science & Technology, 35(2), 295–302.
[46] Xu, Z. B., Lv, X. W., Wang, X. Y., Feng, J. A., Liu, X. L., Peng, J. F., & Wu, H. (2024). Electrochemically assisted laser surface microtexture preparation and tribological properties research. Journal of Manufacturing Processes, 122, 54–64.

7 Mechanical and Tribological Behavior Analysis on Laser-Surface-Textured Metals/Ceramics/Polymers

S. Sathish, S. Vijayakumar, V. Mugendiran, and A. Riyaz Ahmed

7.1 INTRODUCTION

The mechanical and wear behavior of various metals, polymers, and ceramics is discussed extensively in this chapter. Within the dimension of mechanical behavior, laser surface texturing (LST) has been found to be a promising method to fabricate dissimilar materials by joining of metals and polymers. These dissimilar materials play a major role in enabling weight reduction in automotive and aerospace industries. The texturing process is done in the surface of the metal and the textured metal is then joined with a polymer by means of appropriate joining process such as welding, brazing, or hot pressing. Then the mechanical behavior of the metal–polymer composites is evaluated using a tensile testing machine in which the joint strength is evaluated. Also an interesting observation is the drastic increase in surface roughness (SR) of the samples post LST. This increase in SR plays a crucial role by reducing the wear rate and coefficient of friction (COF) of surfaces.

7.2 EFFECT OF TEXTURING ON MECHANICAL BEHAVIOR

7.2.1 LST of Metals

Stainless steels are widely preferred in industries as they provide a combined property of good strength along with excellent corrosion resistance at ambient conditions. LST is a promising process to sustain the corrosion resistance of steels at high operating temperatures. The surface of *AISI 301 LN austenitic grade stainless steel* was textured by a nanosecond pulsed laser beam, which had a power of 50 W. The beam was wobbled over the surface of the material in an oscillating mode, as represented in Figure 7.1. SR of all the samples increased post texturing. The maximum SR

DOI: 10.1201/9781003542087-7

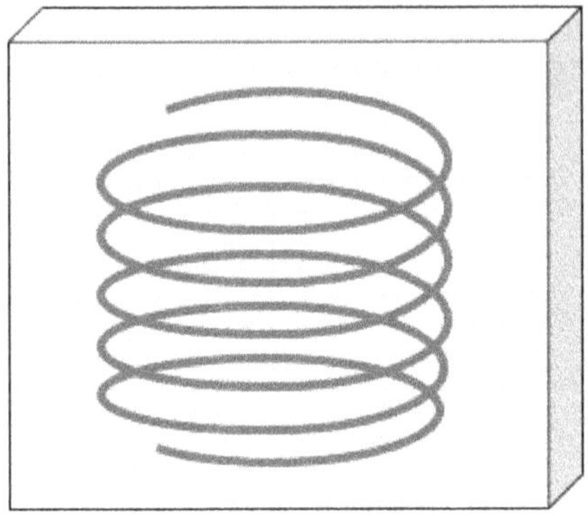

FIGURE 7.1 Wobble texture [1].

obtained was 3.17 μm when textured with an amplitude of 30 μm and a frequency of 300 Hz, compared to non-textured sample which had an SR of 0.85 μm. All the textured samples showed micro hardness higher than the non-textured sample. The highest hardness of 290.14 ± 11 HV was obtained when the laser beam was wobbled at an amplitude of 30 μm and a frequency of 600 Hz, which was 34% more than the non-textured sample, which had an average hardness of 216.42 HV [1]. Parallel linear patterns were textured with a power of 60W on the surface of *AISI 301LN steel*. The non-textured sample had a micro hardness of about 212 HV and there was no significant variation in the microhardness on increasing the depth from the surface. The textured specimen had a maximum microhardness of 428.7 HV at the surface and the microhardness started to decrease with increasing depth and attained stability at 220 HV. The significant point to note is an increase in hardness of 102% post texturing. Another interesting outcome of texturing is the increase in tensile strength from 734 ± 16 MPa in non-textured sample to 821 ± 12 MPa in textured sample. At the same time the increase in tensile strength is also accompanied by a reduction in elongation at fracture from 72% ± 3% to 58% ± 2%. The failure mechanism analyzed by scanning electron microscope (SEM) showed a 45° shear fracture, which had originated from the sub surfaces [2].

Micro grooves were textured on the surface of *5A06 Al*, with a nanosecond pulsed laser, as shown in Figure 7.2. The effect of treatment factors such as energy density, pulse frequency, and scanning speed on the hardness was experimented. It was found that the scanning speed has the maximum effect, and energy density had the minimum effect on the hardness post texturing. The hardness of the surface increases with increase in energy density and pulse frequency and decreases with increase in scan speed. A maximum hardness of 355.08 HV was obtained at an energy density of 30 J/cm^2, pulse frequency of 50 KHz, and scanning speed of 2.5 m/min. The minimum hardness of 106.29 HV was obtained at an energy density of 30 J/cm^2, pulse frequency of 10 KHz, and scanning speed of 2.5 m/min [3].

LST was performed in Ar and N_2 atmosphere on *Ti6Al4V alloy*. Texturing process increased the SR from 0.29 μm in non-textured surface to 3.86–5.38 μm in the

FIGURE 7.2 Micro groove texture [3].

textured surfaces. There was also formation of tensile residual stress of about 303–497 MPa upon texturing and it was found that the residual stresses increased with the increase in laser power. The micro hardness increased with increase in laser power in both Ar and N_2 atmospheres. The micro hardness in Ar-atmosphere-textured samples reached a maximum of 630 VHN at 900 W laser power. The micro hardness in N_2-atmosphere-textured samples reached a maximum of 1327 VHN at 1100 W laser power. However, in both the cases, the VHN reduced with increasing depth from the surface. The nano-hardness measurement curves show a leftward shift, indicating an increase in hardness in the melt zone. Here also the hardness was found to increase with increase in laser power. The maximum nano-hardness of samples in Ar and N_2 was 6.14 GPa and 19.14 GPa, respectively, compared to non-textured sample, which had a nano-hardness of 4.38 GPa. There was also a significant increase in the Young's modulus of the sample, where the non-textured sample had a modulus of 126.74 GPa and samples textured in Ar and N_2 atmosphere had modulus of 149.11 GPa and 256.11 GPa, respectively [4]. An interesting application of LST in metal–metal-welded joints is provided next. Grooves with different depths were textured in *Q355 steel*, which is depicted in Figure 7.10 by different laser processing times, 5 times, 10 times, and 20 times and achieved 6 μm, 26 μm, and 56 μm groove depths, respectively. The textured steel was then joined with *Al 6A01* by welding brazing process. A maximum joint tensile strength of 111 MPa was obtained for 20 processing time. It was also found that the strength increased with increase in processing time. SEM analysis on the fractured region indicated tear fracture in Al [5]. Dimple, grid, and chaotic structures were textured on *Ti6Al4V* by a nanosecond pulsed laser source. A lap joint was made by joining the two textured Ti6Al4V samples by means of an epoxy adhesive. The maximum joint strength of 32.3 MPa was exhibited in the grid textured sample compared to a strength of 3.9 MPa of the non-textured sample. The failure mode by SEM revealed that the samples had both adhesive and cohesive failure. Also, the best mechanical interlocking effect was found in the grid patterned sample, justifying its high strength [6].

7.2.2 LST of Metal–Polymer Composites

Plastic–metal *acrylonitrile butadiene styrene* (ABS) was ultrasonically welded to *aluminum alloy Al 5052*. LST was performed in the surface of Al 5052 in the shape of line grid 0.2 mm, 0.3 mm apart and circular paths of 0.2 mm radius with 0.6 mm spacing between them, as shown in Figures 7.3 and 7.4. The texturing was done at laser powers of 8.3 W, 20 W, and 18.2 W for the three cases, respectively. Now the Al5052 was ultrasonically welded to ABS. The joint strength of the welded parts was evaluated by tensile testing machine. A maximum joint strength of 17.87 MPa was obtained in the line grid textured samples having 0.2 mm spacing, when welded for 1.7 s at a pressure of 0.16 MPa. A minimum strength of 2.03 MPa was obtained in the circular grid textured samples, when welded for 1.1 s at a pressure of 0.1 MPa. SEM analysis revealed the samples failed due to interfacial breakage [7].

Al 6061 was textured with line patterns using a nanosecond pulsed laser beam with a power of 70 W, as shown in Figure 7.5. The average SR increased with increase in texture depth and reached a maximum of 28.13 μm at a depth of 215.4 μm. There was a notable decrease in SR on increasing the texture interval and the maximum SR

FIGURE 7.3 Line texture [7,21,35].

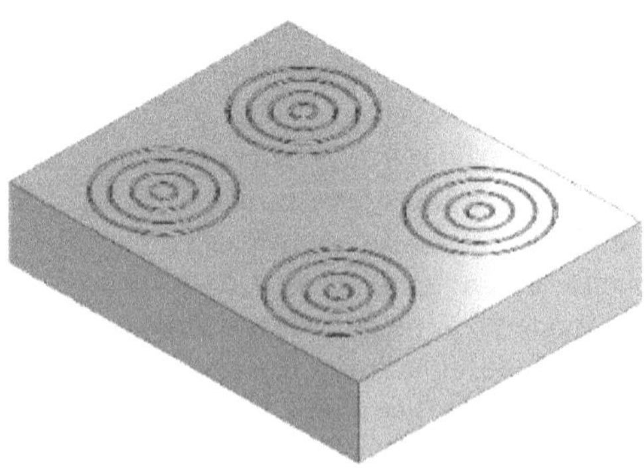

FIGURE 7.4 Circular texture [7].

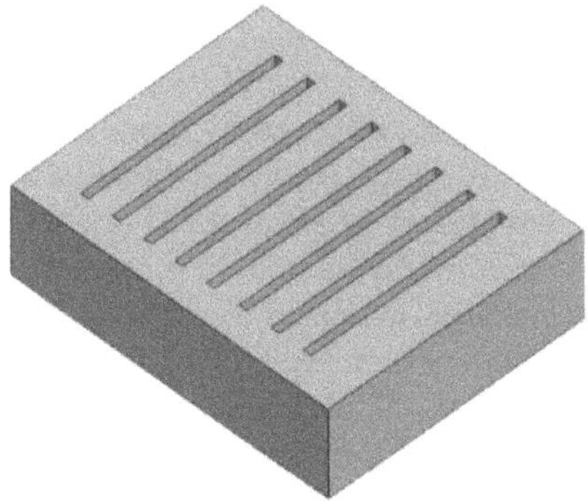

FIGURE 7.5 Line grid texture [8,35].

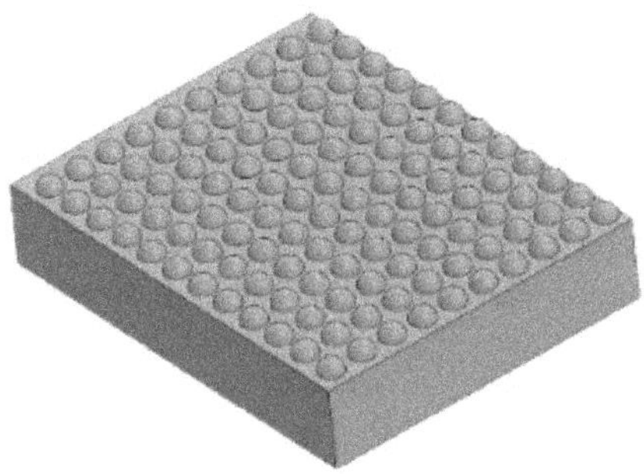

FIGURE 7.6 Papilla textures [9].

of 25.32 μm was obtained at 0.5 mm interval. The effect of SR on the texture width showed an increase up to 0.3 mm width, which showed an SR of 31.84 μm, after which it decreased. Overall, all the textured samples displayed a prominent increase in SR compared to non-textured samples. The textured Al 6061 was joined with *carbon-fiber-reinforced Polyamide 66* by a high speed friction stir lap joining process. The hybrid joints were subjected to tensile shear strength tests. A maximum shear strength of 25.47 MPa was obtained at a texture depth of 179.4 μm, a texture interval of 0.5 mm, and a texture width of 0.2 mm. The failure of the composite occurred at the region of fiber reinforced polymer, thereby indicating the efficacy of strong joining [8].

Biomimetic papilla structures of diameters ranging from 35 μm to 50 μm were textured on *Al 6061* by a femtosecond laser having a power of 52.40 W (Figure 7.6). The textured surfaces were joined with *carbon fiber reinforced PEEK* by friction stir spot welding process. The tensile shear strength of the composite was found to increase with increase in texture diameter up to 45 μm and reduces thereafter, which was attributed to weak mechanical interlocking. The maximum strength obtained at 45 μm was 37.06 MPa, which is 157.18% higher than the non-textured sample. Also papilla structures textured at 45 μm produced a joint efficiency of 69.92%, better than any other texturing features [9]. Nanosecond laser texturing of rectangle, rhombus, triangle, and shapes was performed on *Al 6061* with a 70 W power laser, as shown in Figures 7.7a, 7.7b, 7.7c, respectively. Rectangular-textured samples displayed the highest SR of 1.981 μm, followed by rhombus- and triangle-shaped texture surfaces. The SR of all the textured surfaces increased significantly compared to non-textured surfaces, which had an SR of 0.283 μm. The textured Al 6061 was joined with *glass fiber reinforced Polyamide 66* through hot pressing method. The tensile shear strength experiments indicated that among the textured samples, rectangular texturing resulted in highest strength of 10.90 MPa and triangular texturing displayed a low tensile strength of 7.82 MPa. However, all the textured samples had a strength higher than that of non-textured sample, which had a strength of only about 1.59 MPa. The failure modes were analyzed by SEM analysis, which unveiled a failure in the interfacial region in the non-textured sample and mixed mode of failure in textured samples [10].

U-, V-, and I-shaped grooves were textured on Al 6061 by nanosecond pulsed laser beam. The textured *Al 6061* was joined with *Polyamide-6* by friction stir welding process. Tensile shear strength investigation revealed that the V-shaped grooves withstood the maximum load of 1196 N followed by I-shaped groove and U-shaped groove, which withstood 800 N and 600 N, respectively [11]. *Polybutylene terephthalate* was joined with *textured surface of Al 5052* through injection molding machine. Dimple diameters of size 40 μm and 80 μm were textured by picosecond, nanosecond, and continuous wave lasers, as depicted in Figure 7.8. When the dimple

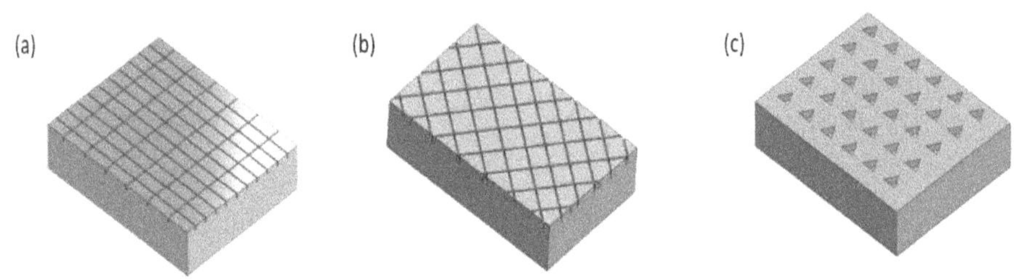

FIGURE 7.7 Textured (a) rectangle, (b) rhombus, and (c) triangle [10].

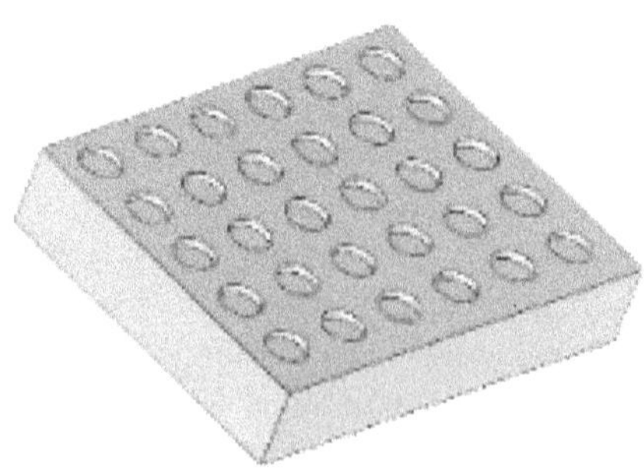

FIGURE 7.8 Circular grid textures [12,25,27,28,31,33–36,39–41].

size was smaller, more dimples can be textured in a given fixed area leading to a joint strength of about 19 MPa. Increasing the dimple diameter leads to formation of lesser number of dimples in a particular fixed area, leading to a reduction in strength to about 12 MPa. The effect of using different laser types on the failure modes was studied by SEM analysis. Texturing with nanosecond laser led to the failure at the polymer side, texturing with picosecond laser led to cohesive failure, and texturing with continuous wave laser led to the polymer being pulled out of the dimples [12].

Grid shaped structures were textured *on cold rolled steel* at a power of 25 W, as shown in Figure 7.9. The cold rolled steel was textured at various hatch distances and laser repetitions. The steel textured with 150 μm hatch distance and a laser repetition of 12 times showed the highest SR of 19.8 μm compared to non-textured surface, which had an SR of 0.447 μm. The textured steel was joined with *flax fiber reinforced vitrimer* by hot pressing method. The cross-tensional strength increased with increase in hatch distance and attained a maximum of 1680 N at a distance of 150 μm and then started to decrease. A similar increasing tendency in strength was noticed with an increase up to 9 laser repetitions, which had a maximum strength of 1680 N, beyond which it started to reduce [13]. Square-shaped texturing of sizes 200 μm and 300 μm was performed on Al 7075 with a nanosecond pulsed laser working at a power of 30 W, which is depicted in Figure 7.9. The textured *Al 7075* was joined with *Polyetheretherketone* (PEEK) by a friction assisted joining process. The sample, which was cleaned and textured with size of 200 μm, showed a maximum tensile shear strength of 25 MPa. The tensile shear strength of all cleaned textured samples was higher than the strength of textured samples which were not cleaned. Also the cleaned and 200-μm-textured samples absorbed a maximum energy of 13.2 J. The failure mode analyzed by SEM showed that the samples failed due to multiple reasons like pull out, teeth failure, and cohesion failure [14]. Another work studied the maximum load carried by Al 7075-PEEK by studying the effect of square grid size and the laser repetitions, which is shown in Figure 7.9. The peak force increases with increase in laser repetitions and it was found to decrease with increase in texture size of the square. A maximum peak load of around 5.7 KN was obtained in the sample having size 200 μm and repetition of 20. The fracture analysis showed that the sample having texture size of 200 μm failed due to pull out of the metal, whereas the sample having texture size of 300 μm failed due to a combination of pull out and teeth failure [15].

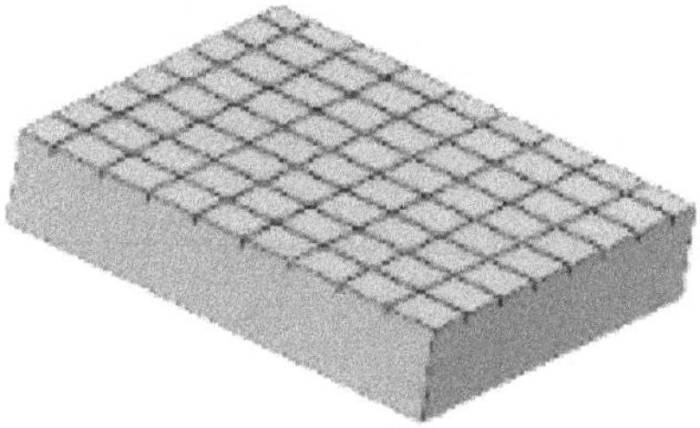

FIGURE 7.9 Square grid textures [13–15,17,26,30].

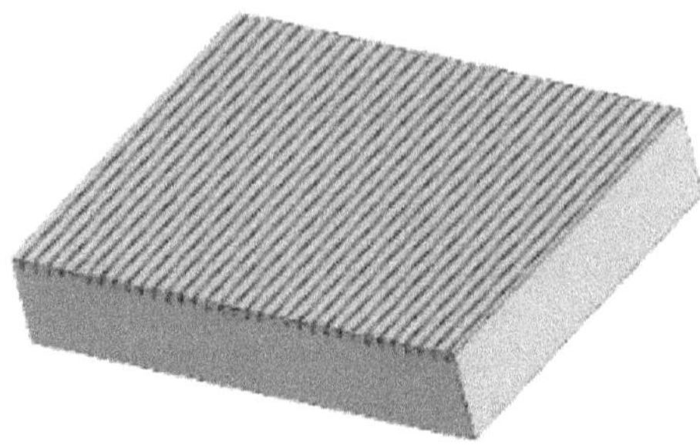

FIGURE 7.10 Linear groove textures [16,5,29,46].

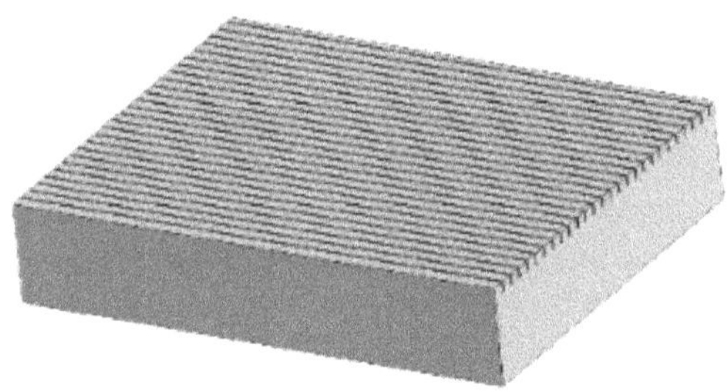

FIGURE 7.11 Perpendicular groove textures [16].

Groove texturing was performed on *low alloy steel* by a nanosecond pulsed laser as shown in Figures 7.10 and 7.11. The textured low alloy steel was joined with *glass-fiber-reinforced polyamide* with the help of a continuous wave laser beam. The variations in groove depth, groove width, groove distance, groove orientation, and laser beam angle on the joint strength of T joints were studied. The samples textured with grooves having a minimum spacing of 200 μm had the highest joint strength of about 11 MPa than the samples having a groove spacing of 600 μm [16]. Nanosecond laser was used to texture the surface of *Ti6Al4V*, which was then joined with *carbon fiber reinforced PEEK* by hot press joining. Square grids of side length 1 mm were textured, as shown in Figure 7.9. The line width was varied from 0.1 mm to 0.4 mm to study the effect of the variation in line width on the tensile shear strength. It was observed that the tensile shear load increased to 5286 N at line width of 0.2 mm, beyond which it started to decrease. The key takeaway is the increase in load-carrying capacity in post textured samples compared to non-textured sample, which carried only a load of 1312 N. The fracture region analyzed by SEM showed a combination of interfacial and cohesive failure [17].

7.2.3 LST of Ceramics

Alumina-toughened zirconia surface was textured by femtosecond-based laser texturing method. The ceramic was subjected to a pre-sintering process prior to LST and was also post-sintered after LST. The same material surface was also modified by sand-blasting method. All the samples were subjected to flexural test, which showed that the sample which was sand blasted had the maximum flexural strength of 1236 MPa. The sample which was textured after the pre-sintering process exhibited a flexural strength of 720–750 MPa compared to the sample textured without pre-sintering, which had a strength of 500–600 MPa. The key takeaway is that sintering the ceramics prior to LST can result in the improvement of flexural property [18]. Straight lines with line gap 0.1 mm were textured in *tungsten carbide cobalt* surface with a nanosecond pulsed laser. The SR increases with increase in energy density and a maximum roughness of about 9 μm was obtained at a scanning speed of 150 mm/s and an energy density of 35.37 J/cm^2. A maximum hardness of 2800–2900 HV was obtained when textured with an energy density of 35.37 J/cm^2 and a scanning speed of 50 mm/s. A minimum hardness of 1700 HV was obtained when textured with an energy density of 5.89 J/cm^2 and a scanning speed of 50 mm/s. The key takeaway is the increase in hardness of textured surfaces in comparison to non-textured surface, which had a hardness of 2332 HV [19].

Squares of 50 μm length and 80 μm spacing were textured on *silicon carbide (SiC) substrates* by nanosecond type pulsed laser, as depicted in Figure 7.12. Silicon hafnium oxide was bond coated over the textured surface by atmospheric plasma spraying method. SR and tensile shear strength of the textured surface was compared with grit blasted surface. The SR of textured SiC was 13.8 ± 0.2 μm compared to an SR of 9.5 ± 0.5 μm for the grit blasted surface. After the bonding process, the textured part had a high strength of 27.31 ± 3.34 MPa compared to the grit-coated part which had a strength of 8.12 ± 0.70 MPa [20]. Microcolumn grid arrays having 100 μm gap were textured on *zirconia* by a nanosecond pulse based laser, as shown in Figure 7.3. Post-processing by heat treatment and silicone-oil-assisted heat treatment revealed an increase in corrosion resistance of the textured samples [21]. Ytria-stabilized tetragonal *zirconia polycrystal (Y-TZP)* surfaces were subjected to nanosecond and picosecond laser treatment. Square trenches were textured on the surface of Y-TZP and the 4-point bend test revealed that a maximum strength of 1340 MPa was obtained when utilizing picosecond pulsed laser treatment [22]. *Zirconia* surfaces were textured with nanosecond-pulsed-based laser treatment. It was found that the flexural strength decreases from 1347 ± 64 MPa in non-textured sample to 1190 ± 81 MPa in

FIGURE 7.12 Square groove textures [20,27,28,37,45].

textured sample. Further the nano indentation hardness and Young's modulus are also found to decrease in the samples which were textured [23]. *Ytria-stabilized tetragonal zirconia* polycrystal (Y-TZP) surfaces were subjected to nanosecond pulse based LST. A maximum SR of 3.89 μm was obtained when textured at a density of 424.5 J/cm^2 and scanned with a velocity of 15 mm/s. The textured samples were also subjected to heat treatment. Among the textured and textured+heat treated samples, a maximum flexural strength of about 1400 MPa was obtained when textured with a density of 254.70 J/cm^2, scanning speed of 50 mm/s, and heat treated at 500° C [24]. A compilation of the mechanical behavior of textured materials is shown in Table 7.1.

TABLE 7.1
Mechanical Behavior of Metals/Polymer/Ceramics

Serial		Micro Hardness (HV)		Joint Strength (MPa)	
Number	Material	Non-Textured	Textured	Non-Textured	Textured
	Metals				
1	AISI 301 LN steel [1]	216.42	290.14 ± 11	–	–
2	AISI 301 LN steel [2]	212	428.7	734 ± 16	821 ± 12
3	5A06 Al [3]	73.975	355.08	–	–
4	Ti6Al4V [4]	278	630 (Aratm) 1327(N_2 atm)	126.74 GPa	149.11 GPa (Aratm) 256.11 GPa (N_2 atm)
5	Q355 steel-Al6A01 [5]	–	–	–	111
6	Ti6Al4V [6]	–	–	3.9	32.3
	Metal–polymers				
7	Al5052–ABS [7]	–	–	–	17.87
8	Al6061–CFRPEEK [8]	–	–	6.3	25.47
9	Al–CFRPEEK [9]	–	–	14.41	37.06
10	Al6061–GFRPA [10]	–	–	1.59	10.90
11	Al6061–PA [11]	–	–	–	1196 N
12	Al5052–PBT [12]	–	–	–	19
13	Steel–flax fiber reinforced vitrimer [13]	–	–	–	1680 N
14	Al 7075–PEEK [14]	–	–	–	25 13.2J
15	Al 7075–PEEK [15]	–	–	–	5.7 KN
16	Steel–GFRPA [16]	–	–	–	11
17	Ti6Al4V–CFRPEEK [17]	–	–	1312	5786
	Ceramics				
18	Al_2O_3 toughened Zr [18]	–	–	–	720–750
19	WC–cobalt [19]	2332	2800–2900	–	–
20	SiC–SiHfO$_2$ [20]	–	–	8.12 ± 0.70	27.31 ± 3.34
21	YTZP [22]	–	–	–	1340
22	Zr [23]	–	–	1347 ± 64	1190 ± 81
23	YTZP [24]	–	–	–	1400

7.3 EFFECT OF TEXTURING ON TRIBOLOGICAL BEHAVIOR

7.3.1 LST of Metals

Dimple shapes were textured on *SS 316*, *NiTi*, *Ta2*, and *Ti6Al4V* with a laser having power 16 W, as depicted in Figure 7.8. The texture density was varied from 0% to 11%. From the wear rate experiments, the lowest wear rate was found at a texture density of 5% for SS 316, 7% for NiTi, 5% for Ta2, and 11% for Ti6Al4V. For SS 316, the COF was found to increase with the increase in texture density from 0.18 for non-textured sample to 0.48 for the sample textured with 11% density. Ta2 and Ti6Al4V followed the same trend. However, due to the elastic property of NiTi, the COF reduced after an initial increase. An interesting observation is that the COF of all textured samples was higher than the COF of non-textured sample [25]. Cross hatch grids with various spacing width and scanning speeds were textured in *20CrMnTi steels* with a laser having a power of 20 W, as shown in Figure 7.9. The SR of all textured samples were higher than the SR of non-textured samples. It was found that the SR increased with a decrease in scanning speed. The spacing width did not have any effect on the SR. Texturing with a scanning speed of 100 mm/s and a space width of 50 μm had the highest SR of 0.9 μm compared to non-textured sample, which had an SR of 0.224 μm. Ball-on-disc tribometer testing with lubrication was conducted to study the tribological behavior. Texturing at 100 mm/s having width of 20 μm gap produced lowest wear quality of 10.8 mg compared to non-textured sample having a wear loss of 17.2 mg. The lubricant forms a hydrodynamic film due to which all the textured samples exhibit COF lesser than non-textured samples. The lowest COF was exhibited by the sample textured at a scanning speed of 100 mm/s having a gap of 20 μm. The efficacy of the results was confirmed by experimenting the same in gear samples, which proved that textured gears were more efficient than non-textured gears [26].

Circle, square, rhombus, and hexagon shapes were textured in *AISI 316L steel* by laser powder bed fusion method (LPBF), as shown in Figures 7.8, 7.12, 7.13, and 7.14, respectively. Non-textured samples were also prepared by LPBF and casting for comparison. The SR of the textured samples was found to be 4.5–4.7 compared to casted sample, which had an SR of 1.1–1.4 μm. The micro hardness also increased from 240 HV in casted samples to 270 HV in textured samples. The wear behavior was experimented in dry, aqueous, and oil lubricated conditions. Circle-shaped

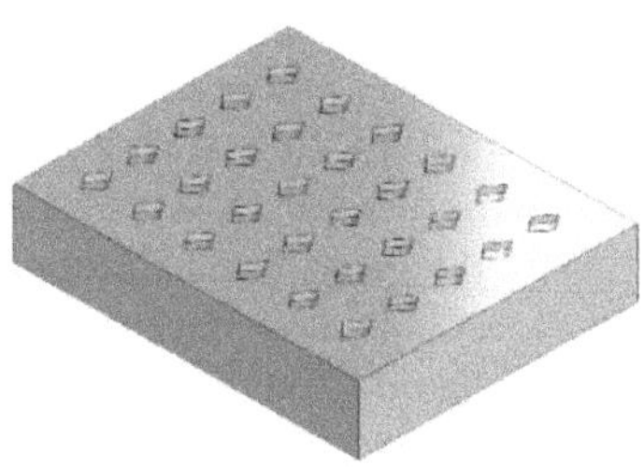

FIGURE 7.13 Rhombus groove textures [27].

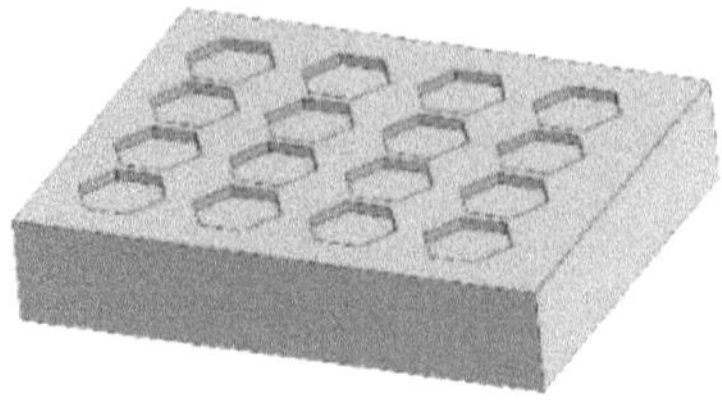

FIGURE 7.14 Hexagon groove textures [27].

texturing exhibited a lowest COF of 0.37, 0.32, and 0.22 in dry, aqueous, and oil lubricated conditions, respectively, compared to a COF of 0.58, 0.48, and 0.39 for the non-textured sample operating in the same condition. The circle-shaped textured surface also had the lowest wear loss of 1×10^{-3}g, 0.3×10^{-3}g, and 0.1×10^{-3}g in dry, aqueous, and oil lubricated conditions, respectively, compared to a wear loss of 2.6×10^{-3} g, 1.8×10^{-3} g, and 1.2×10^{-3} g for the non-textured sample operating in the same condition [27].

Circle, square, and triangle shapes were textured on the surface of *Al 7075* by a femtosecond laser, as represented in Figures 7.8, 7.12, and 7.15, respectively. The texturing was performed at varying spacings of 100 μm, 150 μm, and 300 μm. The COF was found to increase with increase in the space between the textures. The circle textured sample had the lowest average COF of about 0.12, followed by square and triangle textured samples, compared to non-textured sample, which had the highest COF of 0.65. The circle textured samples also had the lowest wear of 0.007 mm^3/Nm compared to non-textured sample, which had a wear rate of 0.033 mm^3/Nm. Among the textured samples, those having triangle shape had the most severe wear [28]. Square, parallel, and perpendicular texturing was performed in *AISI 304 steel* by a femtosecond pulsed laser, as shown in Figure 7.10. The square textured sample exhibited the maximum SR of 90 nm compared to 20 nm in non-textured sample. The COF of square textured sample was 0.07, followed by perpendicular and parallel textured samples, having COF of 0.13 and 0.14, respectively. The COF is found to be drastically reduced upon texturing operation, compared to non-textured sample, which had a high COF of 0.22. The lowest wear rate was also found in square textured sample, having a value of 0.2×10^{-4} mm^3/Nm compared to non-textured sample, which had a wear rate of 0.45×10^{-4} mm^3/Nm [29]. Square grids were textured on *AZ31B magnesium alloy* by direct laser shock surface patterning method using a nanosecond pulsed laser, as shown in Figure 7.9. The micro hardness increased with increase in laser intensity and reached a maximum of 88 VHN when textured with 1.92 GW/cm^2. The COF of the textured surface was 0.124, lesser than that of non-textured surface having a COF of 0.151. Also, the COF was found to decrease with an increase in laser intensity [30]. Circular and ellipsoid patterns were textured on *AISI 52100 chrome steel* with a 20-W-powered laser, as shown Figures 7.8 and 7.16, respectively. The SR of the ellipsoid textured surface increased drastically to 23.6 μm compared to non-textured surface, which had an SR of 3.21 μm. The wear rate was found to increase with increase in load. The circle textured surfaces had the highest wear rate of 1.2 mm^3/m when operating at speeds of 750 r/min and 1400 r/

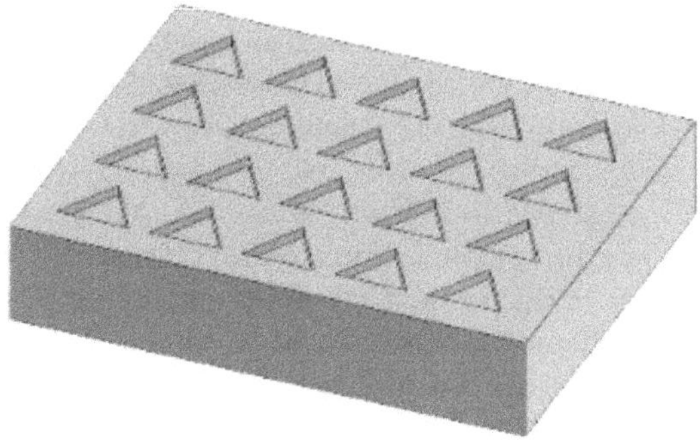

FIGURE 7.15 Triangular groove textures [28].

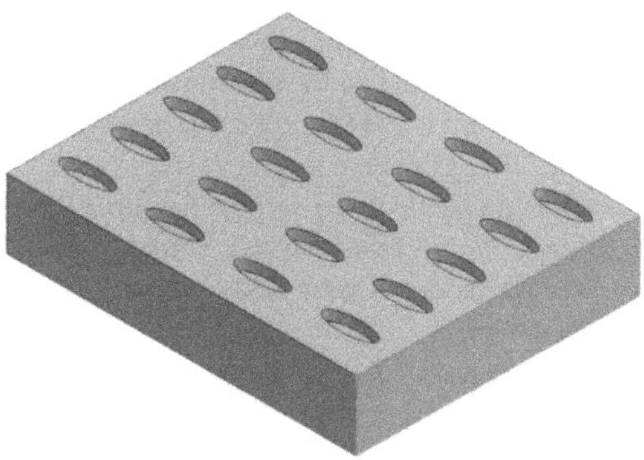

FIGURE 7.16 Elliptical groove textures [31,32].

min. For the ellipsoid textured surfaces, there was no notable change in COF on increasing the loads. For the circle textured surfaces, the COF reduces drastically on increasing the load [31]. Ellipse-shaped patterns were performed on the surfaces of *HT 250 cast iron* block, as represented in Figure 7.16. The COF was found to increase with increase in load reaching a maximum of 0.05 at 889 N. However, it was lesser than that of non-textured sample, which had a COF of 0.07. The depth and width of the wear track also increased with an increase in load. Interestingly, the wear loss for the textured samples was found to be greater than that for the non-textured samples [32]. Circular arrays in the shape of square arrays were textured in the surfaces of *cast iron* by a nanosecond pulse based laser, as depicted in Figure 7.8. This study mainly focused on the lubrication parameters on the friction behavior. It was found that the textured surfaces had the ability to maintain elastohydrodynamic lubrication (EHL) for a long time compared to non-textured samples [33]. Micro dimple circular arrays were textured on *copper* surfaces with a nanosecond pulse based laser, as shown in Figure 7.8. The texture density was varied: 5%, 13%, and 35%. A minimum COF of 0.1 was obtained when textured with a density of 13% and a load rate of 1.27 N/s. The COF was found to reduce drastically in comparison with a non-textured sample having a COF of 0.317. The experiments on the effect of rotational speed on the COF revealed that the COF decreased with an increase in the rotational speed. A minimum COF of 0.07 was exhibited at a maximum rotational speed of 900 rpm and when texturing at a density of 13%. The mechanism of wear was found to be abrasive and adhesive [34]. Micro dimple, grooves, and grid patterns were textured on *chromium alloys* by LST, as shown in Figures 7.5 and 7.8. A highest microhardness of 315 HV was obtained in the sample textured with grid pattern. The studies on friction revealed that it took 900 s for the COF to become stable. The lowest COF was achieved on texturing with grid patterns, whose width was 100 μm and were spaced 200 μm apart. The lowest wear rate was also found in the grid textured samples, which was 0.194×10^{-4} mm^3/Nm. This was very much lower than the wear rate of non-textured sample, which was 0.452×10^{-4} mm^3/Nm [35]. Micro crater patterns were textured on *diamond nanocomposite* by femtosecond pulse laser, as shown in Figure 7.8. The wear testing was done in oil lubricating conditions. The COF reduced from 0.22 in non-textured samples to 0.12 in textured samples. The depth of wear also reduced from 30 nm in non-textured samples to 20–25 nm in textured samples [36]. Groove and square shaped patterns were textured in *CVD diamond films* by a nanosecond pulsed laser, as shown in Figure 7.12. It was found that the square textured sample exhibited a COF higher than groove textured sample. Subsequently, the wear rate is also lowest for square textured sample having 3.95×10^{-5} mm^3/Nm. This is very low when compared to non-textured sample having a wear rate of 1.52×10^{-4} mm^3/Nm [37]. Micro pores were textured on *T8 steel* by a nanosecond pulsed laser system. The effect of COF on speed and load was experimented. The COF of textured samples was lower than the non-textured samples in all loads and speeds. A minimum COF of 0.07 was obtained at a load of 500 N and a minimum COF of 0.05 was obtained at a speed of 1.4 m/s [38]. *GCr15 bearing steel* was coated with babbit alloy wire and mask layer. Micro dimple of diameters ranging from 120 μm to 150 μm was textured on the surface by a picosecond pulse based laser system, as depicted in Figure 7.8. COF stabilization took place after 1000s. Textured layer showed a COF of 0.034, which was lower than the COF of non-textured sample. The wear rate also followed the same trend, the textured having a low wear rate of 2.08×10^{-7} mm^3/Nm compared to non-textured one. The rotation speed was found to have a profound influence on the wear rate. The wear mechanism was a combination of adhesive, abrasive, and oxidative wear [39]. Dimples with varying densities—1.1%, 7.1%, and 11.2%—were textured on a *nickel composite*, which was prepared by powder metallurgy, as shown in Figure 7.8. The lowest COF was 0.03–0.1, which was found in the surface of the sample textured with a density of 11.2% and when the temperature was increased up to 400°C. The wear rate was also found to be lowest in the same sample [40]. TiN was coated on *AISI M2 tool steel* and the surface was textured with rectangular micro dimple arrays using a nanosecond pulse based laser, as shown in Figure 7.8. The effect of load on the COF was studied in depth. The COF increased with an increase in load and this was attributed to continuous transition of the lubrication regime from hydrodynamic to boundary condition. Thereby the least COF of 0.076 was obtained at a load of 50 N when textured with a density of 5%, which is lesser than non-textured sample having a COF of 0.096. The optimal condition for texturing suggested was to use a laser power of 8 W and irradiating at 0.1 s to obtain a dimple having diameter 50 μm and depth 20 μm [41]. Moat structures with square and circular pillars were textured on *Ti6Al4V* by a nanosecond based pulsed laser, as depicted in Figures 7.17 and 7.18 respectively. The square moat textured sample exhibited the highest hardness of 384 HV, which was way better than the non-textured sample, which had a hardness of 344 HV. The tribological behavior was studied both in dry and in lubricated conditions. In dry testing at a density of 50%, the square moats showed the lowest COF of 0.35 compared to non-textured surfaces, which had a COF of 0.456. In lubricated testing at a density of 50%, the square moats exhibited a decrease in COF to 0.257. In dry testing at a density of 50%, the square moats showed the lowest wear depth of 104.65 μm compared to non-textured surfaces, which had a wear depth of 159.91 μm. In lubricated testing at a density of 60%, the square moats exhibited a decrease in wear depth to 88.6 μm [42].

Bionic textures on the surfaces of *Ti6Al4V* was found to play a significant role in reducing the wear. The bionic textures are represented in Figures 7.19 and 7.20. Ripple-shaped bionic textures reduced the COF prominently from 0.19 ± 0.03 in non-textured surfaces and to 0.05 ± 0.02 in textured surfaces. The wear resistance was found to be more prominent when the texturing was done in a spacing of 0.2 mm [43].

FIGURE 7.17 Square moat textures [42].

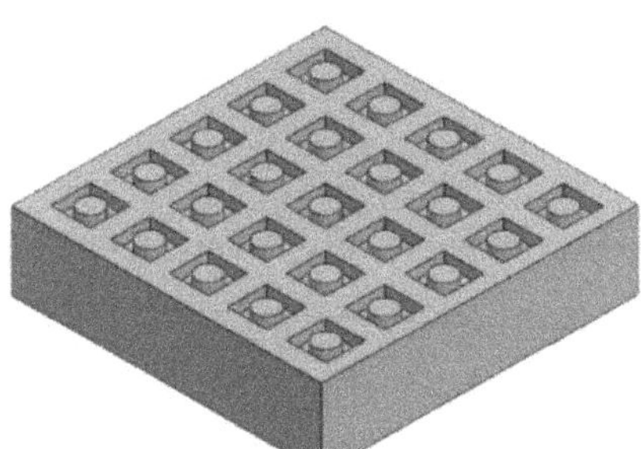

FIGURE 7.18 Circular moat textures [42].

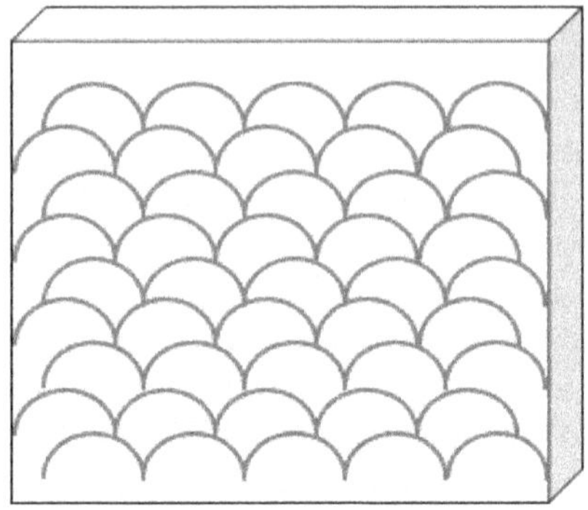

FIGURE 7.19 Fish bionic textures [43].

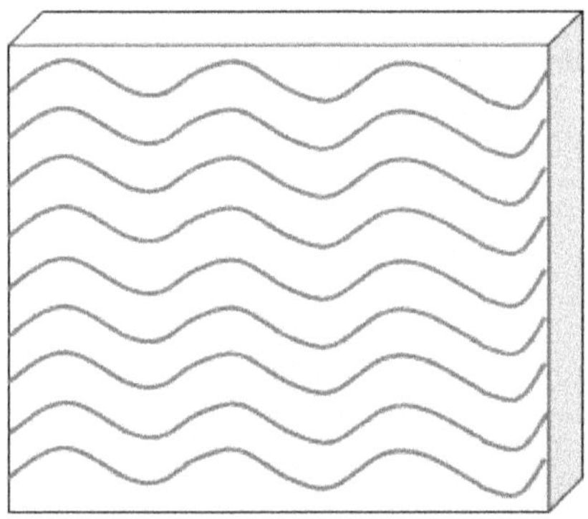

FIGURE 7.20 Shell bionic textures [43].

Dimple textures with 70 μm diameter and 50 μm depth were done on the surface of *Ti6Al4V* by nanosecond pulse based laser. Friction testing was performed without and with solid lubricants. Increasing the sliding speed was found to reduce the COF. Textured surfaces wear tested with lubricants exhibited the lowest COF of 0.1 up to 1 minute, which only increased up to 0.37 [44].

Skew surfaces were fabricated on *AISI 1045 carbon steel* by indirect laser shock surface patterning, as shown in Figure 7.12. Increasing the laser intensity increased the COF, which reached a maximum of 0.76 at an intensity of 0.89 GW/cm^2. A minimum COF of 0.15 was obtained at an intensity of 0.484 GW/cm^2. Texturing process also increased the hardness from 212 VHN in non-textured sample to 238 VHN in textured sample [45]. Parallel lines having a gap of 0.1 mm were textured on *Al-SiC composites* by a nanosecond pulsed laser system, as shown in Figure 7.10. The COF was found to decrease with an increase in energy density and reached 0.12 for a maximum density of 35.37 J/cm^2. The COF increases with increase in scanning speed and had a value of 0.5 at the maximum speed of 200 mm/s. The wear volume was found to decrease up to a maximum of 70% on textured samples compared to non-textured ones. The minimum wear volume was 1 mm^3 obtained when textured at 8.84 J/cm^2 with a speed of 10 mm/s [46].

7.3.2 LST of Ceramics

Chevron-shaped micro textures were patterned on the surface of *cemented carbide* by femtosecond-pulse-based laser, as shown in Figure 7.21. The texturing was done at various densities—5.05%, 9.5%, 13.02%, and 15.2%. It was found that the COF decreased with increase in load. The COF was largest when textured at 15.2% having a value of 0.22 at 40 rpm and 15 N load. The COF was lowest when textured at 9.5% having a value of 0.02 when operated at a load of 300 N. Texturing with a density of 9.5% was found to be optimal for obtaining lower friction and wear characteristics. The mechanism of wear was found to be abrasive and adhesive wear [47].

Circular, groove, and square pit shapes having spacing of 120 μm were textured on the surface of *cemented carbide*. The sample with circular pit had the lowest COF of 0.421 compared to non-textured sample, which had a COF of 0.526. Adhesive wear dominated as the major mechanism of wear. The suggested texturing parameters to reduce the wear is to have a laser energy of 1103.494J/cm^2, and scanning 3 times with a speed of 991.2mm/s [48]. Dimple and groove shapes were textured on the rake of *sintered tungsten carbide tool*, which was used in machining Al2024 alloy. Friction tests were conducted at 10 N load and contact areas—0.5 mm, 1 mm, and 2 mm. The frequency was varied from 0.5 Hz to 50 Hz. EHL was observed at a frequency of 40

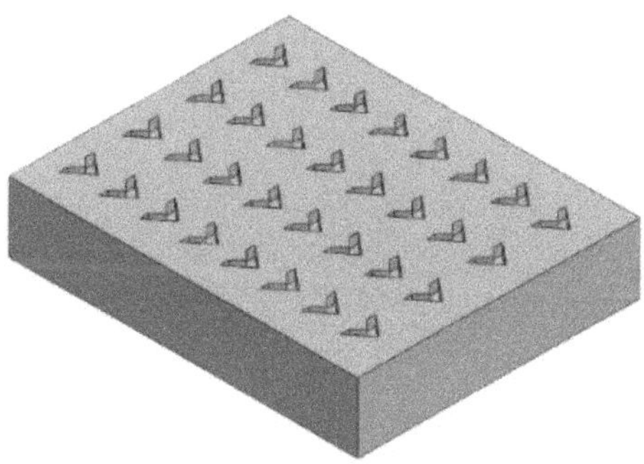

FIGURE 7.21 Chevron-shaped texture [47].

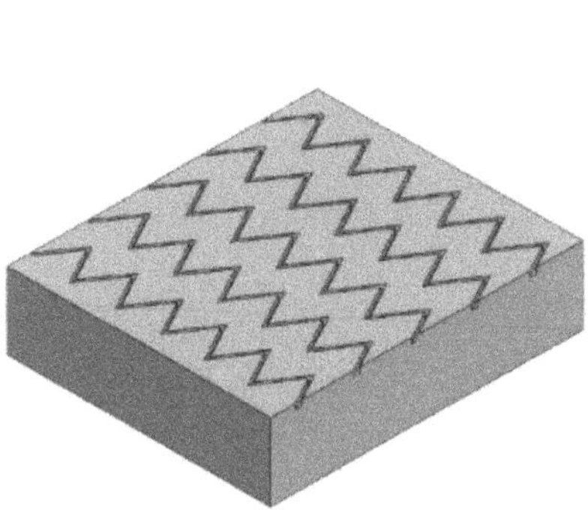

FIGURE 7.22 Weave textures [50].

Hz. For the dimple-shaped and groove-shaped structures, the COF increased 4.1 times and 30.6 times with decrease in pin diameter [49]. Circular-shaped textures in linear and weave arrangements were textured in hot pressed Al_2O_3/TiC discs, as shown in Figure 7.22. The texturing was done at spacings of 100 μm, 150 μm, 200 μm, and 250 μm. Increasing the spacing distance decreases the COF in all samples [50]. A compilation of the tribological behavior of textured materials is shown in Table 7.2.

TABLE 7.2
Tribological Behavior of Metals/Polymer/Ceramics

Serial Number	Material	COF		Wear Rate	
		Non-Textured	Textured	Non-Textured	Textured
	Metals				
1	SS 316 [25]	0.48	0.18	–	–
2	20CrMnTi [26]	–	–	17.2 mg	10.8 mg
3	AISI 316 [27]	0.39	0.32	1.2×10^{-3} g	0.1×10^{-3} g
4	Al 7075 [28]	0.65	0.12	0.033 mm³/Nm	0.007 mm³/Nm
5	AISI 304 [29]	0.22	0.07	0.45×10^{-4} mm³/Nm	0.2×10^{-4} mm³/Nm
6	AZ31B [30]	0.151	0.124	–	–
7	AISI5200 chrome steel [31]	–	–	6.3	1.2 mm³/m
8	QT600 cast iron [32]	0.07	0.05	–	–
9	Cu [34]	0.317	0.07	–	–
10	Chromium [35]			0.45×10^{-4} mm³/Nm	0.194×10^{-4} mm³/Nm
11	Diamond nanocomposite [36]	0.22	0.12	30 nm	20–25 nm
12	CVD diamond film [37]	–	–	1.52×10^{-4} mm³/Nm	3.95×10^{-5} mm³/Nm
13	T8 steel [38]	–	0.05	–	–
14	GCr15 steel [39]	–	0.034	–	2.08×10^{-7} mm³/Nm
15	Ni composite [40]	–	0.03–0.1	–	–
16	AISI M2 tool steel [41]	0.096	0.076	–	–
17	Ti6Al4V [42]	0.456	0.35 (dry), 0.257 (lub)	159.91 μm	104.65 μm, 88.6 μm
18	Ti6Al4V [43]	0.19 ± 0.03	0.05 ± 0.02	–	–
19	Ti6Al4V [44]	–	0.1	–	–
20	AISI 104 carbon steel [45]	–	0.15	–	–
21	Al–SiC [46]	–	0.12	–	–
	Ceramics				
22	Cemented carbide [47]	–	0.02	–	–
23	Cemented carbide [48]	0.526	0.421	–	–

7.4 SUMMARY

The mechanical and wear behavior of laser surface textured metals, polymers, and ceramics have been discussed extensively. Nanosecond- and femtosecond-based lasers were utilized to perform the texturing operation. Circular, square, groove, rhombus, hexagon, triangular, moat shaped, chevron, wobble, and elliptical shapes were textured on the surfaces, of which circular, square, and groove shapes were the most preferred ones. Apart from this, nature-inspired fish and shell-bionic textures and biomimetic papilla textures were also performed. The surface roughness and joint strength of textured materials were evaluated, showing an increase compared to non-textured materials. In terms of tribological behavior, the wear rate, wear volume, and coefficient of friction decreased for the textured materials compared to non-textured materials. The future scope of LST is extending the study of mechanical and tribological behavior toward emerging composite materials.

REFERENCES

[1] Rezayat M, Morales M, Moradi M, Mateo A. Laser wobbling surface texturing of AISI 301LN steel for enhancement of the corrosion resistance at high temperature. Opt Laser Technol 2024;171:110375. https://doi.org/10.1016/j.optlastec.2023.110375.

[2] Rezayat M, Roa JJ, Mateo A. Effect of laser surface texturing on Schmid factor and plastic deformation mechanisms on AISI 301LN steel. Metals (Basel) 2023;13:1789. https://doi.org/10.3390/met13101789.

[3] Jia Z-Y, Ye T, Ma J-W, Cao X-K, Liu W, Yu W-J, et al. Effect of process parameters on the hardness of laser surface textured 5A06 aluminum alloy. J Mater Eng Perform 2021;30:5858–5867. https://doi.org/10.1007/s11665-021-05840-y.

[4] Das B, Srivastava SK, Manna I, Majumdar JD. Mechanically tailored surface of titanium based alloy (Ti6Al4V) by laser surface treatment. Surf Coat Technol 2024;479:130560. https://doi.org/10.1016/j.surfcoat.2024.130560.

[5] Yang B, Li H, Sun H, Xu W, Xia H, Su X, et al. Towards enhanced mechanical performance of Al/steel welded-brazed joints via laser surface texturing modification. J Mater Res Technol 2023;27:5278–5290. https://doi.org/10.1016/j.jmrt.2023.11.029.

[6] Maressa P, Anodio L, Bernasconi A, Demir AG, Previtali B. Effect of surface texture on the adhesion performance of laser treated Ti6Al4V alloy. J Adhes 2015;91:518–537. https://doi.org/10.1080/00218464.2014.933809.

[7] Tan X, Hu Y, Zhi Q, Wu J, Li Y, Liu J, et al. Effect of laser texturing on mechanical strength and microstructure of ultrasonically welded ABS to aluminum alloy. J Mater Res Technol 2024;31:3425–3435. https://doi.org/10.1016/j.jmrt.2024.07.057.

[8] Liu Y, Wang X, Zhou L, Zhao H, Han X, Tan C, et al. Achievement of high-strength Al/CFRP hybrid joint via high-speed friction stir lap joining and laser texturing pretreatment parameters variation. Thin-Walled Struct 2024;199:111762. https://doi.org/10.1016/j.tws.2024.111762.

[9] Bi X, Liu H, Li Y, Xu M, Wang Z. Biomimetic papilla texture by femtosecond laser for high-strength CFRTP/A6061-T6 FSSW hybrid structures. Compos B Eng 2024;281:111500. https://doi.org/10.1016/j.compositesb.2024.111500.

[10] Zhang X, Su J, Feng Z, Liu H, Xia H, Zhang Z, et al. Effect of laser textured patterns on the interfacial characteristics and mechanical properties of 6061Al/GFRTP joints via hot-pressing. Thin-Walled Struct 2023;189:110916. https://doi.org/10.1016/j.tws.2023.110916.

[11] Wang W, Wang S, Zhang X, Xu Y, Tian Y, Huang H. Enhanced aluminum alloy-polymer friction stir welding joints by introducing micro-textures. Mater Lett 2021;295:129872. https://doi.org/10.1016/j.matlet.2021.129872.
[12] Enami K, Kimura F, Yokoyama K, Murakami T, Kajihara Y. Experimental and simulative investigation of the effects of laser-structured metal surface on metal-polymer direct joining. Precis Eng 2020;62:273–281. https://doi.org/10.1016/j.precisioneng.2019.12.011.
[13] Jeon JH, Choi JY, Kim GY, Hong S, Callanga J, Lee C, et al. Effect of laser texturing and thermomechanical joining parameters on bond strength of steel-flax fiber reinforced vitrimer composites. J Manuf Process 2024;124:212–225. https://doi.org/10.1016/j.jmapro.2024.06.009.
[14] Leone C, Paoletti A, Babu Yanala P, Lambiase F. Improving bonding strength of aluminium-PEEK hybrid metal-polymer joints by two-step laser surface treatment. Opt Laser Technol 2024;170:110304. https://doi.org/10.1016/j.optlastec.2023.110304.
[15] Lambiase F, Yanala PB, Leone C, Paoletti A. Influence of laser texturing strategy on thermomechanical joining of AA7075 aluminum alloy and PEEK. Compos Struct 2023;315:116974. https://doi.org/10.1016/j.compstruct.2023.116974.
[16] Rodríguez-Vidal E, Sanz C, Soriano C, Leunda J, Verhaeghe G. Effect of metal micro-structuring on the mechanical behavior of polymer–metal laser T-joints. J Mater Process Technol 2016;229:668–677. https://doi.org/10.1016/j.jmatprotec.2015.10.026.
[17] Liu Y, Su J, Tan C, Feng Z, Zhang H, Wu L, et al. Effect of laser texturing on mechanical strength and microstructural properties of hot-pressing joining of carbon fiber reinforced plastic to Ti6Al4V. J Manuf Process 2021;65:30–41. https://doi.org/10.1016/j.jmapro.2021.03.021.
[18] Han J, Göksel B, Li M, Vleugels J, Braem A, Castagne S. Effects of femtosecond laser surface texturing on mechanical properties and low temperature degradation of alumina toughened zirconia. J Eur Ceram Soc 2024;44:4196–4209. https://doi.org/10.1016/j.jeurceramsoc.2024.01.036.
[19] Vázquez JM, Salguero J, Del Sol I. Texturing design of WC-Co through laser parameter selection to improve lubricant retention ability of cutting tools. Int J Refract Hard Met 2022;107:105880. https://doi.org/10.1016/j.ijrmhm.2022.105880.
[20] Yu S, Wang Y, Wang S, Zhao Q, Li Y, Ren D, et al. Inspired square micro-texture structures designed by nanosecond laser sculpture for enhancing adhesion strength of plasma spraying Si-HfO2 bond coat on SiC substrates. Opt Laser Technol 2024;175:110840. https://doi.org/10.1016/j.optlastec.2024.110840.
[21] Liu C, Fu J, Li L, Wang H, Pei X, Zhang T, et al. Fabrication of antibacterial and anti-corrosive zirconia ceramics with extreme wettability by facile laser-based surface modification. Ceram Int 2024;50:22733–22747. https://doi.org/10.1016/j.ceramint.2024.03.376.
[22] Parry JP, Shephard JD, Hand DP, Moorhouse C, Jones N, Weston N. Laser micromachining of Zirconia (Y-TZP) ceramics in the picosecond regime and the impact on material strength: Laser micromachining of Zirconia ceramics. Int J Appl Ceram Technol 2011;8:163–171. https://doi.org/10.1111/j.1744-7402.2009.02420.x.
[23] Roitero E, Anglada M, Mücklich F, Jiménez-Piqué E. Mechanical reliability of dental grade zirconia after laser patterning. J MechBehav Biomed Mater 2018;86:257–263. https://doi.org/10.1016/j.jmbbm.2018.06.039.
[24] Pereira RSF, Moura CG, Henriques B, Chevalier J, Silva FS, Fredel MC. Influence of laser texturing on surface features, mechanical properties and low-temperature degradation behavior of 3Y-TZP. Ceram Int 2020;46:3502–3512. https://doi.org/10.1016/j.ceramint.2019.10.065.

[25] Yuan S, Lin N, Wang W, Zhang H, Liu Z, Yu Y, et al. Correlation between surface textural parameter and tribological behaviour of four metal materials with laser surface texturing (LST). Appl Surf Sci 2022;583:152410. https://doi.org/10.1016/j.apsusc.2021.152410.

[26] Zhang Y, Jia X, He K, Dong X, Wang Z, Liao K. Lipophilic and friction properties of 20CrMnTi steel with laser-induced texturing. Tribol Int 2024;194:109550. https://doi.org/10.1016/j.triboint.2024.109550.

[27] Türkdayı M, Kavasoğlu YS, Kovacı H. Increasing the friction and wear performance of AISI 316L under dry, aqueous and oil lubricated sliding conditions with surface textures fabricated by laser-powder bed fusion (L-PBF). Tribol Int 2024;200:110135. https://doi.org/10.1016/j.triboint.2024.110135.

[28] Li Q, Wu X, Zhang K, Liu W. Influence of femtosecond laser textured shapes on the friction characteristics of 7075 aluminum alloy. Mater Lett 2024;373:137098. https://doi.org/10.1016/j.matlet.2024.137098.

[29] Seid Ahmed Y, DePaiva MJ, Amorim LF, Torres DR, de Rossi W, Veldhuis CS. Laser surface texturing and characterization of austenitic stainless steel for the improvement of its surface properties. Int J Adv Manuf Technol 2021;115:1795–1808. https://doi.org/10.1007/s00170-021-07284-z.

[30] Zhang X, Mao B, Siddaiah A, Menezes PL, Liao Y. Direct laser shock surface patterning of an AZ31B magnesium alloy: Microstructure evolution and friction performance. J Mater Process Technol 2020;275:116333. https://doi.org/10.1016/j.jmatprotec.2019.116333.

[31] Paul Joshua S, Dinesh Babu P. Effect of laser textured surface with different patterns on tribological characteristics of bearing material AISI 52100. J Cent S Univ 2020;27:2210–2219. https://doi.org/10.1007/s11771-020-4442-7.

[32] Ding S, Hu Y, Wei H, Mu D, Zhan X, Wang J, et al. The tribological behavior of cast iron by laser surface texturing under oil-lubricated initial line contact for rotary compressor. Lubricants 2023;11:486. https://doi.org/10.3390/lubricants11110486.

[33] Kim B, Chae YH, Choi HS. Effects of surface texturing on the frictional behavior of cast iron surfaces. Tribol Int 2014;70:128–135. https://doi.org/10.1016/j.triboint.2013.10.006.

[34] Li K, Yao Z, Hu Y, Gu W. Friction and wear performance of laser peen textured surface under starved lubrication. Tribol Int 2014;77:97–105. https://doi.org/10.1016/j.triboint.2014.04.017.

[35] Lu L, Zhang Z, Guan Y, Zheng H. Comparison of the effect of typical patterns on friction and wear properties of chromium alloy prepared by laser surface texturing. Opt Laser Technol 2018;106:272–279. https://doi.org/10.1016/j.optlastec.2018.04.020.

[36] Pimenov SM, Jaeggi B, Neuenschwander B, Zavedeev EV, Zilova OS, Shupegin ML. Femtosecond laser surface texturing of diamond-like nanocomposite films to improve tribological properties in lubricated sliding. Diam Relat Mater 2019;93:42–49. https://doi.org/10.1016/j.diamond.2019.01.022.

[37] Sui T-Y, Cui Y-X, Lin B, Zhang D-W. Influence of nanosecond laser processed surface textures on the tribological characteristics of diamond films sliding against zirconia bioceramic. Ceram Int 2018;44:23137–23144. https://doi.org/10.1016/j.ceramint.2018.09.122.

[38] Wan Y, Xiong D-S. The effect of laser surface texturing on frictional performance of face seal. J Mater Process Technol 2008;197:96–100. https://doi.org/10.1016/j.jmatprotec.2007.06.019.

[39] Wang W, Yuan W, Guo Q, Wang N, Chi B, Yu J. Effect of picosecond laser surface texturing under Babbitt coating mask on friction and wear properties of GCr15 bearing steel surface. Eng Fail Anal 2024;157:107878. https://doi.org/10.1016/j.engfailanal.2023.107878.

[40] Li J, Xiong D, Dai J, Huang Z, Tyagi R. Effect of surface laser texture on friction properties of nickel-based composite. Tribol Int 2010;43:1193–1199. https://doi.org/10.1016/j.triboint.2009.12.044.
[41] Saetang V, Taweejun N, Zhu H, Jaafar H, Daodon W. Effects of laser texturing on tribological properties and wettability of titanium nitride-coated AISI M2 high-speed steel. Surf Coat Technol 2024;490:131173. https://doi.org/10.1016/j.surfcoat.2024.131173.
[42] Velayuthaperumal S, Radhakrishnan R. A novel nanosecond pulsed laser textured moat configurations for enhancing surface wettability, corrosion and tribology behaviour of Ti6Al4V implant material. Mater Today Commun 2024;39:108701. https://doi.org/10.1016/j.mtcomm.2024.108701.
[43] Li L, Li Z, Xing Z, Guo W, Huang Y, Wang H. Effect of femtosecond laser bionic texture on anti-wear properties of medical Ti-6Al-4 V. Tribol Int 2023;190:109062. https://doi.org/10.1016/j.triboint.2023.109062.
[44] Wu Z, Xing Y, Huang P, Liu L. Tribological properties of dimple-textured titanium alloys under dry sliding contact. Surf Coat Technol 2017;309:21–28. https://doi.org/10.1016/j.surfcoat.2016.11.045.
[45] Mao B, Siddaiah A, Menezes PL, Liao Y. Surface texturing by indirect laser shock surface patterning for manipulated friction coefficient. J Mater Process Technol 2018;257:227–233. https://doi.org/10.1016/j.jmatprotec.2018.02.041.
[46] Del Sol I, Batista M, Mayuet P, Bañon F, Vazquez-Martinez JM. Improvement of the tribologicalbehavior of high reinforcement concentration Al–SiC (63%) metal matrix composite through laser surface texturing. J Mater Res Technol 2024;28:2338–2346. https://doi.org/10.1016/j.jmrt.2023.12.107.
[47] Dan L, Xuefeng Y, Chongyang L, Jian C, Shouren W, Yanjun W. Tribological characteristics of a cemented carbide friction surface with chevron pattern micro-texture based on different texture density. Tribol Int 2020;142:106016. https://doi.org/10.1016/j.triboint.2019.106016.
[48] He C, Yang S, Zheng M. Analysis of synergistic friction reduction effect on micro-textured cemented carbide surface by laser processing. Opt Laser Technol 2022;155:108343. https://doi.org/10.1016/j.optlastec.2022.108343.
[49] Butler-Smith P, Esfahani RN, Chrostek-Mroz A, See T. On the tribological and machining performance of laser textured sintered carbide cutting tools in turning of Al2024. Procedia CIRP 2022;108:358–361. https://doi.org/10.1016/j.procir.2022.04.072.
[50] Xing Y, Deng J, Wu Z, Wu F. High friction and low wear properties of laser-textured ceramic surface under dry friction. Opt Laser Technol 2017;93:24–32. https://doi.org/10.1016/j.optlastec.2017.01.032.

8 Hydrophobicity and Corrosion Behavior Analysis on Laser Surface Textured Materials

T. Sathies, Kumaran M., and Jayanth N.

8.1 INTRODUCTION

The surface properties of materials, especially wettability characteristics, play a predominant role in various engineering applications. Wettability refers to the ability of a liquid to spread across a solid surface due to intermolecular interaction. The wetting regime can be categorized as hydrophilic, super hydrophilic, hydrophobic, and superhydrophobic. The schematic representation of different wetting regimes is shown in Figure 8.1. The wettability of the surface could be measured using a sessile technique or by measuring the contact angle (CA) of a fluid drop at rest. The surfaces that are difficult to wet (low surface wettability) are called superhydrophobic. Metallic surfaces with superhydrophobic surfaces exhibit corrosion resistance and there is a huge demand for techniques that can fabricate superhydrophobic texture. A superhydrophobic surface is one in which the water droplet sliding angle is less than 10° while the contact angle of a water droplet on the solid surface is greater than 150°. Air is readily trapped by the valley present on the superhydrophobic surface and it helps in protecting the metallic surface against corrosion by preventing the direct contact of the metallic surface with the external corrosive media [1,2]. A few examples of naturally occurring superhydrophobic surfaces are rose petals, water strider legs, butterfly wings, cicada wings, lotus leaves, and rice leaves. The observed hydrophobicity was attributed to the presence of micro/nanoscale structures on the surface along with fitting chemistry. Generally, the wettability of the solid surface is a function of surface topography and chemical structure [3]. The wettability regime of a surface can be modified by developing microscale or nanoscale textures on the surface. The free energy of the surface also controls the wettability. The solvents containing fluorine-based functional groups, such as CF_2, CF_3, and fluoroalkyl silanes (FAS), are excellent choices for decreasing the free energy of solid surfaces. The three fundamental models governing the wettability of the surface with different topographies are Wenzel's model, Cassie-Baxter's model, and Young's model, schematically represented in Figure 8.2. The Wenzel model is applicable when complete contact is established between a liquid droplet and a rough surface.

DOI: 10.1201/9781003542087-8

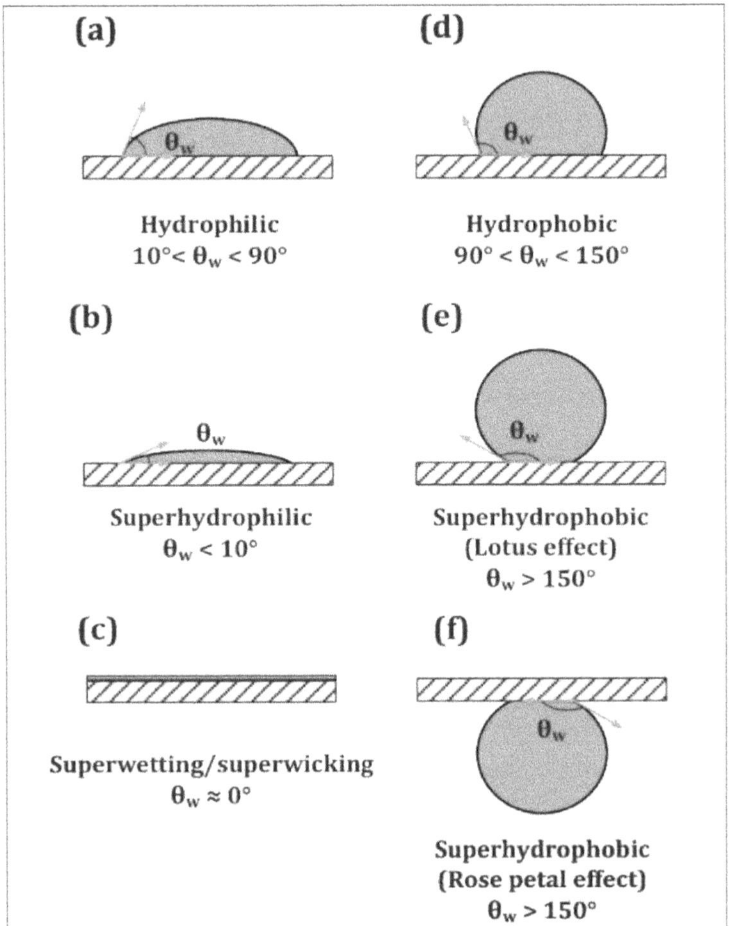

FIGURE 8.1 Schematic representation of different wetting regimes (Reproduced with permission from Ref. [1]).

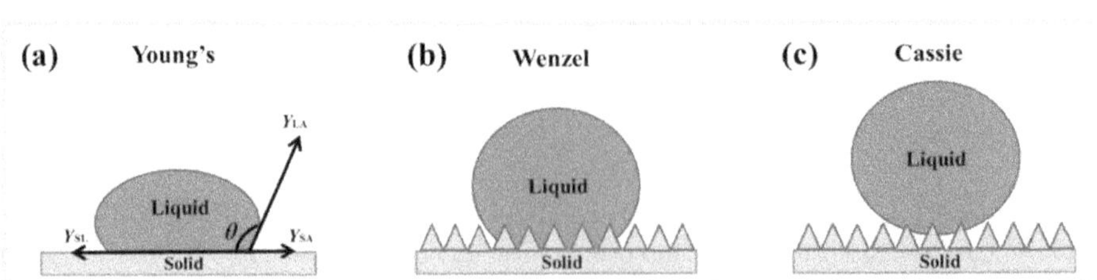

FIGURE 8.2 Schematic representation of different models governing wetting regimes: (a) Young's model, (b) Wenzel model, (c) Cassie model (Reproduced with permission from Ref. [4]).

The Cassie-Baxter model is used in places where air pockets get entrapped between the texture and liquid droplet and it prevents the liquid from wetting the surface [4].

Superhydrophobicity can be easily achieved on the surfaces of the functional component by creating periodic, three-dimensional micro/nano-patterned topography via laser texturing and subsequent surface treatment [5]. Laser-based texturing techniques are an attractive platform for creating micro- and nano-structures, such as laser-induced periodic surface structures or hierarchical structures on a metallic substrate. However, a laser-textured surface lacks desired wettability and corrosion resistance and the textured materials are subjected to subsequent surface chemistry modification techniques. The laser texturing method generally produces super hydrophilic surfaces, which may subsequently convert into superhydrophobic surfaces through various post-processing techniques. Soon after laser texturing, oxide or hydroxide layers form on the top surface of textured material and these layers make the surface more hydrophilic by increasing the surface energy. Superhydrophobicity can be attained on laser-textured material by a variety of surface chemistry modification methods, including low-temperature heat treatment, exposing the nanostructured surface to atmospheric conditions for sufficient duration, and attaching low surface free energy groups like fluorinated groups through chemical modifications techniques [1]. The substrate material properties (specific heat, thermal conductivity, bandgap, etc.) and laser processing parameters such as pulse energy, repetition rate, pulse duration, scanning speed, wavelength, processing environment, and polarization determine the surface structure of the laser-textured material.

Numerous laser surface texturing techniques, including picosecond (shorter pulse duration), femtosecond (ultra-short pulse duration), and nanosecond (longer pulse duration) laser texturing, have been used to modify surfaces of engineering materials and the effect of different texturing patterns with varying sizes and forms on corrosion resistance, antifouling properties, anti-icing characteristics, drag reduction, lubrication, and wear resistance is reported in the literature. Figure 8.3 schematically shows the laser texturing process and the effect of different lasers on heat affected zone. Before laser selection for an indented application, it is essential to have a clear idea about the surface morphological behavior, pulse repetition frequency, depth of texturing, laser fluence, texture size, and accuracy level required [6]. The effect of different post-processing techniques on the hydrophobicity and corrosion resistance of the laser-textured materials are discussed in the following section.

8.2 CHEMICAL TREATMENT OF LASER TEXTURED MATERIALS

Chemical treatment is a well-established method for inducing superhydrophobicity in laser-textured materials. Fluorinated groups ($-CF_2$, $-CF_3$) and alkyl $(CH_2)_n$ groups can be anchored on the surface of textured material to lower surface energy and cause superhydrophobicity. Ma et al. fabricated a circular-shaped texture with superior hydrophobicity on carbon steel through a nanosecond laser texturing technique and chemically modified it with 1 w% 1H,1H,2H,2H perfluorodecyltriethoxysilane/ethanol solution to improve its corrosion resistance. Samples with lower texture pitch (p)/diameter (d) (1.0) possessed lower surface energy and exhibited higher water contact angle (161.5°), better corrosion resistance, durability, and abrasion resistance

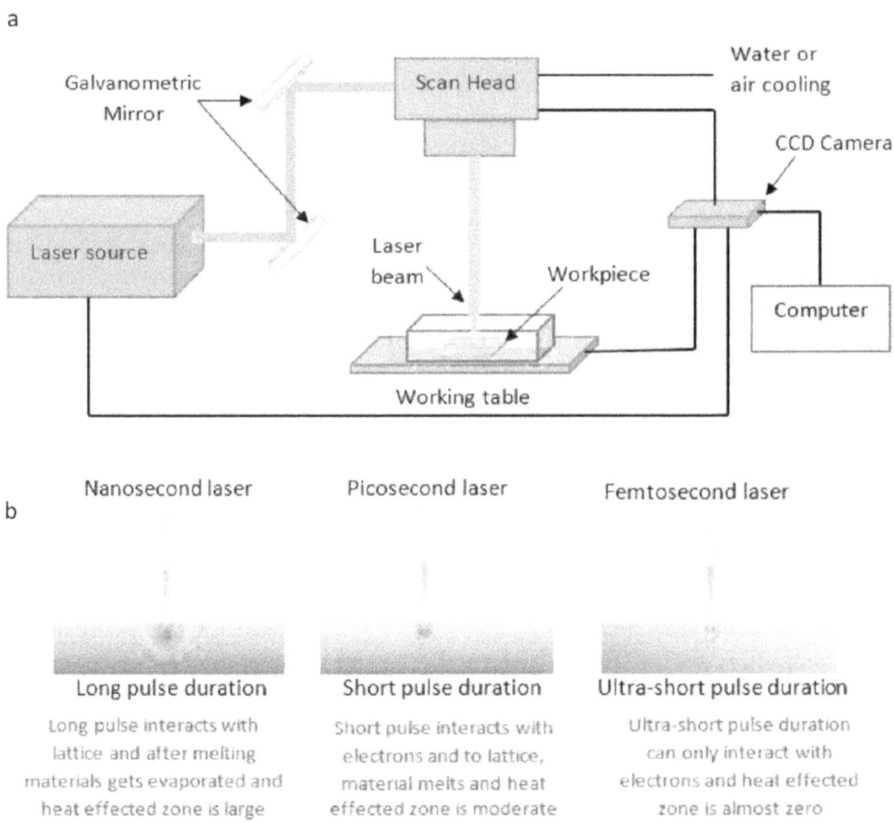

FIGURE 8.3 (a) Laser texturing process, (b) different types of lasers used for laser texturing (Reproduced with permission from Ref. [6]).

[7]. Yang et al. developed grid patterns on a 1060 aluminium surface through laser ablation technique and silanated the textured surface with (heptadecafluoro-1, 1, 2, 2-tetradecyl) triethoxysilane (AC-FAS) to improve the super-hydrophobicity and corrosion resistance of aluminium material. The treated surfaces showed excellent chemical stability in both alkaline and acidic solutions, with a maximum water contact angle of 160.6° ± 1.5° and a rolling angle of 3.0° ± 1.0°. The surface free energy of the chemically modified textured samples was reduced by the $-CF_2$, $-CF_3$, and cross-linked Si–O–Si functional group present on the sample surface [8]. Boinovich et al. fabricated a superhydrophobic layer on aluminium-magnesium (AMG) alloy by combining laser texturing and chemisorption techniques. The hydrophobic agents are induced on the surface of the components by subjecting laser-textured components to methoxy-{3-[(2,2,3,3,4,4,5,5,6,6,7,7,8,8,8-pentadecafluorooctyl)oxy]propyl}-silane vapours for 1 hour followed by drying the samples in the oven at 130°C for 1 hour. Surface compounds such as aluminium oxynitride and aluminium oxide formed on the substrate surface during laser treatment and chemical processing promote the hydrophobicity and corrosion resistance of the material [9]. Numerous laser

texturing process parameters such as laser power, laser scan speed, pulse, number of laser passes, repetition frequency, spot size, and line spacing affect the texture quality of the sample surface. González-Fernández utilized laser texturing as a corrosion mitigation strategy for SS310 samples employed in high-temperature molten salt applications. The impact of laser texturing on the corrosion resistance of SS310 samples was investigated by exposing the samples to the flow of molten Li_2CO_3–K_2CO_3–Na_2CO_3 salts for a duration of 600 hours. Long-chain hydrocarbons adhere to the surface of steel specimens during laser texturing and these hydrocarbons decompose at high temperatures and mitigate corrosion by forming protective carbides and carbonate layers. As a result, the corrosion rate of the laser-textured steel material subjected to harsh environments decreases drastically [10]. Lu et al. created a superhydrophobic surface comprising micro-cracks and brain-like microstructure on SS316L alloy by employing a UV nanosecond laser texturing technique and fluoroalkyl silane as a chemical treatment reagent. Optimum texturing parameters for creating such features were identified as a scanning interval of 30.0 μm and laser fluence of 8.14 J/cm^2. A self-assembled film comprising R-O-Si and Si-O-Si groups was formed on the substrate during chemical modification and it helped in promoting the hydrophobicity and corrosion resistance of the laser-textured material. The textured surface exhibited a maximum static contact angle and sliding angle of $160 \pm 5°$ and $3 \pm 0.5°$ [11]. Zhang et al. improved the super-hydrophobicity and corrosion resistance of laser surface textured AISI 304 stainless steels by decorating the textured material with hexadecyltrimethoxysilane (HDTMS). Micropores with a radius of 10 μm were produced on the substrate through laser texturing and the effect of micropore-micropore distances and chemical modification on corrosion resistance and superhydrophobicity was investigated. Chemically treated samples with lesser micropore distance possessed high surface roughness and lesser surface energy (0.535 mJ/m^2) and demonstrated higher superhydrophobicity (water contact angle of 158.9° and rolling angle of 3°) and corrosion resistance [4].

Wang et al. introduced a hybrid approach comprising laser texturing, carbon ion implantation, and fluoroalkyl silane (FAS) surface modification technique to induce superhydrophobicity and anti-corrosion stability on nitinol material. The nitinol substrate subjected to laser texturing, silanization, and carbon ion implantation contains regular grid patterns with micron-scale ridges, pits, and rosette protrusions. The hierarchical microstructure formed by the hybrid approach encapsulates multistage air pockets and it restrains the corrosive medium from permeating into the substrate. Consequently, the thermal stability, abrasion resistance, hydrophobicity, and corrosion resistance of NiTi alloy increases drastically [12]. Kumthekar et al. used a femtosecond pulse width laser to create controlled hydrophobic metasurfaces on Ni-Mn-Ga-based alloy. Micro-nano features and grooves as deep as 100 μm were produced on the material through multistage laser texturing and hydrocarbon-rich lubricant was applied over the textured region to achieve a superhydrophobic surface with static contact angle (SCA) of water droplets as high as 150° [13]. Madapana et al. analysed the effect of laser surface structuring on the contact angle, corrosion resistance, hardness, roughness, and bioactivity of Ti6Al4V alloy. LSS improved the surface hardness, roughness, corrosion resistance, and contact angle against simulated body fluid. When the samples were immersed in simulated body fluid, the deposition of calcium phosphate on textured material was higher than the as-received material. In addition to improved surface texture, the organic species were absorbed by the textured material from the environment and it helped in reducing the polarity and surface energy of the sample [14]. To improve the corrosion resistance of Ti6Al4V alloy in HF and HCl/HNO_3 solutions, Tang et al. laser textured hierarchical micro-/nanostructure surface comprising radiating flocculent nanostructures and parallel linear peak-like microscale structures on titanium alloy. The textured surface upon silanization exhibited excellent superhydrophobicity (contact angle of 158.4° and sliding angle of 4.7°) and corrosion resistance. The presence of self-assembled fluorosilane film and air packets between the textures impedes the corrosion. No appreciable changes in surface topography were observed when the textured material was immersed in 4% HF solution at room temperature and 5% HNO_3 solution at 80°C for 24 hours [15].

Lorenz et al. improved the hydrophobicity and chemical stability of the laser-textured stainless steel by functionalizing the textured surface with (3,3,4,4,5,5,6,6,7,7,8,8,9,9,10,10,11,11,12,12,12-heneicosafluorododecyl) phosphonic acid (F21)/tetrahydrofuran (THF) solution mixture. The chemical structure of a hydrophobic self-assembled monolayer formed on a steel sample during chemical treatment is shown in Figure 8.4. The F21 self-assembled layer endures the textured surface with superhydrophobicity and prevents the degradation of textured material in hydrogen peroxide (H_2O_2) for more than 2 weeks [16]. Cobalt–chromium–molybdenum alloy is one of the preferred materials for the fabrication of implants and dental fixtures. To improve the biocompatibility, hydrophobicity, and corrosion resistance of $Co_{28}Cr_6Mo$ alloy, Zhang et al. created a bionic 3D micro nanocomposite structure (microgrid grooves and layered nanoparticles) on $Co_{28}Cr_6Mo$ alloy through laser texturing and provided slippery coating over the textured surface by using polydimethylsiloxane solution modification and silicone oil treatment. Functional groups such as C–O, C=O, and Si–O were observed on the processed samples and it augments the hydrophobicity. Using the aforementioned technique, a slippery surface with less adhesion, high contact angle (155.236° ± 1.5°), good abrasion resistance, and corrosion resistance was developed by the authors [17].

Numerous applications exist in the marine industry for amorphous alloy coatings with exceptional anti-fouling and anti-corrosion properties. Wei et al. developed micro-nano hierarchical structures on $Fe_{43}Cr_{20}Mo_{10}W_4C_{15}B_6Y_2$ (at%) amorphous alloy coating and treated the surface with 2.0 wt% poly-perfluoroacrylate to induce fluorine-containing organic molecules on the textured coating. During the laser texturing, trench and dimple structures containing granular deposits and nanoparticles were created on the coating surfaces. After grafting low surface free energy organic compounds onto LT surfaces, coated surfaces showed a minimum sliding angle of 3.6° and a maximum contact angle of 166.5° ± 0.6° and it results in better corrosion resistance. Various steps involved in the chemical treatment of laser-textured samples are shown in Figure 8.5 [18]. Qiu developed a superhydrophobic micro-nanostructure on Inconel 718 by subjecting it to a spatially shaped femtosecond Bessel beam and fluorination treatment. Compared to the unprocessed surface, the created superhydrophobic surface demonstrated improved dynamic anti-icing performance and steady static-droplet delay icing performance; it indicates that droplet condensation time was prolonged by the produced structure [19].

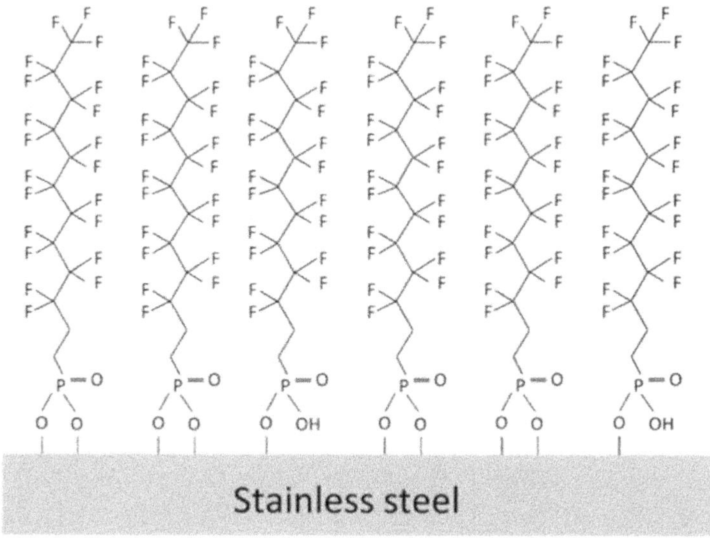

FIGURE 8.4 Self-assembled monolayer (F21) formed on stainless steel sample (Reproduced with permission from Ref. [16]).

Liang et al. employed the femtosecond (fs) laser direct writing method to create protective armour with micro-nano structures from magnesium alloy. Grain-refined layers with nanocrystals were also deposited on the material surface during laser direct writing. The laser-processed magnesium alloy exhibited remarkable corrosion resistance and it was attributed to the combined action of the nanocrystal layer and the micro-nano structure [20]. Tang et al. introduced a hybrid femtosecond laser-chemical method to fabricate superhydrophobic surfaces with better wear resistance and corrosion resistance on 316L stainless steel. Surfaces with different textures (grooves, mesh, and honeycomb) were created by femtosecond laser texturing. Subsequently, the surfaces with texture patterns were subjected to laser-assisted stearic acid decomposition to improve the hydrophobicity. For laser structuring, laser power of 4.5 W, repetition rate of 800 kHz, and scanning speed of 750 mm/s were adopted, whereas for laser-assisted chemical decomposition, laser parameters involving 1 W, 150 kHz, and 50 mm/s were used. Of all the texture designs produced by the authors, the honeycomb structure with a 0.1 mm depth and 0.4 mm side length had the best superhydrophobicity, with a contact angle of 154.2° and a sliding angle of 8.9° [21]. To prepare a superhydrophobic surface on a copper substrate, Shu et al. combined laser texturing and sol–gel methods. The micro-nano structure (pyramidal micro-pit array) fabricated on the copper substrate was coated with SiO_2@PDMS to improve the corrosion resistance, anti-icing property, and self-cleaning ability. The developed structure had a water contact angle of 159.5°, higher delayed freezing time (3.33 times more than normal surface), and lesser ice adhesion strength (8.12 kPa) [22]. Yayoglu et al. fabricated pillar-shaped microstructure patterns on magnesium alloy and modified the surface of the textured surface with stearic acid. The textured surface remained hydrophilic while the stearic acid–modified surface had hydrophobic properties. The gas layer and alkalinity-induced oxide layer helped

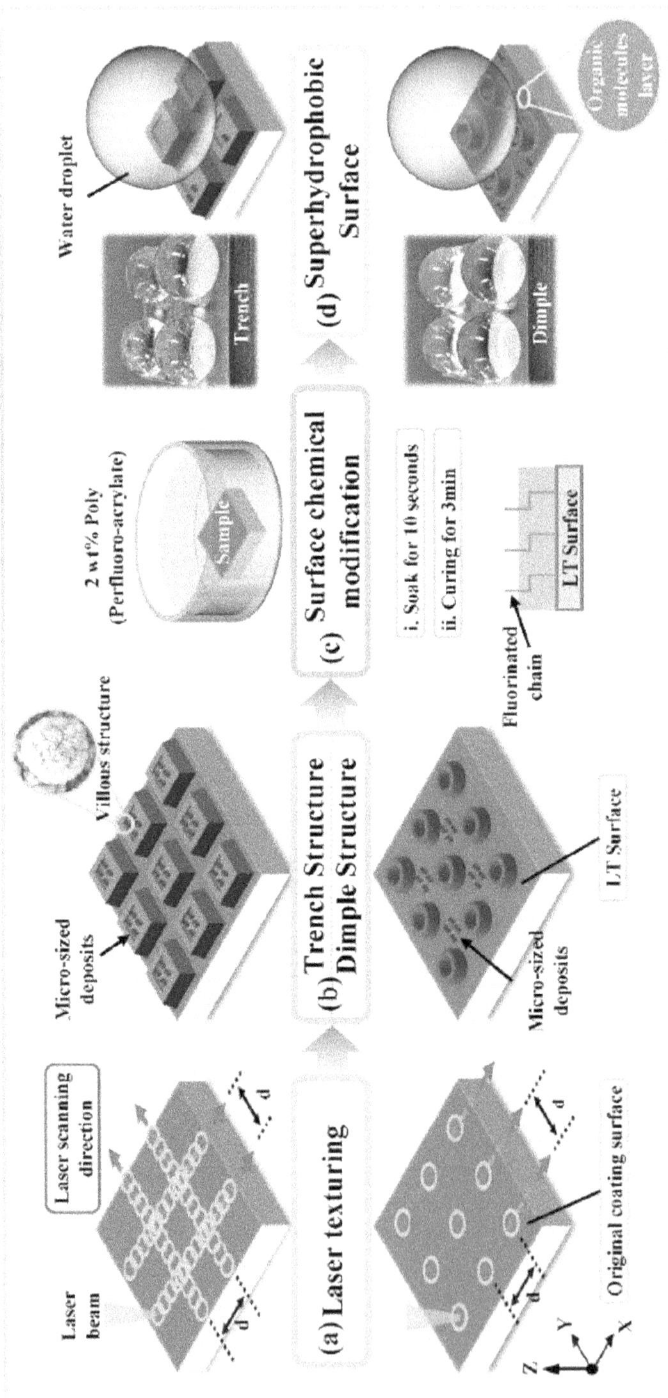

FIGURE 8.5 Schematic illustration of the fabrication of superhydrophobic laser-textured surfaces: (a) laser texturing; (b) generation of hierarchical trench and dimple micro-/nano-structures; (c) fluorinated surface chemical modification; and (d) resulting superhydrophobic behavior of the textured surfaces.

in increasing the corrosion resistance of chemically modified textured magnesium alloy [23]. Table 8.1 summarizes the effect of different chemical modifications on the wettability of laser-textured material. The superhydrophilic characteristics of laser-textured material are successfully converted to superhydrophobic through chemical modification. The chemical used for the modification varies between materials and the duration of chemical treatment affects the resultant properties.

8.3 LONG-TIME EXPOSURE OF TEXTURED MATERIAL TO AIR

When a laser-textured metal surface is exposed to atmospheric conditions for sufficient duration, its surface wettability state can shift from hydrophilic to hydrophobic. The adsorption of airborne organics with aliphatic groups on the textured surface causes a transition in the material's wettability state and the time taken for this kind of transition varies between materials. Trdan et al. detected a transition from superhydrophilic to superhydrophobic states after exposing laser-textured SS316L alloy to atmospheric conditions for one month. This trend helped in diminishing the corrosion rate of the particular alloy. Laser texturing prevents inter-granular corrosion of SS316L alloy by creating a homogeneous microstructure with fewer galvanic cells, resulting in Cr-depleted grain boundaries and $M(Fe, Cr)_{23}C_6$ carbides. The transition from hydrophilic to hydrophobic might have happened through (i) the development of hydrophobic functional groups on the material surface, (ii) partial surface deoxidation, (iii) airborne hydrocarbon contamination and organic matter absorption on the textured material [24]. Meng et al. modified the surface of 7075 aluminium alloy with a femtosecond laser and analysed the effect of laser modification on corrosion behavior. Deformed layers and amorphous oxidative layers were introduced during and after laser modification. These layers along with the textured surface mitigate the diffusion of corrosive medium through the samples and improve the corrosion resistance [25]. To prevent the premature failure of aluminium alloy in the marine atmosphere, Cholkar et al. utilized a femtosecond laser system to create corrosion-resistant surface textures with biofouling characteristics on aluminium alloy 7075 T6 alloy. The aluminium alloy 7075 samples were textured with five different designs namely crosshatched, rhombus, parallel, triangular, and star. The sample immediately after texturing was hydrophilic but the contact angle of the samples varied over some time. Interestingly, the samples with star and triangle texture designs showed a progressive transition to extremely hydrophobic states, with contact angles of 157° and 150°, respectively. In comparison to pristine samples, aged samples with a star texture pattern showed a 79% reduction in biofilm formation and a 25% reduction in corrosion rate [26]. Exposing the laser-textured material to atmospheric conditions to modify the wettability is a simple and low-cost technique. However, the time taken for transition from superhydrophilic to superhydrophobic will be more and the environmental conditions influence the transition.

8.4 HEAT TREATMENT OF LASER-TEXTURED MATERIAL

Heat treatment is another important post-processing technique that can affect the surface wettability of laser-textured material. The transition from superhydrophilic state to superhydrophobicity depends on the heat treatment temperature. Low-temperature annealing usually results in a textured surface with superhydrophobic properties whereas textured surface with superhydrophilic characteristics could be obtained through high-temperature annealing conditions. The wettability state of a textured surface usually changes because of changes in the surface chemistry that happen during or after heat treatment. Li et al. produced a corrosion-resistant superhydrophobic H59 brass substrate by subjecting it to laser ablation and heat treatment. The observed superhydrophobicity was attributed to the synergistic interaction between the micro/nano hierarchical structure developed by laser ablation and the adsorption of C–C and C–H organic groups from air during heating treatment. The rough grooves trap air pockets and improve corrosion resistance by reducing the solid-electrolyte contact area [27]. The scanning interval followed during laser texturing affects the wettability of the material. As the scanning interval increases from 100 μm to 600 μm, the superhydrophobicity and corrosion resistance of laser-textured H59 brass substrate decreases. The superhydrophobic surface with less adhesion properties (scanning interval 100 μm) had more corrosion resistance than the superhydrophobic with more adhesion properties and plain brass substrate [28]. Liu et al. fabricated a superhydrophobic Ti alloy surface with a periodic micro-bump structure by combining laser texturing and a heat treatment process. After laser texturing, the Ti alloy surface exhibited super hydrophilicity. Post heat treatment (150°C for 2 hours), the textured surface quickly transformed from superhydrophilic to superhydrophobic and it was attributed to the reduction in surface energy of material because of the deposition of carbon-containing hydrophobic groups [carboxyl (–COOH alkyl and (C–C/C–H)] groups. The surface erosion–corrosion resistance of the textured material was boosted by 18% as a result of the reduced solid-liquid contact area [29].

The laser powder bed fusion technique is widely adopted by industries for the fabrication of customized NiTi alloy components. However, LPBF-built NiTi alloy does not exhibit sufficient corrosion resistance. Hence, Yang et al. utilized laser texturing and low-temperature heat treatment to improve the hydrophobicity and corrosion resistance of LPBF-built NiTi alloy. Micro-nano papillary structures were constructed on the material surface by employing a grid-like scanning strategy and a large number of carbides and C–C/C–H functional groups were anchored on the material surface during heat treatment. Post-heat treatment, corrosion resistance improved drastically and the corrosion current density of the textured material decreased from 3.87×10^{-7} A/cm^2 to 2.92×10^{-8} A/cm^2 [30]. Similarly, Jiang et al. combined laser surface texturing and post-heat treatment techniques to prepare bronze surfaces with superhydrophobic and superoleophilic properties. Micro-pores of different depths were produced on the substrate through laser texturing and the uneven absorption of energy and redeposition of the ablated nano-size particles resulted in the formation of hierarchical micro-nano structures on the surface. A maximum water contact angle of 158° was achieved when the textured surface was heat treated at 150°C in ambient conditions for 12 hours. The C element in the form of the C–C(H) group was attached to the material surface during heat treatment and it enhanced the superhydrophobicity by reducing surface energy [31]. Ma et al. obtained a super-hydrophobic surface on a copper substrate by combining laser texturing and heat treatment. The laser-textured copper surface with grid pattern exhibited super-hydrophilic properties and upon the heat treatment at 150°C for 2 hours, the textured surface characteristics changed from super-hydrophilic to super-hydrophobic (water contact angle of 155.6°). During heat treatment, numerous organic compounds were adsorbed on the oxidized copper surface, consequently lowering surface polarization and favouring wettability modification [32]. Fan et al. created a sinusoid pattern and grid pattern on 5051 aluminium alloy by using a nanosecond laser and heat-treated the textured specimen at 150°C. The contact angle of samples increases gradually as the heat treatment duration increases and a maximum contact angle of 134.88° and 136.97° was noticed on samples with sinusoid and grid patterns that were heat treated for 72 h [33]. Misyura et al. reported the effect of heat treatment (100°C for 6 hours) on the corrosion resistance of laser-textured aluminium alloy (AlMg6). Low-temperature heat treatment accelerated the absorption of polar functional groups from the air by creating terminal oxide groups and removing water vapours. Heat treated sample had a water droplet contact angle of 157° and a roll-off angle of 4.5°. A denser oxide layer was formed on the sample during texturing and it helps in increasing the corrosion resistance of the textured samples. Micro cracks were detected on heat-treated textured specimens and led to a slight reduction in corrosion resistance compared to textured samples without heat treatment [34]. Wang et al. achieved hydrophobicity on laser-textured Ti-6Al-4V alloy by heat treating the textured sample. Heat treatment created more active sites for the adsorption of functional groups such as $-CH_2-$ and $-CH_3$ and it helped in the transition of titanium samples with microbumps from a superhydrophilic state to superhydrophobic state [35]. Table 8.2 summarizes the effect of heat treatment on the wettability of different laser-textured materials. Better hydrophobicity on laser-textured material could be attained through heat treatment. The heat treatment duration affects the final wettability and it differs between materials and textures.

8.5 HYBRID APPROACH

To garner the benefits rendered by different post-processing techniques, several researchers combined different laser texturing and post-processing techniques to fabricate surfaces with better hydrophobicity and corrosion resistance. Yang et al. combined femtosecond and nanosecond laser texturing techniques to produce multilayered micro/nanostructures on NiTi shape memory alloy. Further, the textured samples underwent a silanization process to improve hydrophobicity. The air pockets trapped by the multilayered micro/nanostructure improved the superhydrophobicity and protected the material against corrosion. The presence of fluorinated functional groups with low surface energy on the textured material was also responsible for the creation of superhydrophobic surfaces [36]. The laser ablation technique in combination with the laser-guided melt stacking method was used by Cao et al. to fabricate a terracotta warrior pit superhydrophobic structure on magnesium alloy. The proposed structure consists of a wall structure comprising a terracotta warrior-like pit and a lotus-like surface papillae structure. The Cassie–Baxter-based design protects the texture and the surface demonstrated excellent mechanical durability and repairability. The grain size of magnesium alloy was refined during femtosecond laser processing and the combined effect of grain refinement and superhydrophobicity leads to increased corrosion resistance of magnesium alloy [37]. Yang et al. integrated nanosecond and femtosecond laser processing and silanization treatment to increase the stability and corrosion resistance of aluminium substrates. The superhydrophobic surface produced through the proposed approach has a smaller wetting area and a higher concentration of chemisorbed FAS molecules. The surface modifications happening on the surface of the aluminium sample during the hydrolysis and chemisorption process are schematically shown in Figure 8.6. As a result, surface modification delayed the creation of wetting defects and improved corrosion resistance by partially resisting the erosion by chloride ions in the NaCl solution [38]. Tran et al. amalgamated three techniques, namely laser texturing, boiling water treatment, and silicone oil heat treatment, to produce superhydrophobic and corrosion-resistant aluminium surfaces. The hierarchical micro-nanoscale composite structure formed on the surface of the sample during the different stages of post processing of laser-textured material is shown in Figure 8.7. The laser texturing generated nano-micro hierarchical composite structures whereas boiling water treatment introduced pseudo-boehmite nanostructure on the textured material. The subsequent silicon oil treatment bolstered the absorption of the hydrophobic group (–CH_3) on the textured surface leading to a superhydrophobic aluminium surface with excellent corrosion resistance [39]. Wang et al. generated a hierarchical surface structure (grid pattern) on a copper substrate through laser texturing and modified the surface chemistry of the textured surface by involving silicone oil dripping and heat treatment techniques. Pure laser texturing leads to a superhydrophilic copper surface but the simultaneous application of silicone oil dripping and heat treatment to laser-textured surface resulted in a superhydrophobic surface with a contact angle as high as 159.2 ± 2.1° [40].

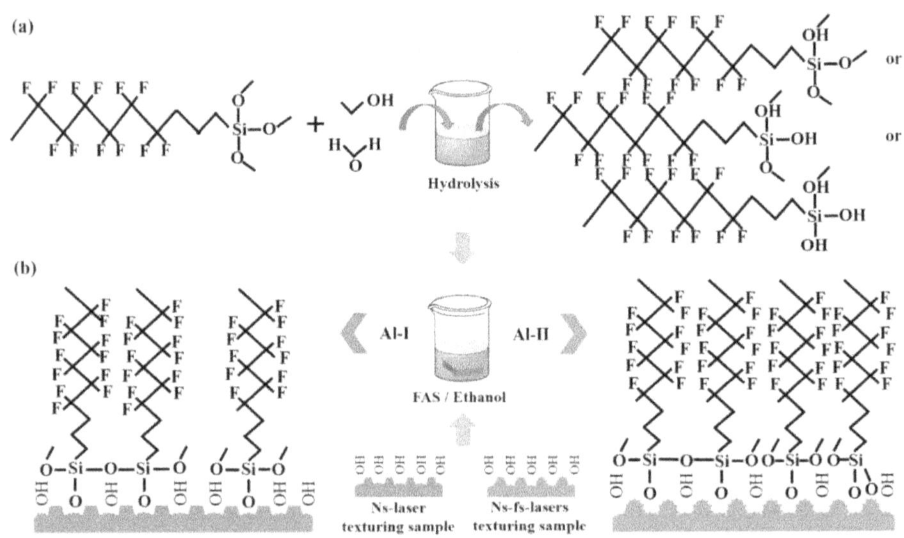

FIGURE 8.6 Hydrolysis and condensation of FAS molecules on aluminium sample [38]. (a) Hydrolysis of FAS precursor in ethanol/water solution to generate reactive silanol (Si–OH) groups. (b) Condensation and grafting of hydrolyzed FAS molecules onto laser-textured aluminium surfaces, forming a low-surface-energy fluorinated siloxane coating.

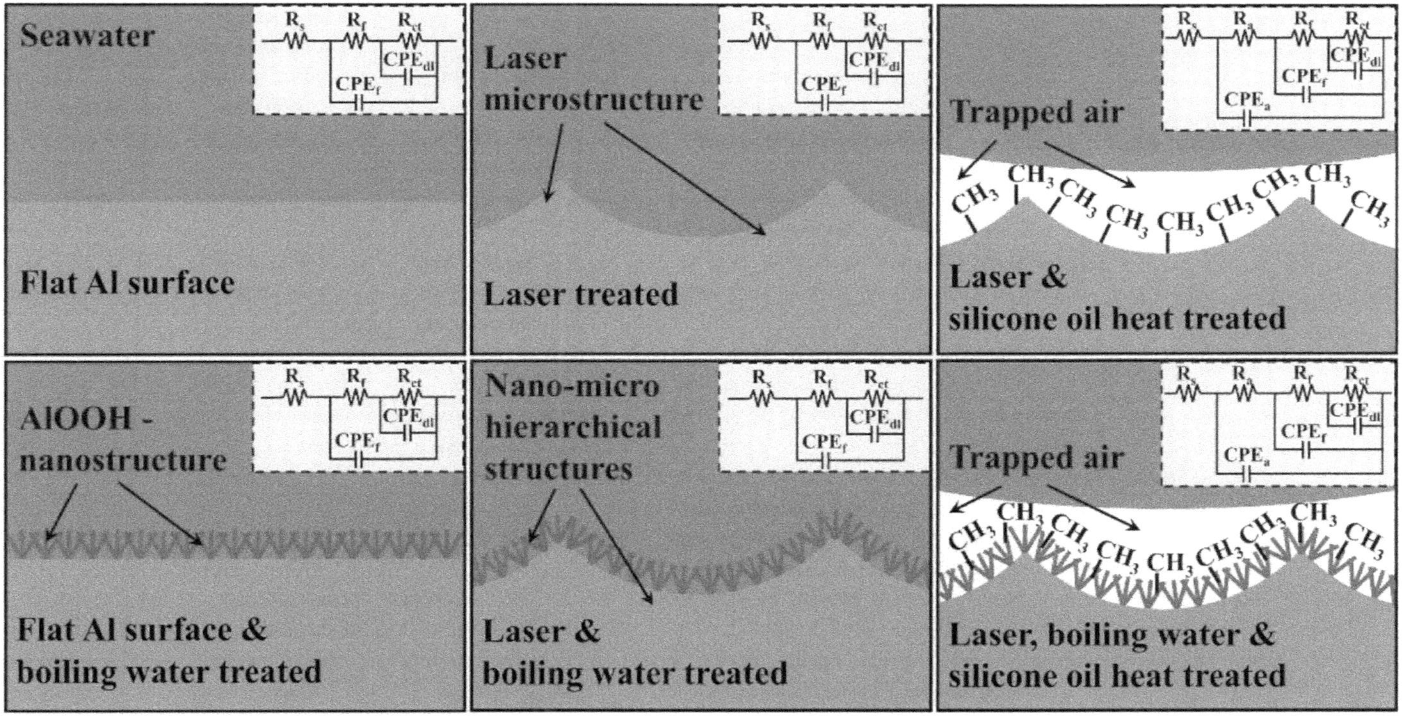

FIGURE 8.7 Effect of laser texturing, boiling water treatment, and silicone oil heat treatment on surface morphology of aluminium sample and their corresponding electrochemical equivalent circuit for EIS [39].

TABLE 8.1
Summary of Chemical Treatment Effect on Hydrophobicity of Laser-Textured Material

Material	Laser Type	Texture Type	Chemical Type	Water Contact Angle and Sliding Angle	Reference
S45C steel plates	Nanosecond IR laser fiber machine	Circle-shaped texture patterns (diameter = 100 μm)	1.0 wt% FAS-17 (1H,1H,2H,2H-perfluorodecyltriethoxysilane)/ethanol	161.5°	[7]
1060 aluminium	Nanosecond Ytterbium pulsed fibre laser	Grid-pattern	AC-FAS [CF_3–$(CF_2)_7$–$(CH_2)_2$–Si–$(OC_2H_5)_3$]/ethanol	160.6° ± 1.5° and 3.0° ± 1.0°	[8]
AMG aluminium alloy	IR ytterbium fiber laser	Grooves	Methoxy-{3-[(2,2,3,3,4,4,5,5,6,6,7,7,8,8,8-pentadecafluorooctyl) oxy]propyl}-silane	165.2 ± 0.9° & 2.9 ± 0.4°	[9]
316 L SS steel	Nanosecond laser	Micro-cracks and brain-like microstructures	Fluoroalkylsilane (FAS, 1H, 1H, 2H, 2H-perfluorooctyltriethoxysilane) ethanoic solution	160 ± 5° and 3 ± 0.5°	[11]
304 SS	Fiber laser	Micropore-micropore distances of 50 μm	Hexadecyltrimethoxysilane ($CH_3(CH_2)$ 15$Si(OCH)_3$)	158.9	[12]
NiTi alloy	Ytterbium nanosecond pulsed fiber laser	Macro-nano structures	1H, 1H, 2H, 2H-per-fluorodecyltrichlorosilane; solvent: ethanol and distilled water	158.7° ± 1.2° and 1.6° ± 0.4	[13]
3 wt% TiC/Ti6Al4V	YAG laser	Hierarchical micro-/nanostructures	1.5 wt% 1H,1H,2H,2H-perfluorodecyltrichlorosilane	158.4° and 4.7°	[15]
SS316 SS steel	Ultrashort pulse infrared laser	Hierarchical micro- and nanostructures	0.8 mmol/l (3,3,4,4,5,5,6,6,7,7,8,8,9,9,10,10,11,11,12,12,12-heneicosafluorododecyl) phosphonic acid (F21)/tetrahydrofuran (THF) solution	153° ± 5°	[16]
Co28Cr6Mo alloy	Nanosecond laser	Microgrid and nanoparticle composite structure	PDMS solution and silicone oil	155.236° ± 1.5°	[17]
$Fe_{43}Cr_{20}Mo_{10}W_4C_{15}B_6Y_2$ (at%) amorphous alloy	Picosecond pulsed laser	Trench and dimple structure	2.0 wt% poly-perfluoroacrylate	166.5° and 3.6°	[18]
Inconel 718	Spatially shaped bessel-beam and gaussian beam	3D micro-nano-structures	Trichlorosilane (1H,1H,2H,2H-perfluoroctyl)	154.3°	[19]
316L SS steel	Femtosecond laser	Honeycomb	Stearic acid	154.2° and 8.9°	[21]
Copper	Nanosecond laser	Micro-nano surface structures	SiO_2@PDMS	159.5° and 0.5°	[22]
Magnesium	Picosecond	Micro-pillars	Stearic acid	135°	[23]

TABLE 8.2
Various Heat Treatment Conditions Used for Modifying Laser-Textured Material

Material	Texture Pattern	Heat Treatment Condition	Contact Angle	Reference
H59 brass	Micro-scale protrusions and micro-scale pits	150°C, 4 hours	162°	[27]
Ti-6Al-4 V alloy	Periodic micro-bump structure	150°C, 2 hours	156.2°	[29]
NiTi shape memory alloy	Micro-nano structure	150°C, 1 hour	150.3°	[30]
Bronze	Pore arrays	150°C, 12 hours	158°	[31]
Copper	Grid pattern	150°C, 2 hours	155.6°	[32]
5051 aluminium	Grid pattern	150°C, 72 hours	136.97°	[33]
Aluminium alloy (AlMg6)	Dimple	100°C, 6 hours	157°	[34]
Ti-6Al-4V	Micro-grooves	150°C, 2 hours	151.2°	[35]

8.6 DISCUSSIONS

Laser texturing successfully induces superhydrophobicity on the material surface by developing periodic three-dimensional micro/nanostructure patterns with lesser surface energy on the material surface. Laser type, laser pulse frequency, and laser beam energy decide the morphology achievable through laser texturing. Lasers types used for creating textures on material surfaces are classified as millisecond, nanosecond, picosecond, and femtosecond lasers. Most of the textured surface upon texturing exhibit superhydrophilic characteristics due to the formation of oxides and the superhydrophilic surface is converted into superhydrophobic through different post-processing techniques. Some of the commonly used post-processing techniques are (i) chemical modification (ii) heat treatment, and (iii) long-term exposure to atmospheric contaminants. Exposure of textured material to atmospheric conditions resulted in the adsorption of organic contaminants rich in $-CH_2-$ and $-CH_3$ groups and it helps in improving the hydrophobicity. Chemical treatment techniques increase the hydrophobicity of textured surfaces by lowering the surface energy of the samples. Fluoro-silane molecules with $-CF_3$ and $-CF_2-$ groups are anchored to the textured material surface during chemical treatment and it helps in transition from superhydrophilic to superhydrophobic. Similarly, lower temperature heat treatment achieves super hydrophobicity by favouring the adsorption of organic molecules with less free energy. Various patterns, namely, nano protrusions, micro-grating, micro-grooves, micro-hole arrays, micro-pits, micro-bumps, sub-micron ripple, nano-ripples, hierarchical textures, conical peaks, dimples, crossed grooves, concentric rings, grid, and line pattern, are generated by research fraternity on various materials for inducing superhydrophobic behavior. The air pockets trapped by the superhydrophobic surfaces reduce the contact area of the surface with the corrosive medium and improve the corrosion resistance. The published work confirms that laser texturing is an effective technique for altering the surface characteristics of the material. Desired hydrophobicity and corrosion resistance can be anchored on laser-textured material by subjecting it to different post-processing techniques. Laser texturing could act as a tool to improve the functionality of material in several applications.

REFERENCES

[1] A. Samanta, Q. Wang, S. K. Shaw, H. Ding, Roles of chemistry modification for laser textured metal alloys to achieve extreme surface wetting behaviors, Materials & Design, 192 (2020), 108744.
[2] Z. Liu, T. Niu, Y, Lei, Y. Luo, Metal surface wettability modification by nanosecond laser surface texturing: A review, Biosurface and Biotribology, 8(2) (2022), 95–120.
[3] A. Cholkar, R. McCann, G. Perumal, S. Chatterjee, M. Swayne, D. Kinahan, D. Brabazon, Advances in laser-based surface texturing for developing antifouling surfaces: A comprehensive review, Applied Surface Science Advances, 18 (2023), 100513.
[4] L. Zhang, N. Lin, J. Zou, X. Lin, Z. Liu, S. Yuan, et al., Super-hydrophobicity and corrosion resistance of laser surface textured AISI 304 stainless steel decorated with Hexadecyltrimethoxysilane (HDTMS), Optics & Laser Technology, 127 (2020), 106146.
[5] A. O. Ijaola, E. A. Bamidele, C. J. Akisin, I. T. Bello, A. T. Oyatobo, A. Abdulkareem, . . . & E. Asmatulu, Wettability transition for laser textured surfaces: A comprehensive review, Surfaces and Interfaces, 21 (2020), 100802.
[6] V. Kumar, R. Verma, S. Kango, V. S. Sharma, Recent progresses and applications in laser-based surface texturing systems, Materials Today Communications, 26 (2021), 101736.
[7] Q. Ma, Z. Tong, W. Wang, G. Dong, Fabricating robust and repairable superhydrophobic surface on carbon steel by nanosecond laser texturing for corrosion protection, Applied Surface Science, 455 (2018), 748–757.
[8] Z. Yang, X. Liu, Y. Tian, Hybrid laser ablation and chemical modification for fast fabrication of bio-inspired super-hydrophobic surface with excellent self-cleaning, stability and corrosion resistance, Journal of Bionic Engineering, 16 (2019), 13–26.
[9] L. B. Boinovich, E. B. Modin, A. R. Sayfutdinova, K. A. Emelyanenko, A. L. Vasiliev, A. M. Emelyanenko, Combination of functional nanoengineering and nanosecond laser texturing for design of superhydrophobic aluminum alloy with exceptional mechanical and chemical properties, ACS Nano, 11(10) (2017), 10113–10123.
[10] L. González-Fernández, A. Anagnostopoulos, T. Karkantonis, S. Dimov, M. Chorążewski, Y. Ding, Y. Grosu, Laser-texturing of stainless steel as a corrosion mitigation strategy for high-temperature molten salts applications under dynamic conditions, Solar Energy Materials and Solar Cells, 257 (2023), 112380.
[11] Y. Lu, Y. Guan, Y. Li, L. Yang, M. Wang, Y. Wang, Nanosecond laser fabrication of superhydrophobic surface on 316L stainless steel and corrosion protection application, Colloids and Surfaces A: Physicochemical and Engineering Aspects, 604 (2020), 125259.
[12] M. Wang, C. Yang, Z. Yang, Y. Tian, The effects of carbon ion implantation on wettability, abrasion, thermal and anti-corrosion stabilities of laser ablated super-hydrophobic Nitinol surface, Journal of Materials Research and Technology, 28 (2024), 1784–1793.
[13] A. Kumthekar, V. Laitinen, K. Ullakko, Characterization of hydrophobic metasurfaces fabricated on Ni-Mn-Ga-based alloys using femtosecond pulsed laser ablation, Materials & Design, 244 (2024), 113128.

[14] D. Madapana, R. Bathe, I. Manna, J. D. Majumdar, Effect of process parameters on the corrosion kinetics and mechanism of nanosecond laser surface structured titanium alloy (Ti6Al4V), Applied Surface Science Advances, 20 (2024), 100580.
[15] M. Tang, L. Zhang, Facile preparation of hierarchical micro-/nanostructure superhydrophobic surface with excellent corrosion resistance on Ti6Al4V-based composite for nuclear fuel reprocessing, Advanced Engineering Materials, 25(23) (2023), 2300923.
[16] P. Lorenz, J. Zajadacz, J. Bauer, M. Ehrhardt, G. Hommes, S. Peter, K. Zimmer, Chemical stability of superhydrophobic stainless steel surfaces realized by laser-texturing and chemical functionalization, Surfaces and Interfaces, 37 (2023), 102685.
[17] L. Zhang, Z. Tan, T. Wu, L. Zhang, B. Hao, Investigation on the corrosion resistance properties of cobalt–chromium–molybdenum alloy artificial human body components with robust biomimetic superhydrophobic and slippery surfaces based on laser texturing, Langmuir, 39(42) (2023), 14996–15013.
[18] X. Wei, Z. Liao, Y. Liang, L. Zhang, L. Wang, B. Chen, J. Shen, Superhydrophobic Fe-based amorphous alloy coatings with excellent anti-fouling and anti-corrosion properties by picosecond laser texturing, Applied Surface Science, 642 (2024), 158612.
[19] Z. Qiu, W. Chen, D. Li, H. Wang, J. Zhang, Y. Zhang, . . . & B. Dai, Preparation of large-area superhydrophobic and anti-icing 3D micro-nano-structures using femtosecond Bessel beams with fluorination treatment, Applied Surface Science, 664 (2024), 160239.
[20] M. Liang, Y. Yuan, X. Li, J. Cheng, X. Zhang, K. Zhang, R. You, Enhanced corrosion resistance of Mg and Mg alloys with fs-laser printed hydrophobic and periodically microrippled surface for armor application, Applied Surface Science, 639 (2023), 158156.
[21] Y. Tang, Y. Cai, L. Wang, X. Luo, B. Wang, Q. Song, Z. Liu, Fabrication of superhydrophobic stainless steel via hybrid femtosecond laser-chemical method with wear-resistance and anti-corrosion properties, Optics & Laser Technology, 164 (2023), 109474.
[22] Y. Shu, X. Lu, W. Lu, W. Su, Y. Wu, H. Wei, et al., Mechanically robust superhydrophobic copper surface with self-cleaning, anti-icing, and corrosion resistance, Surface and Coatings Technology, 455 (2023), 129216.
[23] Y. E. Yayoglu, R. G. Toomey, N. B. Crane, N. D. Gallant, Laser machined micropatterns as corrosion protection of both hydrophobic and hydrophilic magnesium, Journal of the Mechanical Behavior of Biomedical Materials, 125 (2022), 104920.
[24] U. Trdan, M. Hočevar, P. Gregorčič, Transition from superhydrophilic to superhydrophobic state of laser textured stainless steel surface and its effect on corrosion resistance, Corrosion Science, 123 (2017), 21–26.
[25] S. Meng, Y. Yu, X. Zhang, L. Zhou, X. Liang, P. Liu, Investigations on electrochemical corrosion behavior of 7075 aluminum alloy with femtosecond laser modification, Vacuum, 221 (2024), 112911.
[26] A. Cholkar, S. Chatterjee, C. Richards, É. McCarthy, G. Perumal, F. Regan, . . . & D. Brabazon, biofouling and corrosion protection of aluminum alloys through ultrafast laser surface texturing for marine applications, Advanced Materials Interfaces, 11(6) (2024), 2300835.
[27] X. Li, Y. Jiang, Z. Jiang, Y. Li, C. Wen, D. Zhang, . . . & Z. Zhang, Improvement of corrosion resistance of H59 brass through fabricating superhydrophobic surface using laser ablation and heating treatment, Corrosion Science, 180 (2021), 109186.
[28] X. Li, Y. Jiang, X. Tan, Z. Zhang, Z. Jiang, J. Lian, et al., Superhydrophobic brass surfaces with tunable water adhesion fabricated by laser texturing followed by heat treatment and their anti-corrosion ability, Applied Surface Science, 575 (2022), 151596.

[29] C. Liu, S. Tong, Y. Yue, H. Wang, J. Song, Y. Li, . . . & Z. Wang, Laser-based fabrication of superwetting titanium alloy with enhanced corrosion and erosion-corrosion resistance, Colloids and Surfaces A: Physicochemical and Engineering Aspects, 688 (2024), 133648.
[30] Y. Yang, Y. Ma, W. Li, L. Ma, Z. Li, B. Jia, . . . & P. Liang, Anti-corrosion superhydrophobic surface of LPBF-NiTi alloy fabricated by nanosecond laser machining, Optics & Laser Technology, 158 (2023), 108858.
[31] C. Jiang, W. Zhang, M. Wang, T. Sun, X. Zhang, Y. Wu, H. Zheng, Attaining superhydrophobic and superoleophilic bronze surface by picosecond laser texturing and post-heat treatment, Optics & Laser Technology, 177 (2024), 111139.
[32] L. Ma, L. Wang, C. Li, J. Guo, P. Shrotriya, C. Deng, J. Zhao, Hybrid nanosecond laser processing and heat treatment for rapid preparation of super-hydrophobic copper surface, Metals, 9(6) (2019), 668.
[33] C. Fan, X. Wang, Y. Liu, C. Li, X. Liu, Rapid fabrication of near superhydrophobic aluminium surface through nanosecond laser texture and heat treatment, In Journal of Physics: Conference Series, 2174(1) (2022), 012050. IOP Publishing.
[34] S. Y. Misyura, D. V. Feoktistov, V. S. Morozov, E. G. Orlova, G. V. Kuznetsov, A. G. Islamova, Effect of heat treatment on corrosion of laser-textured aluminum alloy surfaces, Journal of Materials Science, 56(22) (2021), 12845–12863.
[35] Q. Wang, H. Wang, Fabrication of textured surface with controllable wettability via laser-thermal hybrid processing, Materials Letters, 315 (2022), 131954.
[36] X. Yang, C. Yang, Z. Yang, D. Zhang, A study on strengthening the corrosion resistance of NiTi SMA by composite femtosecond and nanosecond laser-induced hierarchical micro/nanostructures, Optics & Laser Technology, 163 (2023), 109339.
[37] J. Cao, B. Ma, Y. Yang, L. Li, X. Li, Femtosecond laser preparation of terracotta warrior-like pit superhydrophobic structure on magnesium alloy with mechanical durability and corrosion resistance, Langmuir, 39(29) (2023), 10230–10239.
[38] X. Yang, C. Yang, Z. Yang, D. Zhang, Corrosion resistance and stability of superhydrophobic surfaces with hierarchical micro/nanostructures induced by nanosecond and femtosecond lasers, Optics & Laser Technology, 169 (2024), 110179.
[39] N. G. Tran, D. M. Chun, A. G. Abd-Elrahim, Superhydrophobic aluminum surfaces with nano-micro hierarchical composite structures: A novel and sustainable approach to corrosion protection, Journal of Alloys and Compounds, 960 (2023), 170907.
[40] Q. Wang, K. Yin, Z. Bai, J. Liu, L. Huo, H. Wang, Fabrication of robust superhydrophobic copper surface via highly efficient nanosecond laser-based surface functionalization, Optik, 276 (2023), 170690.

9 Laser Surface Texturing on the Piston Top Surface

Enhancing Surface Modification, Reducing Residual Stresses, and Improving Surface Quality

Kumaran M., S. Mohan, S. Ravi, and T. Sathies

9.1 INTRODUCTION

9.1.1 Overview of Laser Surface Texturing

A contemporary high-energy laser-based surface treatment technique called laser surface texturing (LST) is used to precisely regulate material surface texture features at the micro or nanoscale in order to improve or generate new useful capabilities. In order to create dimple, groove, or pore surface topographies that contribute to the material's tribology, wettability, and mechanical durability, material is selectively removed during the ablation process. By reacting with the material through localized melting, vaporization, or plasma creation, lasers—which can be short or ultrashort-pulsed systems—can produce controlled surface textures. New opportunities have been created by precisely altering the surface in this manner without compromising the bulk material qualities, as opposed to the present methods of mechanical machining, shot peening, or coatings. In the automotive and aerospace industries, laser texturing is widely used to enhance the functionality of vital components including turbine blades, bearings, and pistons. LST can increase lubricant retention, decrease wear and friction, and improve engine efficiency and longevity in piston applications. The formation of micro-dimples on the piston's top surface aids in the formation of an oil reservoir and dramatically reduces the likelihood of direct metal-to-metal contact, which results in large frictional energy losses. The ability to fine-tune surface roughness and topography to maximize the adherence of functional coatings and hence influence the general mechanical and thermal properties of the treated component is the other benefit of LST. When contrasted with conventional

DOI: 10.1201/9781003542087-9

surface finishing methods, LST doesn't require extra consumables like abrasives or chemical etchants; it is extremely reproducible, contactless, and ecologically benign. It is highly adjustable to various engineering applications since it allows for the precise setting of process parameters including laser power, pulse duration, spot size, and scanning speed to create the necessary surface characteristics. The development of laser technologies, such as fibres and ultrafast lasers, in recent years has improved the efficiency and viability of LST and made it possible to integrate them with high-volume industrial manufacturing processes. It incorporates advancements in real-time monitoring and computer modelling approaches that improve prediction and control over the laser–material interaction, enabling the optimization of texturing patterns for particular applications. Interest in LST has also grown as a result of the growing need for high-performance and energy-efficient mechanical systems, particularly in the automotive industry where strict technical rules govern emissions and fuel efficiency. Textured surface treatment for enhanced heat dissipation, corrosion resistance, and biomedical implants is another field in which LST is being investigated. Laser surface texturing is a crucial enabling technology in the development of materials for the next generation of engineering applications because it allows for the customization of surface characteristics at the micro and nanoscale. In fact, new developments in hybrid LST approaches that combine laser texturing with coatings or chemical treatments are anticipated to keep up with advances in research on laser processing techniques. To put this in a broader context, laser surface texturing is one of the many advanced production technologies that offer tremendous advantages in terms of precise surface modification for enhanced mechanical, thermal, or tribological qualities of a wide range of industries. Because of its ability to precisely control surface microstructures, the process has become practically essential in modern manufacturing, particularly for high-performance parts with extremely specific needs like a piston's top, where better lubrication, friction, and wear resistance are essential to an engine's performance and lifespan [1–4].

9.1.2 Significance of Piston Top Surface Modification

One of the most crucial components of an engine that burns gasoline internally is the piston's facing-up section, which has a direct impact on the engine's performance, energy consumption, and lifespan. The piston surface is subjected to continuous cycles of heat and motion due to its operation in extremely hot and pressurized circumstances, as well as mechanical strain. This makes it vulnerable to wear and tear, friction loss, carbon buildup, and the accumulation of residual stress. By addressing these issues and enhancing the piston's sliding and heat characteristics, high-tech techniques like LST can significantly improve engine performance by altering the piston's top surface. Reducing friction between the piston and the cylinder wall and improving lubricant retention are two of the primary goals of piston surface modifications. Even though they may be polished or covered, standard piston surfaces typically don't maintain adequate lubrication under high load conditions, which increases wear and energy loss. In order to reduce direct metal-to-metal contact, a stable lubricating coating can be formed by creating dimples, grooves, or craters on the surface that effectively capture and reserve more oil. By preventing scuffing and adhesive wear, this will further minimize friction and increase the piston's service life. In addition to these facts, the piston's surface texturing can have an impact on its heat dissipation properties, which is also a critical factor in thermal fatigue and overheating. Proper transfer of heat from the piston to the environment will be facilitated by patterning the laser texturing for increased thermal conductivity. Additionally, this could help prevent hotspots that could cause banging or pre-ignition and regulate combustion chamber temperatures. Another significant area that surface change at the piston top influences is residual stress management. Changes in the casting, forging, and machining processes leave the piston material with residual stress and may weaken its mechanical strength, which could cause cyclic loading to fail too soon. LST has been shown to alter the distribution of stress in the surface layer, reducing tensile stresses and creating advantageous compressive stresses that are linked to increased fatigue resistance. This type of application is very beneficial for heavy-duty and high-performance engines when the pistons are subjected to prolonged high mechanical and thermal loading. Additionally, modifications to the piston surface may concurrently enhance resistance to fouling and carbon deposition, which are often linked to full combustion and oil degradation. By decreasing surface roughness and changing the energy characteristics on the surface, an effective surface can reduce carbon adhesion. Otherwise, carbon buildup could negatively impact engine efficiency and emissions. Surface modification techniques are used into modern piston design in response to the increased demand for fuel economy and pollution reduction. In order to comply with new, stricter environmental standards, improved combustion qualities can result in significantly better fuel atomization, lower levels of unburned hydrocarbons, and reduced particle emissions. Warmth treatment and other surface engineering techniques are very useful in hybrid electric vehicle applications, where the components need to be made with minimal energy consumption and excellent durability. In the meantime, hybrid techniques including chemical modification, nitriding, or laser surface treatment in conjunction with coating are being explored in the research to further improve the pistons' performance. LST is a potent enabler of the next generation of high-performance engines in terms of enhancements in lubrication efficiency, wear resistance, thermal stability, and general durability because of its capacity to customize surface features at the micro/nanoscale. In conclusion, one of the necessary techniques to improve engine efficiency and reduce wear while prolonging life is piston top surface modification. By adding functional surface patterns that aid in lubrication, enable the suppression of residual stresses, and improve thermal management, laser surface texturing offers a very efficient and controllable means to achieve these advantages. Advanced surface modification techniques like LST will play an increasingly important part in the automobile industry's transition to more sustainable and energy-efficient technologies, spurring innovation in piston design and helping to create high-performance, low-emission engines [5–8].

9.2 FUNDAMENTALS OF LASER SURFACE TEXTURING

9.2.1 Mechanisms of Laser–Material Interaction

The term "laser surface texturing" refers to the precise interaction of high-energy laser beams with material surfaces to modify their properties and add useful textures. The basic processes that underlie how laser beams interact with materials are dependent on a number of factors, including the material's absorption coefficient, thermal conductivity, and reflectivity, as well as the laser's parameters, including wavelength, pulse duration, energy density, and scanning speed. The material absorbs the laser beam's energy when it is focused on the piston's upper surface. Rapid heating and associated thermal effects, including melting, vaporization, and plasma production, are part of the process. In pulsed laser systems, the length and strength of each pulse dictates the amount of material removed and the creation of micro-textures such as dimples, grooves, or crater-like patterns. The majority of the material bulk is preserved while providing great resolution in surface modification thanks to short and ultrashort pulsed lasers, such as femtosecond and picosecond lasers, which allow for very high precision ablation with little heat-affected zone surrounding the ablation point. The most crucial factor in obtaining the intended surface morphologies is the phase changes brought on by the laser. For instance, surfaces can be restructured through controlled melting and resolidification, improving their mechanical qualities. For durability and fatigue resistance, laser treatment may also create compressive residual stresses in the layer. Furthermore, the optical properties of the material in terms of light absorption also influence relative efficiency; for example, different laser wavelengths are needed for steel and aluminium to absorb the most energy. To guarantee consistent texturing without undue heat damage or associated surface damage, the laser parameters in the piston top surface region must be optimized. One of the most intriguing effects of LST is laser-induced surface structuring, which occurs when incident laser light interferes with dispersed waves to create self-organized nano- and micro-patterns that give the surface additional characteristics. The depth and spacing of the surface textures are also determined by the laser fluence and repetition rate, which have a direct effect on wear resistance, friction reduction, and lubricant retention. The efficiency of the texturing process is partly determined by the interaction between heat diffusion and material removal mechanisms; lower energy densities enable regulated surface patterning, while greater energy densities produce significantly deeper textures. By avoiding oxidation and guaranteeing clean ablation, shielding gases like argon or nitrogen can further affect the interaction between the laser and the material. In addition, laser-induced shock waves and rapid cooling rates during LST can alter the microstructural properties of the surface layer, grain size, and hardness. Developing complex surface textures for highly particular tribological and thermal requirements in automotive applications is made possible by the ability to precisely manage the laser–material interaction to within a fraction of a second. Real-time laser processing parameter management is becoming increasingly feasible and practical with greater repeatability in enhanced surface quality as computer modelling and in situ monitoring get more sophisticated. Thus, a complex interplay of thermal, optical, and mechanical phenomena results in the creation of designed surface patterns through the laser–material interaction mechanisms involved in laser shock texturing. The optimization of laser processing conditions with higher surface performance in piston applications that provide lower friction, increased wear resistance, and improved thermal stability would be based on a proper understanding of these interactions [9, 10].

9.2.2 Influence of LST on Tribological Properties

The piston top surfaces' tribological properties are significantly impacted by LST. To improve engine performance and prolong its life, it lowers friction, improves wear resistance, and increases oil retention. By altering the surface's interaction with lubricants, micro-textures—such as dimples, grooves, and craters—reduce direct metal-to-metal contact and increase load-bearing capacity. Most of the time, the micro-dimpled textures created by LST would effectively serve as lubricant reservoirs, guaranteeing a sufficient supply of oil even under high-load and high-temperature situations—a crucial function in internal combustion engines. Additionally, it has been discovered that these textures help to maintain a stable hydrodynamic lubrication regime, which reduces boundary friction and the energy loss caused by piston-cylinder interaction. In addition, by enhancing the distribution of contact stresses, LST would increase surface roughness and load-bearing capacity, making wear even more uniform and less noticeable. It is feasible to lower the coefficient of friction while maintaining an adequate layer thickness for lubrication by optimizing the depth diameter and spacing of the laser-textured features, which have a major impact on frictional behavior. In certain engine operating conditions, research has already demonstrated that well-crafted laser texturing can lower the coefficient of friction by as much as 50%. Wear resistance is also another crucial component of enhancing tribology with LST. Abrasive and adhesive wears are decreased because the development of microtextures regulates the contact interface between the piston and cylinder liner. Longer engine component life and improved performance are anticipated because the LST surfaces do not have a tendency to become more rough over time. The material's resistance to plastic deformation and fatigue is strengthened by the laser irradiation's localized heating and quick cooling effects, which result in improved wear characteristics. The same texturing affects the stability and production of tribo-films, which are protective layers that are frequently oxide or carbon layers that form during sliding. Wear rates are further reduced by these layers. In real-world applications, pistons having laser texturing on their surface exhibit increased oil retention, reduced blow-by losses, and enhanced scuff resistance. LST also has the advantage of improving tribological qualities by reducing pollutant and combustion byproduct adhesion, which prevents deposits from accumulating and impairing performance. For extended durations of operation, the piston surface's functional integrity is maintained by the microtextures' self-cleaning. LST can also be used in conjunction with coatings such as ceramic or carbon diamond-like layers to improve tribological performance and provide good wear and friction resistance. LST is a particularly flexible technique in tribology improvement because of the features that may be customized by adjusting the laser parameters towards textures based on the unique needs of engines. The combination of experimental validation and computational modelling has enabled highly accurate texturing pattern creation for optimal performance improvements. In summary, LST enhances the piston surface's tribological behavior in terms of uniform stress distribution, reduced frictional losses, improved wear resistance, and optimized lubrication dynamics. As a result, the engine is more durable and efficient and uses less fuel. LST will become increasingly important in future automotive and aerospace industrial tribological

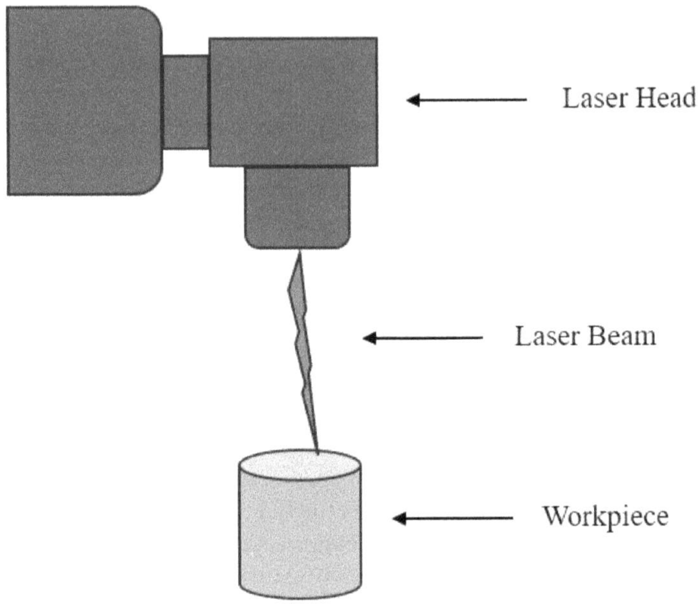

FIGURE 9.1 Layout for laser surface texturing.

applications as a result of continued advancements in laser technology and precise manufacturing techniques [11, 12]. Figure 9.1 shows the layout for laser surface texturing.

9.3 RESIDUAL STRESS ANALYSIS AND MITIGATION

9.3.1 Origin and Impact of Residual Stresses in Pistons

Residual material stresses are caused by a variety of production processes, including casting, forging, machining, and thermal treatments, as well as by operating conditions, high temperatures, and mechanical loads (Figure 9.2). The mechanical behavior and structural integrity of the piston are influenced by these stresses, which are the equilibrium forces that are retained inside the material after external loads or temperature gradients are removed. The material undergoes differential contraction as a result of casting, forging, and uneven cooling rates. This causes tensile and compressive residual stresses in different areas of the piston portion. The distribution of stresses is also altered by machining operations, particularly finishing operations like grinding and honing, which cause plastic deformation of the material close to the surface. This typically results in surface zones with tensile stresses, which can act as sites for the initiation of cracks. Repeated heating and cooling of the piston during engine operation, or thermal cycling, also starts the buildup of stresses in the material, which may eventually result in the formation of possible material degradation pathways. Particularly because it promotes fatigue failure and crack propagation with decreased load-bearing capacity, the contribution from tensile residual stress is harmful. As a result, under high-stress circumstances, the piston is much more likely

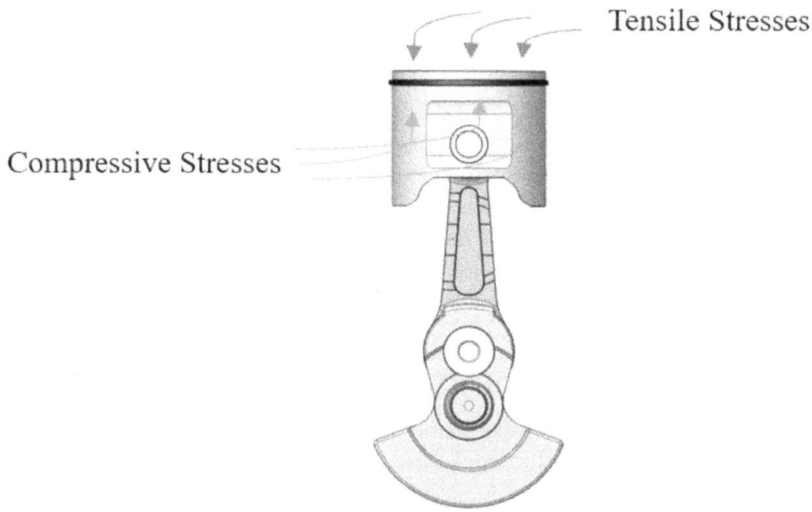

FIGURE 9.2 Residual stress visualization in piston design.

to fail catastrophically. Conversely, compressive residual stresses would be advantageous since they enhance fatigue resistance and postpone the onset of cracks. For these reasons, surface treatments like as laser surface modification and shot peening are used to create advantageous stress states. The residual stresses have a very big impact on the piston's performance. Tensile tensions encourage thermal fatigue, premature wear during operation, and are bad for durability. Excessive residual stresses can lead to dimensional instability, which can impact piston-cylinder clearances and ultimately reduce engine efficiency. Furthermore, stress-induced modifications to the piston's shape may result in excessive friction, an uneven distribution of pressure, and increased oil consumption, all of which would further impair performance. Proper residual stress management is even more crucial in high-powered engines when pistons operate under demanding circumstances in order to ensure strength and longevity. Unchecked stress accumulation can accelerate the development of cracks where stress concentrates, particularly in close proximity to thin spots where heat and tension are greatest, such as ring grooves, sharp corners, and the piston's top. Excessive stress accumulation affects the material's structure at the microscopic level, changing its hardness, strength against brittle fracture, and resistance to gradual deformation—all of which are critical for preserving the integrity of the piston after extended usage. The distribution of residual stresses in pistons can be evaluated using sophisticated measurement techniques such hole-drilling, neutron diffraction, and X-ray diffraction (XRD). They offer priceless insights into the mechanics of stress relaxation and the effectiveness of stress-reduction techniques. In order to maximize piston design, material selection, and surface treatment procedures and enhance service life and performance, residual stresses and their causes and consequences are considered from a wider angle. By selectively altering surface characteristics, spreading internal stresses, and adding advantageous compressive stresses, laser surface texturing essentially offers a novel method of regulating residual stresses. Pioneers can modify the surface residual stress distribution to improve the overall mechanical performance in terms of fatigue resistance, fracture initiation life, and the piston's overall mechanical behavior by carefully adjusting the laser parameters. In conclusion, the impacts of persistent residual stresses in pistons are likely more related to manufacturing and application parameters, and they have a significant impact on thermal stability, fatigue life, and structural integrity. For the pistons to operate at their best and for engines to be reliable and efficient over the long run, new surface engineering methods like the one discussed must effectively handle the residual stresses that now exist in such crucial components [13, 14].

9.3.2 Role of LST in Residual Stress Modification

With LST, residual stresses on the piston top surfaces can be optimally controlled because the process changes the microstructure at the surface and redistributes the stress fields by creating some advantageous compressive stresses that improve fatigue resistance and overall mechanical performance (Figure 9.3). Typically, the pistons have residual stresses from manufacture, which are created during processes like forging and casting, machining, and additional thermal and mechanical loading while the engine is running. The detrimental effects of these stresses are most pronounced when they are in the form of high tensile residual stresses initiating and furthering the process of cracking, which in turn causes preliminary fatigue failure and hence reduction in the service life. All these can be effectively controlled by LST through

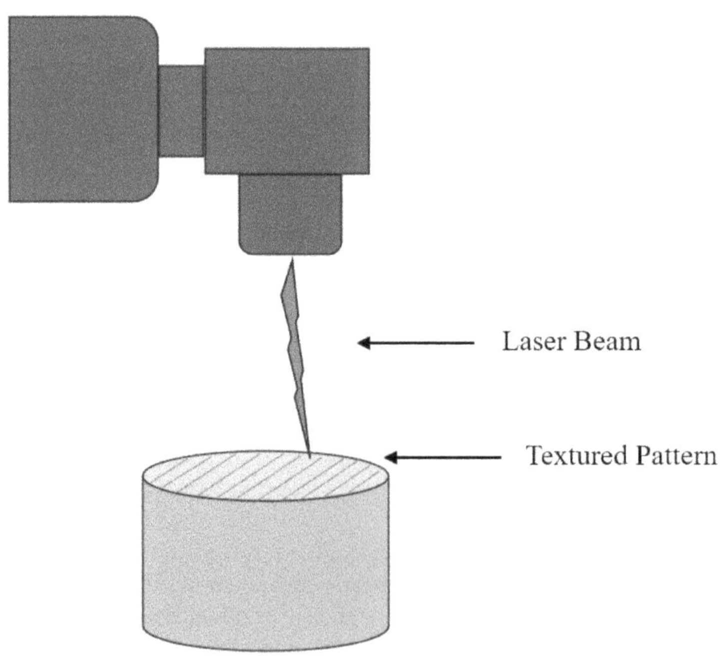

FIGURE 9.3 Process of surface modification.

the creation of effects of localized thermal nature such that, with the interaction between the high-energy laser beam and the piston surface, rapid heating followed by rapid cooling then takes place thereby leading to localized plastic deformation and phase transformations; hence an ultimately refined microstructure with enhanced mechanical properties is created. By creating compressive residual stresses on the surface layer, this technique significantly increases fatigue strength by counteracting previously created tensile stresses. The components' cyclic loading durability is increased by compressive residual stresses, which are advantageous because they prevent crack start and slow down crack propagation. A number of variables affect how well LST manages residual stresses, including laser power, scanning speed, spot overlap, and pulse duration, which all affect how deeply and intensely stress is modified. Manufacturers can design the piston surface's stress profile to have better resistance to mechanical wear, thermal fatigue, and crack propagation by optimizing these characteristics. Furthermore, because LST can produce periodic surface patterns, grooves, or micro-dimples, it will alter the surface topography. These characteristics transfer localized loads more evenly throughout the surface, which supports stress alleviation in addition to lubrication retention. In relation to this point, the redistribution of stress aids in reducing the concentrations of stress, particularly in regions like the ring grooves and the piston crown. Hence, it improves the chances of failure. Yet another benefit of LST regarding residual stress management is that it alters the grain size of the surface layer. Smaller grains generally enhance hardness and, hence, resistance to plastic deformation. The locally laser-induced melted and solidified structure would be expected to yield a finer grain structure, better suited for the extremes of thermal and mechanical loading that act on the piston in the engine. The laser-induced modifications enhance the adhesion of thermal barrier coatings (TBCs) or wear-resistant coatings. Thus, the overall performance and longevity of the piston can be improved. Apart from direct effects on stress management, LST is a contactless environmentally benign process. It eliminates the traditional treatment of mechanical peening or chemical treatment; therefore, it can be used effectively and sustainably by industry. Real-time monitoring and computational modelling techniques can be merged into the LST process for perfect control over modification of the mentioned stresses. Therefore, repeatability and consistency in industrial applications can be achieved. In short, LST in the piston domain is a very strong device of controlling residual stresses for the introduction of favourable compressive stresses, redistribution of stress concentrations, microstructure refinement, and fatigue resistance. By mitigating the adverse effects of tensile residual stresses and enhancing mechanical durability, LST can contribute to prolonged piston lifespan as well as increased efficiency of engines; hence, reduced costs of maintenance. It is for this reason that LST is presumed to be a considerably effective technique for the applications of modern engines that are high performing and energy-efficient [15, 16].

9.4 SURFACE QUALITY ENHANCEMENT VIA LST

9.4.1 Surface Roughness and Texture Control

As the highly effective method to improve piston crown surface by directly changing microstructure, topography, and mechanical properties of the material, it can be seen in a better perspective. One part of an internal combustion engine that is subject to severe mechanical, thermal, and tribological stresses is the piston crown. Therefore, surface quality has a significant impact on the engine's performance, longevity, and efficiency. The surface characteristics required for maximizing lubrication, minimizing wear, and carbon deposition are frequently not produced by conventional techniques including machining, honing, and coating applications. Surface topography appears to be one of the main ways that LST enhances surface quality by efficiently managing surface roughness and microstructural flaws, in contrast to the high-intensity laser beam micro-texturing of engineered textures on the piston surface to improve such functional characteristics as reduced friction, the lubricant's trapping capacity, and thermal stability. The material undergoes rapid solidification and localized melting, which helps to smooth out flaws brought on by machining of the grain structure and promote a consistent surface texture. One can achieve micro-textures with the ideal depth, shape, and spacing to guarantee a constant and flawless surface finish by properly controlling the laser's parameters, which include pulse duration, energy density, scanning speed, and spot overlap. These brought about by the laser alterations reduce microcracks, fatigue failure, and material delamination, which are frequent occurrences in the traditional piston production method and lead to increased wear resistance. Additionally, by producing micro-dimpled textures that can serve as lubricant reservoirs, LST increases lubrication efficiency and provides stable lubrication even under high-load and high-speed circumstances. These surface textures aid in keeping the oil in place, lowering the coefficient of friction and metallic-to-metal contact between the cylinder liner and the piston. In addition to reducing adhesive and abrasive wear, this thereby prolongs the life of engine components and helps to retain energy losses from friction. LST is also very crucial to help with surface quality, since it can alter the piston's wettability feature. The oil film's stability is then converted to combustion efficiency. Changing the laser processing parameters can change the surface energy, making the surface more hydrophilic or hydrophobic. This could have an impact on how lubricant spreads and stays on the surface. This will reduce excessive oil collection and the formation of carbon deposits in the combustion chamber, improving cleanliness and lowering emissions. Additionally, the laser's creation of surface textures that optimize heat transfer properties will increase thermal stability and heat dissipation. Enhancing effective heat conduction to prevent localized overheating and consequently thermal fatigue is one of the purposes of the laser-modified surface. These are the two main issues with engines operating at high performance. Under extended temperature cycling, this changed surface aids in preserving the piston's integrity. Because there is no material deterioration over time, it helps to guarantee that the engine operates consistently. Better long-term durability is facilitated by the enhanced oxidation and corrosion resistance brought about by the laser shock treatment's optimized microstructure. Improved adhesion of protective coatings, such as ceramic coatings, wear-resistant layers, and thermal barrier coatings, can improve surface hardness and mechanical resistance. A new kind of collective treatment for much superior surfaces to endure the conditions of the newest engines is introduced by the best results of laser shock treatment of the piston surface combined with sophisticated coatings. Additionally, laser shock treatment is a highly scalable and repeatable industrial procedure that may offer consistency in mass production together with a high degree of control over texturing and pattern details. The LST technique is not the same as traditional approaches, which occasionally need finishing processes. Since there is no need for chemical treatment and shorter processing time, this non-contact method is safe for the environment and produces less material waste. Computational modelling and real-time monitoring will work well with LST to optimize processing parameters continuously for a textured surface with the fewest possible flaws. LST is essentially a revolutionary engineering method that can significantly improve the piston top surface quality. The surface becomes extremely durable in real-world applications as a result of microstructural refinement, lubrication dynamics optimization, wear and friction reduction, and heat dissipation improvement. A manufacturer will then have the best chance to have an internal combustion engine with significantly better performance, including increased fuel efficiency, decreased emissions, and improved component life cycle, thanks to laser-induced alterations. The next generation of machines, which are renowned for their significantly better performance and lower energy consumption, will undoubtedly find critical specific applications for advancements in lasers and material science [17, 18].

9.4.2 Mechanical and Thermal Properties Improvement

In order to improve the engine's overall performance, LST has been shown to be highly effective in improving the piston's friction, wear, thermal behavior, and durability. Increased surface treatment is necessary due to the high mechanical and thermal demands placed on a piston while it is operating inside an engine. The coefficient of friction (COF) is significantly decreased by the LST treatment of the piston's top face with micro-patterns that are precisely engineered to affect lubrication, combustion, and mechanical stability issues. As a result, energy losses are reduced and fuel efficiency is improved. The LST reduces adhesive wear and direct metal-to-metal contact by forming dimples or reservoirs that provide an adequate oil supply for the rubbing parts. It has a longer engine life since it reduces power consumption and makes the piston move much more smoothly. Additionally, hydrodynamic lubrication is promoted by the optimized surface textures. Here, a stable oil coating is produced even at high speeds and loads, allowing for minimal wear and barely noticeable scuffing. The function of LST in heat dissipation and thermal management is another crucial component of piston function. By providing more surface area and effective conduction, the textured surface improves heat dispersion and prevents thermal fatigue and localized overheating. This is crucial for high-output engines because the high-speed temperature variations the engines face cause material wear and strain on the pistons as well as distortion and decreased efficiency. Additionally, the laser-induced surface alterations significantly change the piston material's microstructural characteristics, improving hardness and grain structure, two elements that improve a component's mechanical stability under cyclic stress. By opposing the existing harmful tensile stresses on fatigue life, LST adds compressive residual stresses that further strengthen the piston. Compressive stresses primarily improve the overall mechanical integrity of the piston and increase its service life by postponing the onset and spread of cracks. Importantly, LST also alters the combustion process by altering the piston's wettability and top surface roughness. According to analysts, textured motifs improve the fuel and air mixture dynamics, resulting in better combustion, lower emissions, and increased power. In order to ensure clean operation and little maintenance, LST can also regulate engine surface energy for deposit and fouling management. The enhanced resistance to oxidation and corrosion that LST-treated pistons exhibit is another significant characteristic that enhances long-term endurance in harsh engine settings. The adaptability of LST as a method for maximizing piston performance in both traditional and cutting-edge powertrain systems is increased by the ability to adjust the laser parameters for application-specific surface textures. Additional performance improvements are made by integrating LST with other surface engineering methods, such as wear-resistant coatings or TBCs, which increase the pistons' resilience to high pressure and temperature conditions. LST's accuracy and scalability make it ideal for today's industry, enabling high-output top pistons with minimal material waste and great quality. Additionally, the process's no-touch approach ensures that surface modifications occur without the need for additional mechanical pressure or flaws. Workers can continuously enhance the process by working on the optimal surface patterns to obtain the greatest performance advantages, thanks to new developments in computational modelling and real-time tracking for LST. In conclusion, LST alters the way pistons function by ensuring that lubricant adheres better, reducing wear and friction, enhancing heat control, strengthening mechanical stability, and increasing combustion optimization. With all of these advantages, LST is positioned as a key technology in the development of strong and energy-efficient internal combustion engines by increasing engine efficiency, reducing fuel consumption, and reducing emissions. LST will play a crucial role in the design and production of pistons in the future as the automotive and industrial sectors seek durability and efficiency from their products. This will improve the performance and dependability of the next generation of engines [19, 20].

9.5 EXPERIMENTAL STUDIES AND CASE ANALYSES

9.5.1 Comparative Analysis of LST-Treated and LST-Untreated Pistons

Since laser technologies have advanced in two areas—material science and improving laser acuity—to meet the demands of ICE durability and efficiency, the future of LST in piston applications is bright. Surface engineering optimizes piston performance. The aerospace and automotive sectors have been requesting longer component lifespans, reduced emissions, and improved fuel economy, which highlights the significance of LST in fulfilling these tribological requirements in piston performance. LST will be used to meet the tribological requirements of piston performance since the automotive and aerospace industries are requesting components with longer lifespans, reduced emissions, and increased fuel efficiency. Every significant future trend in the field investigates ultrafast laser systems, whether they are femtosecond, picosecond, or of any other type that need hitherto unheard-of levels of precision in the surface modification procedure. The next generation of lasers, known as ultrafast laser systems, may be able to provide smaller heat-affected zones without causing any unwanted thermal distortions while also enabling much finer and more controllable micro-texturing. By employing learning laser texture systems that use machine learning (ML) and artificial intelligence (AI) algorithms to optimize a laser's parameters in real time, smart laser texturing technology has changed the landscape. By adjusting parameters like laser intensity, pulse duration, and scanning speed in response to sensor data and operational conditions, these novel intelligent systems can adaptively improve surface texturing. By maintaining the maximum amount of surface modification, this crucial feature upholds quality standards; it resists wear, lowers friction, and retains lubrication, so quality is guaranteed and advantages are maximized. Such AI-enabled LST lowers waste and boosts production efficiency by detecting and fixing flaws throughout the manufacturing process. Another significant prospective trend for pistons in LST is multi-scale surface texturing. Potential benefits of combining texturing at the micro and nanoscales could include better fuel efficiency, decreased frictional losses, and improved lubrication dynamics. Multi-scale texturing can produce hierarchical patterns that have the potential to improve lubricating film retention while producing inherently self-cleaning qualities against carbon accumulation. By doing this, it prevents the potential danger posed by carbon deposits forming on the piston crown of contemporary direct-injection and turbocharged engines. As part of the future development, the next phases in LST development will focus on improving the piston's qualities and making it compatible with sophisticated coatings like diamond-like carbon (DLC) and TBCs. In the controlled application of these coatings, LST can be used to increase surface roughness and anchoring points that improve adhesion. This would boost resistance to high-temperature oxidation, reduce wear, and increase thermal insulation. Functional coatings applied to LST on pistons can help them withstand severe engine conditions while also maximizing engine efficiency and dependability. Research on hybrid and alternative fuel engines is another area in which LST is being used. The role of the piston and combustion efficiency become crucial as the automotive industry shifts to hydrogen-powered and hybrid-electric vehicles. To optimize combustion characteristics and reduce emissions, LST can be reoriented to improve piston surface qualities for various fuel types. For instance, LST can be utilized to produce combustion engine parts for hydrogen combustion engines. Laser machining is more feasible since it allows for fine direct control over thermal distribution and flame propagation, lowering the possibility of pre-ignition and boosting overall efficiency. A significant advancement in in situ monitoring techniques, high-speed imaging, and spectroscopic analysis should be made possible by such implementation. In order to guarantee that each piston obtains the textured structure precisely planned to meet specific performance parameters, the manufacturer will be able to execute the interaction between the laser and the material with previously unheard-of accuracy thanks to the in-line real-time monitoring tools. In addition, further developments in modelling and simulation tools will enable the prediction performance of LST to be studied regarding the design of various textures before the real application is truly considered. Thus, the ability to treat processes as a green technology will henceforward play a significant role in the development of future LST processes. The fact that laser technology is non-contact and essentially non-chemical in nature contributes to the procedure being environmentally friendly; further advancements in energy-efficient sources of lasers and wastage of the material is kept to the minimum then; all these will lead to the creation of the much sustainable manufacturing ecosystem. Furthermore, the capability to apply LST in terms of repair and reconditioning used-up pistons brings a chance for waste reduction and extending lifespans of parts, which is in line with principles of circular economy. In conclusion, future advancements in ultrafast lasers, AI-based process optimizations, multiscale surface textures, integrated functional coatings, hybrid engine applications, and real-time monitoring technologies will all influence the potential of LST in piston applications. These advancements will assist manufacturers in achieving previously unheard-of levels of precision, effectiveness, and longevity in piston design, leading to the development of next-generation, high-performing, low-emission engines. Even if research on tribology and laser processing is progressing, LST will continue to be a crucial component of surface technology and will coexist with the development of internal combustion engines in the future [21, 22].

9.5.2 Industrial Applications and Implementation Challenges

In order to properly utilize LST for piston applications and achieve its full potential in enhancing engine performance, efficiency, and durability, a number of obstacles and constraints must be addressed. Controlling laser parameters with the precision needed to achieve the desired surface textures without introducing flaws or unintended side effects is one of the main challenges. In actuality, a wide range of additional factors that affect the depth, shape, and homogeneity of textures, including laser power, pulse duration, scanning speed, and spot overlap, affect how effective LST is. Uneven surface alterations, such as extremely high surface roughness, asymmetrical texture patterns, or even microcracking that does not improve piston performance, can arise from any parameter variation from the norm. Instead, it erodes it. It is crucial to strike the right balance between texture depth and surface roughness. Generally speaking, excessive surface texturing increases the likelihood of stress concentration and material deterioration, whereas little surface texturing may not improve the intended lubrication and wear resistance.

The heat-affected zone (HAZ) associated with laser processing is another disadvantage of LST. Traditional laser systems still produce localized heating, which can alter the microstructure of the piston material, even if advancements in ultrafast laser systems, such as femtosecond and picosecond lasers, have reduced thermal injury. The mechanical integrity of the piston may be jeopardized if this kind of heat action results in undesirable phase changes, residual tensile stress, and grain enlargement. Any unintentional microstructural change can reduce fatigue resistance and cause premature failure in high-performance engines where pistons are subjected to intense mechanical and thermal loads. Therefore, one of the main challenges in the implementation of LST is optimizing the laser parameters to minimize the HAZ while simultaneously achieving the surface texturing efficiency.

In large-scale industrial applications, cost-effectiveness and scalability are challenging. Despite being a very accurate and regulated process, LST requires rapid laser systems that will operate in high-throughput modes on thousands of pistons in mass production setups. Conventional lasers' slow processing can reduce LST's throughput, which makes it more time consuming and, thus, more expensive than the current surface treatment techniques of chemical etching or mechanical honing. In addition, small and medium-sized manufacturers find it difficult to adopt lasers due to the high expenses of initial investment, maintenance, and operator training. Future research should address this problem by developing high-throughput improved lasers with parallel texturing techniques and multi-beam processing to preserve high precision and consistency while simultaneously accelerating operation.

How laser surface texturing fits into the process of creating pistons and adding finishing touches is another issue. To maintain the rough surfaces functional, the texturing process must be done carefully in conjunction with subsequent cutting, layering, or surface treatment operations. A roughened piston, for example, may lose some of its little rough bits if it undergoes additional final treatment procedures like polishing or coating, which would reduce its ability to aid in lubrication and wear off. Additionally, rust and grime from the laser treatment can impair the adhesion of protective coatings, potentially resulting in premature peeling and short lifespan. A novel method of manufacturing that combines laser surface texturing with additional surface treatments is required to overcome these mixing issues and achieve the best outcomes.

The material is another factor that limits the usability of LST for pistons of different types. Aluminium alloys, steel, and titanium are material types on which LST has been successful; the effect of texturing, however, depends on the thermal and mechanical characteristics of the base material. Higher cracking, oxidation, or microstructural degradation susceptibility that a material exhibits through laser processing requires careful consideration of the material and the optimization of the process parameters. Apart from the aforementioned factors, the shift of the automobile industry towards the use of new kinds of fuel and engine technology shall also bring about different changes in the piston materials; therefore, this also introduces another research prospect to study how LST can be adjusted to such materials, say, for instance, lighter composites or high-entropy alloys.

Long-term stability and durability of laser-textured surfaces under realistic operating conditions are issues. While laboratory testing and simulation prove that LST helps reduce friction, improves the provision of lubricant, and enhances resistance to wear, such textured surfaces deliver an outcome in the long run after extended operations of the engine, high-temperature conditions, and extreme mechanical stresses. The long-term effects of textured surfaces in operation are not well understood. Under eventual interaction of the constant motion of the piston and the lubricant, the surface decay and erosion, as well as the gradual smoothing of the micro-textures, may require retexturing at certain intervals or any alternative strategy to uphold the performance.

The dissemination of the LST method in piston manufacturing is also influenced by environmental and regulatory issues. For example, this technique must demonstrate its effectiveness by comprehensive testing and validation in support of these goals, since executives of automakers require an efficiency test to pass the stringent emission regulations and meet the fuel economy threshold. The environmental impacts of laser processing, including waste production and energy consumption,

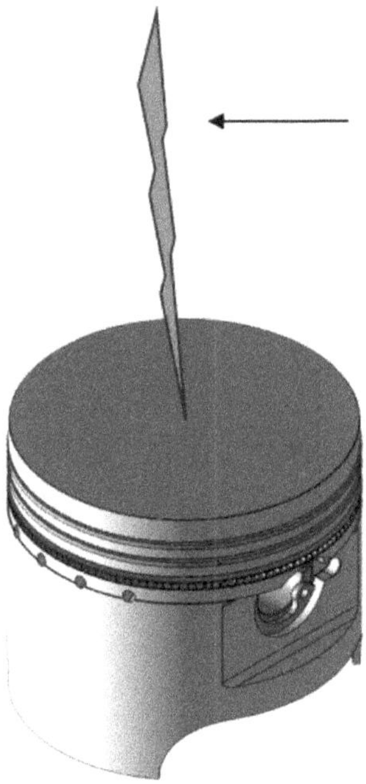

FIGURE 9.4 Laser surface texturing work on the piston.

should guarantee appropriate management in accordance with sustainable manufacturing principles. Even though it causes far less environmental damage than chemical treatment, sustainability improvement necessitates the continued development of more effective energy tools and waste reduction techniques.

In conclusion, even though LST has a lot of potential to improve piston performance, there are a number of problems and restrictions with its application, including long-term durability, regulatory concerns, process control, thermal effects, scalability, material compatibility, and integration into current manufacturing processes (Figure 9.4). Improvements in material science, hybrid manufacturing techniques, process optimization, and additional laser technology development are required to meet these hurdles. By overcoming the aforementioned obstacles, the surface engineering method through LST will be able to advance and become a standard practice, creating the way for internal combustion engines that are more efficient, robust, and capable [23].

9.6 FUTURE TRENDS AND ADVANCEMENTS IN LST

9.6.1 Integration with Industry 4.0 and Smart Manufacturing

LST is regarded as a revolutionary technology that should improve piston durability and performance by maximizing surface characteristics including wear resistance, friction reduction, thermal management, and the overall structural integrity of all materials (Figure 9.5). For instance, LST significantly increases lubricant retention,

FIGURE 9.5 Industry: from the origins to 4.0.

reduces energy loss, and improves piston top surface combustion efficiency, all of which contribute to improved fuel economy and reduced emissions. With decades of advancements in internal combustion engines and tribology, the use of surface texturing in pistons is a relatively new technique. In this instance, LST offers exceptional versatility in creating surface structures at the micro and nanoscales that serve as lubricant reservoirs, guaranteeing dynamic loading without requiring direct metal-to-metal contact. By significantly reducing wear and frictional losses, this method guarantees that the engine can be used farther without experiencing a loss of power. The ability of LST to initiate advantageous compressive residual stresses is another factor. These would provide pistons with the capacity to withstand severe mechanical and thermal loading conditions by acting against the harmful tensile stresses that typically cause fatigue failure. Because of its accuracy and capacity to adjust to various operating conditions, LST is an extremely versatile method that can be used with a variety of engines, including those that run on petrol and diesel fuel, hybrid systems, or other alternative engines. Despite its many benefits, the technique has a number of issues that must be resolved before it can be broadly adopted in the automotive or aerospace industries. Issues include ensuring that the laser operates properly for even and good texturing, minimizing heat-affected areas that could alter material qualities, and making sure it is affordable in large production settings. Fitting LST with existing manufacturing techniques also presents challenges because the textured surfaces' functionality may be altered or diminished by subsequent cutting or covering usage. Additionally, because use, wear, erosion, and carbon buildup could gradually reduce the beneficial advantages of LST over time, research is still ongoing to determine how long laser-textured surfaces persist and remain good in an actual engine's operating environment. Future prospects for LST in piston applications include ultrafast laser technologies, such femtosecond and picosecond lasers, which enable precise and quick texturing while minimizing heat damage. LST will also be altered by the developments in AI and ML. Based on data from in situ monitoring, it can optimize the process in real-time, detect defects, and apply adaptive texturing. Incorporating cutting-edge coatings like TBCs and DLC coatings is one of the future directions for improving engine efficiency, wear resistance, and thermal stability. LST can optimize the piston surfaces to fit improved fuel–air mixture dynamics, combustion stability, and low emissions when it transitions to alternative fuel engines such as hydrogen and biofuel driven combustion systems. LST is a clean, non-use method that produces no waste because its other sustainable aspects do not apply in the contemporary manufacturing setting. In conclusion, LST is a surface engineering technology with a great deal of promise for enhancing the lifespan and performance of internal combustion engine pistons. Notwithstanding these difficulties, further research, laser processing material science, and computer modelling—all of which are involved in the creation of new engine systems—will be necessary to further fine-tune LST and deploy it. The future concept of high-performance and environmentally benign engines will heavily rely on laser surface texturing, making this technology central to the development of automotive and aerospace engineering. Fuel economy, reduced emissions, and increased part life expectancy are all factors that are currently being considered [24].

9.6.2 Emerging Hybrid Techniques and Materials

In order to fully utilize the potential of LST, which is still developing as a crucial surface engineering technique for improving piston performance, more thorough research and industry application techniques are desperately needed (Figure 9.6). The optimization of laser parameters to enable accurate and repeatable surface textures without causing microcracks or undue roughness in the textured surface is a crucial area to be investigated in future studies. More understanding of how laser radiation interacts with piston material would be necessary for this, especially for various alloy compositions utilized in high-performance internal combustion engines. The creation of sophisticated computational models for forecasting surface

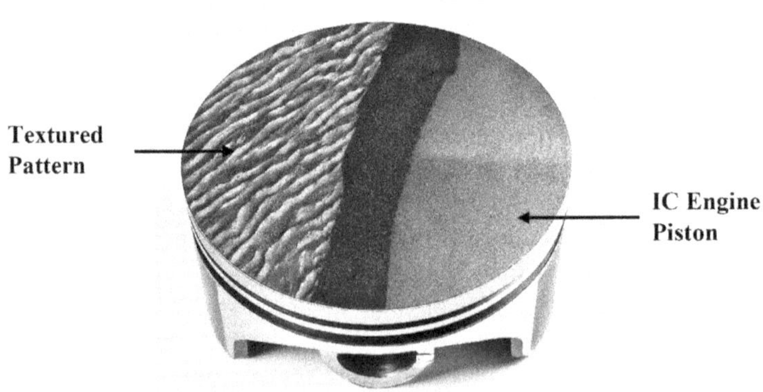

FIGURE 9.6 The LST-piston final component.

integrity, stress distribution, and tribological performance brought on by laser texturing under practical operating circumstances also need further study. Examining how LST works in concert with other surface modification techniques, including as coating deposition and surface hardening procedures, to increase durability and wear resistance and combining LST with functional coatings such solid lubricant coatings, DLC coatings, and TBC may result in hybrid surface engineering solutions that improve oxidation resistance, thermal stability, and friction reduction. For the greatest combination of lubricant trapping, adhesive wear reduction, and load-bearing augmentation, research is also required to explore multi-scaling surface texturing by combining micro- and nano-scale texturing. Last but not least, high-speed laser processing, including multi-beam laser systems and parallel texturing methodologies for higher productivity with high precision, will be envisioned for the practical use of LST in large-scale piston manufacturing. The efficiency of the system will be further increased by automated LST and AI-based LST systems, which will enable real-time process monitoring, adaptive parameter modifications, and defect identification. Machine learning techniques can be used to provide big data analytics about engine performance indicators and laser processing settings, enabling almost constant iterations for empirical data-based design enhancements. In order to ensure quality control and reproducibility in mass production, in situ monitoring techniques such as optical coherence tomography (OCT) and high-speed imaging should eventually be included into industrial LST settings. Future research must also focus on long-term durability and dynamic operation in the real engine in order to validate the laboratory results and move LST closer to mass manufacturing in the aerospace and automotive sectors. Wear, carbon deposits, and surface deterioration mechanisms should be used to detect potential problems that could interfere with the texture's ability to function properly and to devise viable solutions for maintaining the texture at its peak during the piston's service life. Furthermore, since the automobile industry is moving towards engines that run on other fuels like hydrogen, biofuels, and synthetic fuels, further study should be done on how to modify LST in order to optimize new engine architectures for heat management, fuel dynamics, and atomization. Lastly, in order to maximize the advantages of environmental considerations by LST, sustainability aspects should be highlighted. The feasibility of giving worn-out pistons a new texture to extend their service life and reduce waste in accordance with the circular economy's tenets, as well as laser sources with high energy and strategies that minimize material waste, can all be the focus of future research. LST may be more accurately defined and accepted as a mainstream technology in the production of next-generation pistons by tackling these scientific and industrial obstacles. This will ultimately result in increased engine efficiency, reduced emissions, and increased durability in a variety of applications [25].

9.7 CONCLUSION

LST has been able to alter the surface of the piston in the internal combustion engine so as to make that surface more performance- and durability-driven along with high efficiency. LST has raised a new theory in the era of tribology as proper control of lubrication retention, friction, wear resistance, and reduction of residual stresses at such micro/nanostructured surfaces has caused prolonged life of the piston with better performance of the engine. As a result, LST's function in optimizing surface characteristics makes it a highly promising procedure for tribological optimization, fuel consumption efficiency, and emission reduction in the context of the global movement towards more environmentally friendly and fuel-efficient modes of transportation. All of these, however, would only be put into practice after careful evaluation of the process's actual viability on an industrial scale. This means that in order to advance laser surface treatment for industrial applications, hybrid surface modifications with coatings like TBCs and DLC, ultrafast laser systems, and artificial intelligence for process control will be taken into consideration. Future research must improve the laser set-up's specifics, speed up processing, and guarantee long-lasting quality under typical operating conditions. Also, since cars are moving to new fuels and electric power, LST can be changed to help make future engines and energy-saving drives better. Green factors like energy-smart laser work and refreshing old parts should be looked at to match LST with the circle-of-life economy idea. In general, LST is a big change in how to fix piston surfaces, with great possible benefits for better engine work. With more new ideas and work use, LST will have a key part in moving modern car and plane engineering forward to high-power, clean, and fuel-saving engines for the time ahead.

REFERENCES

[1] Lebrun, N., Dupla, F., Bruhier, H., Prudent, M., Borroto, A., Der Loughian, C., . . . & Steyer, P. (2024). Metallic glasses for biological applications and opportunities opened by laser surface texturing: A review. Applied Surface Science, 160617.

[2] Kumaran, M. (2023). Hybrid additive manufacturing technologies. In Handbook of Smart Manufacturing (pp. 251–263). CRC Press. https://doi.org/10.1201/9781003333760-13

[3] Chu, W. S., Shehroze, M. M., Tran, N. G., Dinh, T. H., Hong, S. T., & Chun, D. M. (2024). Green fabrication of superhydrophobic surfaces using laser surface texturing without toxic chemicals: A review. International Journal of Precision Engineering and Manufacturing, 25(5), 1101–1123.

[4] Kumaran, M., Senthilkumar, V., Justus Panicker, C. T., & Shishir, R. (2022). Concept design and analysis of multi-layer and multi-process piston of SS316L and AlSi10Mg by additive manufacturing. In Recent Advances in Materials and Modern Manufacturing: Select Proceedings of ICAMMM 2021 (pp. 441–446). Singapore: Springer Nature Singapore. https://doi.org/10.1007/978-981-19-0244-4_42

[5] Senthilkumar, V., Velmurugan, C., Balasubramanian, K. R., & Kumaran, M. (2020). Additive manufacturing of multi-material and composite parts. In Additive Manufacturing Applications for Metals and Composites (pp. 127–146). IGI Global. https://doi.org/10.4018/978-1-7998-4054-1.ch007

[6] Morales, M., Rezayat, M., & Mateo, A. (2024). Enhancing the corrosion resistance of 2205 duplex stainless steel in molten carbonate salts by laser-surface texturing. Journal of Energy Storage, 78, 110053.

[7] Prabu, G., Yang, C. H., Alnaser, I. A., Jeyaprakash, N. (2025). Analysis of tribological behavior of Laser Powder Bed Fusion (LPBF) CuCrZr alloy through reciprocating loading on the columnar structure. Tribology International, 202, 110301.
[8] Gong, P., Zhang, Y., Wang, C., Cui, X., Li, R., Sharma, S., . . . & Li, C. (2024). Residual stress generation in grinding: Mechanism and modeling. Journal of Materials Processing Technology, 324, 118262.
[9] Krauss, G. (2024). Microstructure, residual stresses and fatigue of carburised steels. Quenching and Carburising, 205–225.
[10] Tekumalla, S., Seita, M., & Zaefferer, S. (2024). Delineating dislocation structures and residual stresses in additively manufactured alloys. Acta Materialia, 262, 119413.
[11] Shen, Z., Dong, R., Li, J., Su, Y., & Long, X. (2024). Determination of gradient residual stress for elastoplastic materials by nanoindentation. Journal of Manufacturing Processes, 109, 359–366.
[12] Kyvelou, P., Huang, C., Li, J., & Gardner, L. (2024, February). Residual stresses in steel I-sections strengthened by wire arc additive manufacturing. In Structures (Vol. 60, p. 105828). Elsevier.
[13] Jeyaprakash, N., Yang, C. H., Ramkumar, K. R. Microstructural, mechanical and wear behaviour of Inconel-718 produced through laser-powder bed-fused additive manufacturing. Materials Science and Technology, 37(3), 326–337.
[14] Zhou, J., Qi, Q., Liu, Q., Wang, Z., & Ren, J. (2024). Determining residual stress profile induced by end milling from measured thin plate deformation. Thin-Walled Structures, 200, 111862.
[15] Duong, V., & Hiep, N. H. (2025). Numerical analysis of piston-cylinder clearances in the firing automotive engine with different piston profiles. Journal of Mechanical Science and Technology, 1–8.
[16] Sharma, M., & Dwivedi, V. K. (2024). Optical microstructural and hardness analysis of an Al–Si alloy-based piston. Journal of The Institution of Engineers (India): Series D, 105(3), 1827–1831.
[17] Dizisaz, M., Parast, M. S. A., Azadi, M., & Dadashi, A. (2024). Plain and fretting fatigue characteristics of 16MnCr5 steel under cyclic bending loading with the application of engine piston pin. Materials Chemistry and Physics, 316, 129110.
[18] Qu, S., Wei, C., Yang, Y., Yao, P., Chu, D., Gong, Y., . . . & Zhang, X. (2024). Grinding mechanism and surface quality evaluation strategy of single crystal 4H-SiC. Tribology International, 194, 109515.
[19] Siddeshkumar, N. G., Suresh, R., Durga Prasad, C., Shivaram, L., & Siddalingaswamy, N. H. (2024). Evolution of the surface quality and tool wear in the high speed turning of Al2219/n-B4C/MoS2 Nano metal matrix composites. International Journal of Cast Metals Research, 37(1), 22–38.
[20] Igwe, N. C., Akhrif, I., El Jai, M., & El Fahime, B. (2025). An experimental investigation of the influence of SLM input factors on the as-built AlSi10Mg surface quality. The International Journal of Advanced Manufacturing Technology, 136(2), 619–674.
[21] Chen, Y., Zhang, Y., Liu, Z., Zhang, W., Pang, B., Zhang, Y., & Wang, D. (2024). Quantitative surface quality evaluation for 3D-printed concrete with coarse aggregate through 3D scanning. Case Studies in Construction Materials, 20, e03077.
[22] Nugroho, A., Kozin, M., Bo, Z., Mamat, R., Ghazali, M. F., Kamil, M. P., . . . & Ashfiya, P. S. (2024). Recent advances in harnessing biolubricants to enhance tribological performance and environmental responsibility: Bibliometric review (2015–2024). Cleaner Engineering and Technology, 100821.

[23] Zhang, G., Tang, J., Yang, K., Wang, R., Chen, Y., Xiong, Y., . . . & Lin, H. (2024). Important contributions of metal interfaces on their tribological performances: From influencing factors to wear mechanisms. Composite Structures, 118027.

[24] Wang, H., Lin, N., Yuan, S., Liu, Z., Yu, Y., Zeng, Q., . . . & Wu, Y. (2024). Structural improvement, material selection and surface treatment for improved tribological performance of friction pairs in axial piston pumps: A review. Tribology International, 109838.

[25] Song, W., Campen, S., Shiel, H., Gattinoni, C., Zhang, J., & Wong, J. S. (2024). Position of carbonyl group affects tribological performance of ester friction modifiers. ACS Applied Materials & Interfaces, 16(11), 14252–14262.

10 Anti-Corrosion and Biomaterial Surface Analyzed Through Laser Surface Texturing

Jeyaprakash Natarajan and T. Sathies

10.1 INTRODUCTION

Corrosion is a pervasive and destructive phenomenon that poses significant challenges to industries worldwide. As an electrochemical process, corrosion occurs when metals and alloys react with their environment, leading to the gradual deterioration and weakening of these materials. This degradation has profound implications for industrial operations, as it affects the integrity, reliability, and safety of equipment, structures, and infrastructure. In fact, the economic impact of corrosion is staggering, with industries across various sectors incurring billions of dollars in direct and indirect costs each year due to maintenance, repairs, replacements, and operational disruptions caused by corrosion-related issues [1–3]. Corrosion is not just a surface-level problem; it is a fundamental threat to the longevity and performance of industrial systems. From the oil and gas industry, where pipelines are susceptible to internal and external corrosion, to the manufacturing sector, where machinery and tools face continuous exposure to corrosive environments, the effects of corrosion are widespread and varied. Additionally, industries such as transportation, construction, and power generation are equally vulnerable, with corrosion impacting everything from bridges and buildings to electrical components and power plants. The underlying causes of corrosion are complex and multifaceted, involving a combination of environmental factors, material properties, and operational conditions. Environmental elements such as humidity, temperature, chemical exposure, and pollutants can accelerate the corrosion process, especially when materials are subjected to harsh or extreme conditions. Furthermore, the mechanical stress and fatigue experienced by materials in industrial settings can exacerbate the effects of corrosion, leading to premature failure and costly repairs. In the oil and gas industry, for example, corrosion can lead to the leakage of hazardous materials, posing severe safety risks and environmental damage. The failure of a corroded pipeline or storage tank can result in catastrophic spills, with long-lasting consequences for both the environment and the industry's reputation. Similarly, in the aviation and maritime industries, corrosion can compromise the structural integrity of aircraft

DOI: 10.1201/9781003542087-10

and ships, endanger lives and leading to significant financial losses. Beyond the direct costs associated with corrosion, such as repairs and replacements, the indirect costs are equally substantial. These include production downtime, loss of efficiency, and the potential for accidents and environmental contamination. In some cases, the financial burden of corrosion can exceed the initial investment in infrastructure or equipment, making it a critical issue that demands ongoing attention and resources. Given the far-reaching effects of corrosion, industries have developed a range of strategies and technologies to combat this persistent problem. These include the use of corrosion-resistant materials, protective coatings, cathodic protection, and the application of corrosion inhibitors. Additionally, regular inspections, maintenance, and monitoring are essential for detecting early signs of corrosion and preventing its progression. The implementation of these measures not only extends the lifespan of industrial assets but also ensures the safety of workers and the protection of the environment. However, despite these efforts, the challenge of corrosion remains significant, requiring continuous research and innovation. Advances in materials science, for instance, are leading to the development of new alloys and coatings that offer enhanced resistance to corrosion. Moreover, the integration of digital technologies, such as sensors and predictive analytics, is enabling more accurate monitoring and early detection of corrosion, allowing for proactive maintenance and reducing the risk of unexpected failures [4,5].

To combat corrosion, surface treatment techniques have become essential, playing a crucial role in enhancing the corrosion resistance of metals. These techniques are designed to modify the surface properties of metals, creating barriers or altering surface chemistry to protect the underlying material from corrosive environments. The importance of surface treatment techniques in enhancing the corrosion resistance of metals cannot be overstated. These techniques offer several key benefits that are critical for the longevity and performance of metal components in various industries.

1. **Extended lifespan of components:** By protecting the metal surface from corrosive agents, surface treatments significantly extend the lifespan of components. This is particularly important in industries such as automotive, aerospace, and construction, where metal structures are exposed to harsh environments. For example, in the aerospace industry, where metal components are exposed to extreme temperatures and corrosive atmospheres, surface treatments ensure the durability and reliability of critical parts [6].
2. **Cost-effectiveness:** Surface treatments are a cost-effective way to enhance corrosion resistance, reducing the need for frequent maintenance, repairs, and replacements. By preventing or slowing down the onset of corrosion, these techniques save industries substantial amounts of money over the long term. Additionally, the cost of applying surface treatments is often much lower than the potential costs associated with corrosion-related failures [7].
3. **Environmental protection:** Surface treatment techniques contribute to environmental protection by reducing the need for mining and processing new materials. By extending the life of existing metal components, industries can minimize their environmental footprint. Moreover, certain surface

treatments, such as the application of eco-friendly coatings, can reduce the release of harmful substances into the environment [8].

4. **Improved safety:** Corrosion can lead to the catastrophic failure of critical infrastructure and equipment, posing significant safety risks. Surface treatments play a vital role in preventing such failures by enhancing the corrosion resistance of metals used in safety-critical applications. For instance, in the oil and gas industry, surface treatments on pipelines and storage tanks help prevent leaks and spills, thereby safeguarding both workers and the environment [9].

From infrastructure to automotive, aerospace, and marine applications, the economic impact of corrosion is substantial, leading to significant costs associated with maintenance, repair, and replacement of metal components [10]. As a result, developing effective methods to enhance the corrosion resistance of metals has become a critical area of research. One such method that has gained prominence in recent years is laser surface texturing (LST), which is a cutting-edge technique that significantly enhances the corrosion resistance of metals by altering their surface properties. LST offers a solution by creating micro- and nano-scale patterns on the metal surface, which can impede these corrosive processes. Figure 10.1 shows a schematic of the LST process. One of the primary ways LST improves corrosion resistance is by reducing the surface area exposed to corrosive agents. The textured patterns can create a barrier that limits direct contact between the metal and its environment.

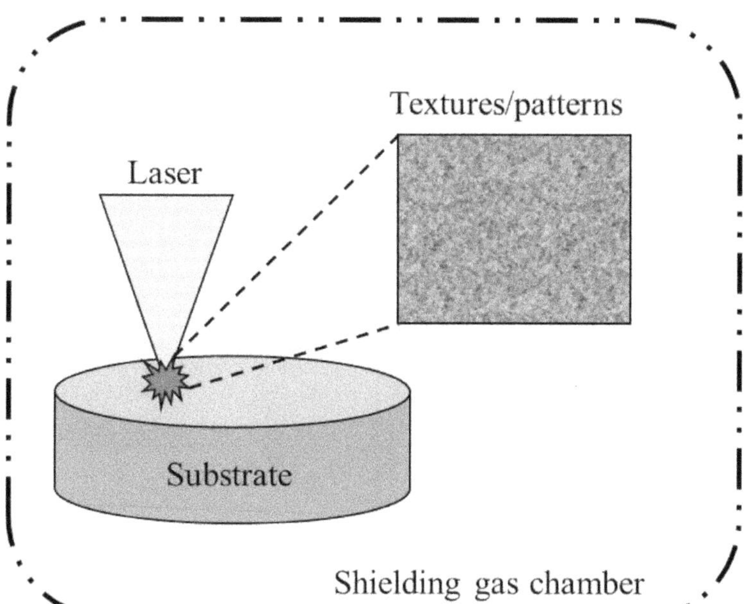

FIGURE 10.1 Laser surface texturing.

For example, laser-induced microstructures can trap air or other protective gases, forming a passive layer that hinders corrosion [11–13]. Additionally, these textures can help in the formation of protective oxide layers, which act as a shield against further corrosion. LST can also enhance the hydrophobicity of metal surfaces, making them less susceptible to corrosion. Hydrophobic surfaces repel water, preventing it from lingering on the metal and reducing the likelihood of corrosion. By carefully controlling the laser parameters, such as power, pulse duration, and scanning speed, it is possible to create surface textures that maximize this water-repelling effect. Furthermore, LST allows for the precise control of surface roughness, which can be optimized to minimize sites where corrosive agents might accumulate. This precision is especially valuable in industries where metals are exposed to harsh environments, such as in marine, aerospace, and chemical processing applications. Thus, laser surface texturing plays a crucial role in enhancing the corrosion resistance of metals by altering their surface properties in ways that reduce exposure to corrosive elements, enhance protective layer formation, and increase hydrophobicity, thereby extending the lifespan of metal components. This chapter focuses on discussing and detailing the mechanism behind enhancing the corrosion resistance via laser surface texturing with relevant literatures.

10.2 TYPES OF CORROSION

Corrosion is a natural, electrochemical process that leads to the deterioration of metals due to their interaction with the environment. It can result in significant economic losses and safety hazards across various industries, making it crucial to understand the different types of corrosion (Figure 10.2). This knowledge helps in selecting the appropriate materials, designing protective measures, and implementing preventive strategies.

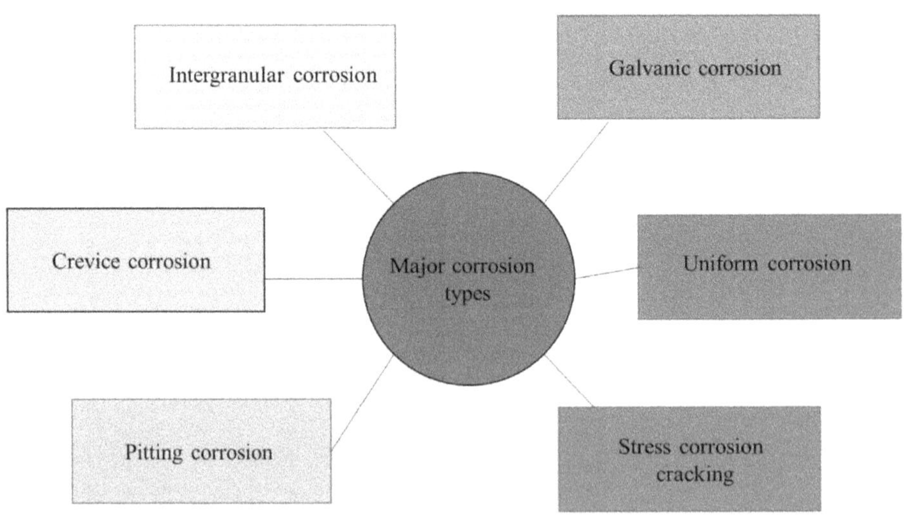

FIGURE 10.2 Major corrosion types.

10.2.1 Uniform Corrosion

Uniform corrosion, also known as general corrosion, occurs evenly across the surface of a metal, leading to a uniform reduction in thickness over time. This type is the most common and predictable form of corrosion, as it typically affects large areas and proceeds at a steady rate. Uniform corrosion is often caused by exposure to environments where the metal reacts uniformly with elements like oxygen, moisture, and acids. For instance, iron rusting in a humid environment is a classic example. Although it results in material loss, uniform corrosion is relatively easy to manage and predict through regular monitoring and protective coatings [14].

10.2.2 Galvanic Corrosion

Galvanic corrosion occurs when two different metals come into electrical contact in the presence of an electrolyte, such as water or moisture. In this electrochemical process, the more reactive metal (anode) corrodes faster, while the less reactive metal (cathode) is protected [15]. This type of corrosion is common in environments where dissimilar metals are used together, such as in piping systems, automotive parts, and marine structures. To prevent galvanic corrosion, it is important to select compatible metals, use insulating materials between them, or apply protective coatings.

10.2.3 Pitting Corrosion

Pitting corrosion is a localized form of corrosion that leads to the formation of small, often deep, pits or holes on the metal surface. It occurs when the protective oxide layer on a metal surface is damaged or penetrated, exposing the underlying metal to aggressive environments. Chloride ions, commonly found in saltwater, are a major cause of pitting corrosion, especially in stainless steels [16]. Pitting is particularly dangerous because it can lead to sudden failure of metal components with minimal overall material loss, making it difficult to detect and monitor. Prevention strategies include using materials resistant to pitting, controlling environmental factors, and maintaining protective coatings.

10.2.4 Crevice Corrosion

Crevice corrosion occurs in confined spaces or crevices, such as those found under gaskets, washers, or joints. These areas can trap moisture, debris, or other corrosive agents, creating a localized environment with limited oxygen flow. As a result, the metal in the crevice becomes anodic and corrodes at a faster rate than the surrounding area. Crevice corrosion is particularly problematic in marine and chemical processing environments [17]. Preventing this type of corrosion involves minimizing crevices in design, using corrosion-resistant materials, and ensuring that joints and fasteners are properly sealed and maintained.

10.2.5 Intergranular Corrosion

Intergranular corrosion occurs along the grain boundaries of metals, often due to the segregation of impurities or the depletion of alloying elements at these boundaries.

This type of corrosion is commonly seen in stainless steels that have been improperly heat treated or welded [18]. In these cases, chromium carbide precipitates at the grain boundaries, depleting the adjacent areas of chromium and making them more susceptible to corrosion. Intergranular corrosion can lead to the disintegration of the metal structure along the grain boundaries, severely compromising the material's strength. Preventive measures include proper heat treatment, the use of low-carbon or stabilized grades of stainless steel, and careful control of welding processes.

10.2.6 Stress Corrosion Cracking

Stress corrosion cracking (SCC) is a form of corrosion that occurs when a metal is subjected to tensile stress in a corrosive environment. This type of corrosion leads to the formation of cracks that can propagate rapidly, resulting in sudden and catastrophic failure of the material. SCC is particularly dangerous because it can occur at stress levels below the material's yield strength and may not show obvious signs before failure [19]. Metals like stainless steels, aluminum alloys, and high-strength steels are particularly susceptible to SCC in environments containing chlorides, ammonia, or caustic substances. Preventing SCC involves reducing stress levels in the material, using stress-relief annealing, selecting resistant materials, and controlling the environmental conditions.

10.2.7 Selective Leaching

Selective leaching, also known as dealloying, is a type of corrosion where one element in an alloy is preferentially removed, leaving behind a porous, weakened structure. A common example is the dezincification of brass, where zinc is leached out, leaving a spongy copper-rich residue. This type of corrosion typically occurs in environments where the selective leaching element is soluble in the corrosive medium. To prevent selective leaching, it is essential to use alloys specifically designed to resist dealloying, apply protective coatings, and control the exposure to corrosive environments [20].

10.2.8 Erosion Corrosion

Erosion corrosion occurs when a corrosive fluid flows over a metal surface at high velocity, combining mechanical wear with chemical attack. This type of corrosion is common in pipelines, pump impellers, and turbine blades, where the fast-moving fluid can remove the protective oxide layer on the metal surface, accelerating the corrosion process [21]. Erosion corrosion often results in a rough, grooved surface and can lead to rapid material loss and failure. To prevent erosion corrosion, it is important to design systems that minimize turbulence, reduce fluid velocity, and use materials with high erosion resistance.

10.2.9 Fretting Corrosion

Fretting corrosion occurs when two metal surfaces are in contact and subject to repeated small oscillatory motions, leading to the removal of protective oxide layers and exposing the metal to corrosion. This type of corrosion is common in joints, bearings, and other assemblies where slight movement occurs between metal parts [22]. Fretting corrosion can result in the formation of pits and debris, which can further accelerate the wear and corrosion process. Preventive measures include using lubrication, increasing contact pressure to reduce relative movement, and selecting materials with good resistance to fretting.

10.2.10 Microbiologically Influenced Corrosion

Microbiologically influenced corrosion (MIC) is caused by the presence of microorganisms, such as bacteria and fungi, which can accelerate the corrosion process by altering the local environment. For example, sulfate-reducing bacteria can produce hydrogen sulfide, leading to the formation of corrosive acids and the subsequent deterioration of metal surfaces. MIC is often found in water systems, pipelines, and storage tanks, where microbial colonies can thrive. Preventing MIC involves controlling microbial growth through biocides, regular cleaning, and maintaining appropriate environmental conditions [23].

10.3 FACTORS AFFECTING THE CORROSION RESISTANCE OFFERED BY METALS

Corrosion resistance in metals is influenced by a complex interplay of factors, both intrinsic to the metal itself and extrinsic environmental conditions. Figure 10.3 shows the corrosion factors in metals. Understanding these factors is essential for selecting appropriate materials and designing effective corrosion prevention strategies [24–26].

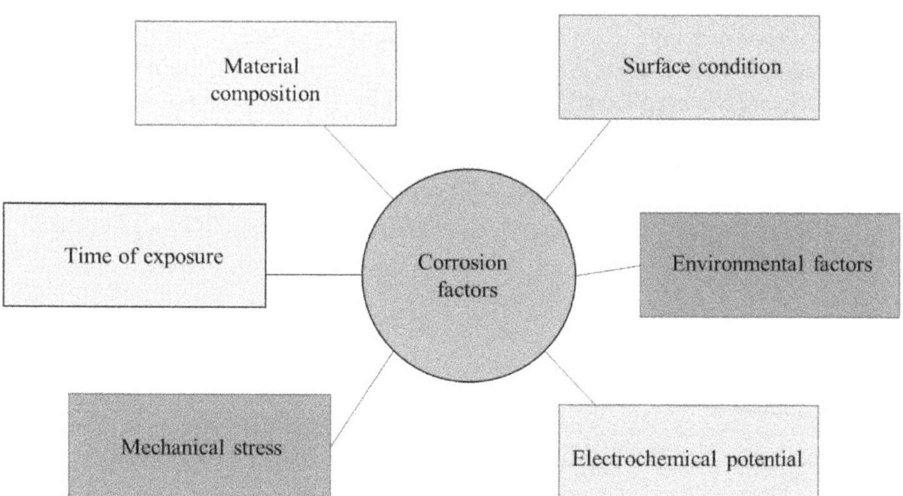

FIGURE 10.3 Corrosion factors.

10.3.1 Material Composition

The chemical composition of a metal or alloy is one of the most significant factors affecting its corrosion resistance. Pure metals tend to corrode more quickly than alloys because they lack the protective qualities that alloying elements provide. For instance, stainless steel, an alloy of iron with at least 10.5% chromium, forms a passive chromium oxide layer on its surface that greatly enhances its resistance to corrosion.

1. **Alloying elements:** Certain elements, when added to metals, can significantly improve corrosion resistance. Chromium, nickel, molybdenum, and copper are common alloying elements that enhance resistance to various forms of corrosion. Chromium, for example, contributes to the formation of a stable, passive oxide layer on stainless steel, protecting it from oxidation. Molybdenum enhances resistance to pitting and crevice corrosion, particularly in chloride-rich environments.
2. **Impurities:** The presence of impurities in metals can reduce corrosion resistance. Elements such as sulfur or phosphorus, when present in small quantities, can form inclusions or segregations that serve as initiation points for localized corrosion, like pitting or intergranular corrosion.

10.3.2 Surface Condition

The condition of a metal's surface has a profound impact on its susceptibility to corrosion. The smoother and cleaner the surface, the more resistant it tends to be to corrosion.

1. **Surface roughness:** Rough surfaces with irregularities, such as scratches or pits, can trap moisture and corrosive agents, creating localized sites where corrosion can initiate. Polishing and other surface treatments that reduce roughness can enhance corrosion resistance by minimizing these sites.
2. **Surface treatments and coatings:** Applying protective coatings, such as paints or plating, or anodizing can significantly improve a metal's corrosion resistance by providing a barrier between the metal and the corrosive environment. Coatings can also be applied to modify the surface chemistry, such as through passivation, which enhances the formation of a protective oxide layer.

10.3.3 Environmental Factors

The environment in which a metal is used is a major determinant of its corrosion behavior. Factors such as humidity, temperature, pH, and the presence of corrosive agents like salts, acids, and gases can all influence the rate and type of corrosion.

1. **Moisture and humidity:** The presence of water is essential for most forms of corrosion, particularly rusting in iron and steel. High humidity levels can

accelerate the formation of an electrolyte on the metal surface, promoting electrochemical reactions that lead to corrosion. In marine environments, where moisture and salts are abundant, metals are particularly susceptible to corrosion.

2. **Temperature:** Higher temperatures generally increase the rate of chemical reactions, including corrosion. Elevated temperatures can also lead to the breakdown of protective coatings or oxide layers, exposing the underlying metal to corrosive agents. However, in some cases, temperature changes can cause condensation, introducing moisture that can initiate corrosion in otherwise dry environments.
3. **pH levels:** The acidity or alkalinity of the environment plays a critical role in corrosion. Most metals are more susceptible to corrosion in acidic conditions, where hydrogen ions can contribute to metal dissolution. However, some metals, like aluminum, form protective oxide layers that are stable in slightly alkaline conditions but can dissolve in acidic environments, making them more vulnerable to corrosion.
4. **Chlorides and other corrosive agents:** Chloride ions, commonly found in saltwater, are highly corrosive, particularly to metals like stainless steel. Chlorides can penetrate and disrupt the passive oxide layer, leading to localized forms of corrosion such as pitting and crevice corrosion. Other corrosive agents, such as sulfur dioxide or hydrogen sulfide, can also contribute to the formation of corrosive acids when combined with moisture.

10.3.4 Electrochemical Potential

The electrochemical potential of a metal relative to other materials in contact with it can drive corrosion through galvanic processes. When two dissimilar metals are in contact in the presence of an electrolyte, the metal with the lower electrochemical potential (anode) will corrode faster, while the metal with the higher potential (cathode) is protected. Metals are ranked in the galvanic series according to their electrochemical potential in a given environment. Metals that are closer together in the series are less likely to experience significant galvanic corrosion when in contact [27]. However, if metals with a large potential difference are coupled, the anodic metal can experience accelerated corrosion.

10.3.5 Mechanical Stress and Deformation

Mechanical factors, such as stress and deformation, can also influence a metal's corrosion resistance.

1. **Stress corrosion cracking:** SCC occurs when a metal is exposed to tensile stress in a corrosive environment, leading to the initiation and propagation of cracks. Metals under high levels of stress, even below their yield strength, can be particularly vulnerable to SCC, especially in the presence of corrosive agents like chlorides or ammonia.

2. **Deformation and work hardening:** Deformation processes, such as bending, rolling, or drawing, can introduce residual stresses into the metal, which may increase its susceptibility to corrosion. Work hardening, while strengthening the metal, can also make it more prone to certain types of corrosion, such as intergranular corrosion, by altering the microstructure [28].

10.3.6 Time of Exposure

The duration of exposure to a corrosive environment is a critical factor in the corrosion resistance of metals. Longer exposure times allow for more extensive corrosion processes to occur, potentially leading to significant material loss or failure.

Initial versus long-term corrosion behavior: Some metals may exhibit good initial resistance to corrosion due to the formation of a protective oxide layer. However, over time, this layer may break down or become less effective, leading to accelerated corrosion. Continuous monitoring and maintenance are essential to manage long-term corrosion risks.

10.3.7 Presence of Microorganisms

MIC is a form of corrosion caused by the presence and activity of microorganisms, such as bacteria and fungi. These organisms can produce corrosive byproducts, like hydrogen sulfide or organic acids, which can accelerate the corrosion process. Microorganisms can form biofilms on metal surfaces, creating localized environments that differ significantly from the bulk environment. These biofilms can concentrate corrosive agents or create differential aeration cells, leading to pitting or other forms of localized corrosion.

10.4 IMPORTANCE OF LASER SURFACE TEXTURING IN CORROSION RESISTANCE ENHANCEMENT

LST offers a novel solution to this problem, providing a way to enhance corrosion resistance through precise surface modifications. It involves the use of high-energy laser pulses to create micro- and nanoscale patterns on a metal surface [29–30]. These patterns can take various forms, such as grooves, dimples, or ridges, depending on the laser parameters and the desired outcome. By altering the surface topography, LST significantly influences the metal's interaction with its environment, particularly with corrosive agents.

One of the key ways LST improves corrosion resistance is by reducing the surface area in direct contact with corrosive substances. The micro-structured surfaces can trap air or other protective gases, creating a barrier that impedes the diffusion of corrosive species like oxygen and chloride ions. Additionally, the surface texturing can promote the formation of a stable passive oxide layer, which further protects the metal from corrosion. Moreover, the precision of LST allows for the creation of surfaces that are hydrophobic, reducing the adhesion of water and other corrosive liquids. This hydrophobicity, combined with the physical barriers created by the textured surface, significantly enhances the corrosion resistance of the metal [31–33].

Compared to traditional corrosion protection methods, LST offers several advantages. First, it is an environmentally friendly process that does not rely on harmful chemicals or coatings. Second, LST is a precise and controlled technique, allowing for the treatment of specific areas without affecting the entire component. This level of precision is particularly useful in complex geometries or where different parts of a component are exposed to varying corrosive environments. Furthermore, LST is a versatile technique that can be applied to a wide range of metals, including aluminum, titanium, and steel. The customization options available with LST allow for the optimization of surface properties based on the specific application, ensuring that the metal's corrosion resistance is maximized for its intended use. Thus, laser surface texturing represents a significant advancement in the field of corrosion resistance for metals. By modifying the surface topography at the micro- and nanoscale [34–36], LST enhances the metal's ability to resist various forms of corrosion through mechanisms such as reduced surface contact with corrosive agents, improved passive film formation, and increased surface hydrophobicity. Its precision, environmental friendliness, and versatility make LST a valuable tool in extending the lifespan and performance of metallic components across diverse industries. As research and development in this area continue, LST is likely to become an increasingly important technique in the battle against corrosion.

10.5 MECHANISMS OF CORROSION RESISTANCE ENHANCEMENT THROUGH LASER SURFACE TEXTURING

As previously mentioned, the LST technique enhances the corrosion resistant characteristics offered by the metals. This enhancement is achieved through several mechanisms (Figure 10.4), as described as follows.

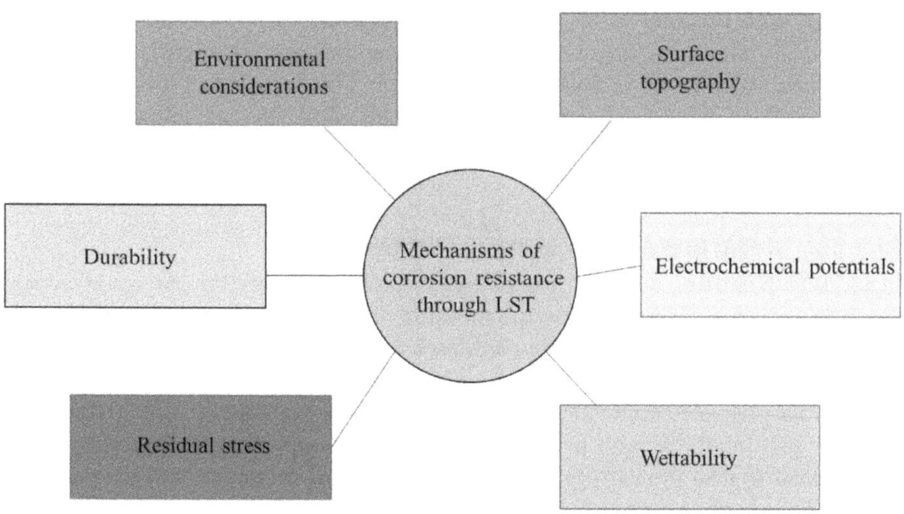

FIGURE 10.4 Corrosion resistance mechanisms.

10.5.1 Surface Topography and Microstructural Modifications

10.5.1.1 Surface Area Reduction

One of the primary mechanisms through which LST enhances corrosion resistance is by altering the surface topography of metals. By creating micro- and nanoscale textures, such as grooves, dimples, or ridges, LST effectively reduces the contact area between the metal surface and corrosive agents. The reduction in exposed surface area minimizes the sites available for corrosive reactions to initiate. This is particularly crucial in environments where localized corrosion, such as pitting or crevice corrosion, is a concern. The micro-textures can trap air or other inert gases within the grooves, further reducing the metal's exposure to moisture, oxygen, and other corrosive substances [37,38].

10.5.1.2 Surface Hardening and Grain Refinement

LST can also induce surface hardening and grain refinement in metals. The rapid thermal cycling caused by laser pulses leads to localized melting and rapid solidification, resulting in a finer microstructure. This refined microstructure typically exhibits enhanced mechanical properties, such as increased hardness and strength, which are beneficial in resisting mechanical wear and corrosion [39]. The grain boundaries in a refined microstructure act as barriers to the diffusion of corrosive species, thereby enhancing the overall corrosion resistance of the material.

10.5.2 Electrochemical Properties and Corrosion Potential

10.5.2.1 Shift in Corrosion Potential

Laser surface texturing can induce changes in the electrochemical properties of metals, which are crucial for corrosion resistance. The microstructural changes, combined with the altered surface chemistry, can shift the corrosion potential of the metal. A more noble (positive) corrosion potential indicates a reduced tendency for the metal to corrode in a given environment. The presence of microstructures and the enhanced passive layer contribute to this shift, as they reduce the active sites available for electrochemical reactions, thereby lowering the overall corrosion rate [40].

10.5.2.2 Reduction in Corrosion Current Density

In addition to shifting the corrosion potential, LST can reduce the corrosion current density. Corrosion current density is a measure of the rate of metal dissolution in an electrochemical cell, and it directly correlates with the severity of corrosion [41–43]. The micro-textures created by LST disrupt the uniformity of the surface, creating areas of varying electrochemical activity. The areas with higher electrochemical stability, such as those with a well-formed passive layer, dominate, leading to an overall reduction in the corrosion current density. This reduction indicates that the metal is less prone to corroding under a given set of conditions.

10.5.3 Wettability and Surface Energy

10.5.3.1 Hydrophobicity and Water Repellence

The wettability of a metal surface plays a significant role in its corrosion resistance. Hydrophobic surfaces, which repel water, are less likely to sustain prolonged contact with corrosive electrolytes, such as saltwater or acidic solutions. LST can induce hydrophobicity by creating micro- and nanoscale roughness on the metal surface [44,45]. The textured surface traps air within its structures, creating a cushion of air that prevents water from spreading across the surface. This air cushion effectively reduces the contact angle of water droplets, making the surface hydrophobic.

Hydrophobic surfaces are particularly beneficial in preventing electrochemical corrosion, as they reduce the amount of water that remains in contact with the metal. By minimizing the presence of water, which acts as an electrolyte, LST reduces the likelihood of corrosion processes such as pitting, crevice corrosion, and galvanic corrosion.

10.5.3.2 Surface Energy Modification

LST also modifies the surface energy of metals, which is directly related to their wettability and interaction with corrosive agents. A lower surface energy corresponds to a reduced affinity for water and other polar substances. The microstructures created by LST, along with the changes in surface chemistry, can lower the surface energy, making the metal less prone to corrosion. The reduction in surface energy also affects the adsorption of corrosive ions, such as chloride ions, thereby inhibiting their ability to initiate corrosion processes [46].

10.5.4 Environmental Considerations and Application-Specific Benefits

10.5.4.1 Environmentally Friendly Corrosion Protection

Unlike traditional corrosion protection methods, such as the application of chemical inhibitors or protective coatings, LST does not involve the use of potentially harmful substances. This makes LST an environmentally friendly alternative for enhancing corrosion resistance. The process is purely physical, relying on the precision and control of laser technology to achieve the desired surface modifications. As a result, there are no chemical waste products or emissions associated with the LST process, making it a sustainable option for industries aiming to reduce their environmental impact.

10.5.4.2 Customization for Specific Applications

One of the key advantages of LST is its ability to be customized for specific applications. The laser parameters, such as wavelength, pulse duration, and energy, can be precisely controlled to create surface textures tailored to the operational environment of the metal component. For instance, in marine applications, where exposure to saltwater is a primary concern, LST can be used to create microstructures that specifically target chloride-induced corrosion. Similarly, in biomedical implants, where biocompatibility and corrosion resistance are critical, LST can be employed to enhance the surface properties of titanium alloys, ensuring long-term stability and performance within the human body.

10.5.5 Long-Term Performance and Durability

10.5.5.1 Resistance to Mechanical Wear

In addition to enhancing corrosion resistance, LST contributes to the overall durability of metal components by increasing their resistance to mechanical wear. The microstructural modifications induced by LST, such as surface hardening and grain refinement, result in a harder, more wear-resistant surface. This is particularly important in applications where metal components are exposed to both corrosive environments and mechanical stress, such as in automotive or industrial machinery. By improving both corrosion resistance and wear resistance, LST extends the service life of metal components, reducing the need for frequent replacements or maintenance [47].

10.5.5.2 Stability Under Aggressive Conditions

The long-term stability of the corrosion-resistant properties induced by LST is another critical factor. The microstructures and passive layers formed through LST are highly stable under a wide range of environmental conditions, including extreme temperatures, pH variations, and exposure to aggressive chemicals. This stability ensures that the corrosion resistance provided by LST is maintained over extended periods, even in harsh environments. As a result, LST-treated metals are better suited for applications where long-term reliability and performance are essential.

10.5.6 Reduction of Residual Stresses

Residual stress refers to the stress that remains within a material after it has been subjected to manufacturing processes such as machining, welding, or surface treatments. These stresses can be either tensile or compressive, depending on the process and the conditions under which the material is treated. Tensile residual stresses are typically undesirable as they can lead to the formation and propagation of cracks, ultimately compromising the material's structural integrity. In contrast, compressive residual stresses are often beneficial because they counteract tensile stresses and inhibit crack formation. In the context of corrosion, residual stress plays a crucial role. Tensile residual stresses can create localized anodic sites on the material's surface, which are more susceptible to corrosion. This localized corrosion can manifest as pitting or stress corrosion cracking, both of which are highly detrimental to the material's performance and longevity [48,49].

The rapid heating and cooling cycles involved in LST cause localized plastic deformation in the material's surface layers. This plastic deformation leads to a redistribution of residual stresses, often converting tensile stresses into compressive stresses. Compressive residual stresses are advantageous because they reduce the driving force for crack initiation and propagation, which are critical factors in the onset of corrosion. By mitigating crack formation, compressive residual stresses effectively inhibit the initiation sites for localized corrosion, such as pitting or stress corrosion cracking. Additionally, compressive residual stresses can improve the material's overall fatigue resistance, which is closely related to its corrosion performance. Fatigue and corrosion often act synergistically, with fatigue cracks serving as pathways for corrosive agents to penetrate the material. By introducing compressive stresses, LST reduces the likelihood of fatigue crack formation, thereby enhancing the material's resistance to corrosion in cyclic loading environments.

Numerous studies have demonstrated the effectiveness of LST in reducing residual stress and enhancing corrosion resistance in various materials. For instance, in stainless steels, LST has been shown to significantly reduce tensile residual stresses and improve resistance to pitting corrosion in chloride-rich environments. Similarly, in titanium alloys, LST has been effective in introducing compressive residual stresses, leading to improved performance in both fatigue and corrosion resistance. The effectiveness of LST in managing residual stresses depends on several factors, including the laser parameters (such as energy density, pulse duration, and scanning speed) and the material's properties. By carefully controlling these parameters, it is possible to optimize the residual stress distribution across the textured surface. For example, lower energy densities and slower scanning speeds tend to induce more compressive residual stresses, which are beneficial for improving corrosion resistance [50].

While the benefits of LST in reducing residual stress and enhancing corrosion resistance are well documented, several challenges remain. One of the primary challenges is the precise control of laser parameters to achieve the desired residual stress profile and surface texture. The complexity of the LST process, combined with the variability in material properties, requires advanced modeling and simulation techniques to optimize the process parameters. Additionally, there is a need for more comprehensive studies to understand the long-term effects of LST on corrosion resistance, particularly in different environmental conditions. Factors such as temperature, humidity, and the presence of aggressive ions can influence the corrosion behavior of laser-textured surfaces, and these factors must be thoroughly investigated to ensure the reliability of LST-treated components.

10.6 PASSIVE FILM FORMATION AND STABILITY

10.6.1 Passive Film Formation: Mechanisms and Importance

A passive film (Figure 10.5) is an oxide layer that forms on the surface of metals when they are exposed to an oxidizing environment. This film is typically very thin, often just a few nanometers thick, but it plays a crucial role in protecting the metal from further corrosion. The formation of a passive film occurs through a process of metal oxidation, where metal atoms at the surface react with oxygen or other oxidizing agents to form a stable oxide compound. The composition, structure, and thickness of the passive film are essential factors in determining its protective qualities. For example, in stainless steel, the passive film is primarily composed of chromium oxide (Cr_2O_3), which provides excellent resistance to corrosion. The chromium in the alloy reacts with oxygen to form the oxide layer, which adheres strongly to the metal surface and serves as a protective barrier [51–53].

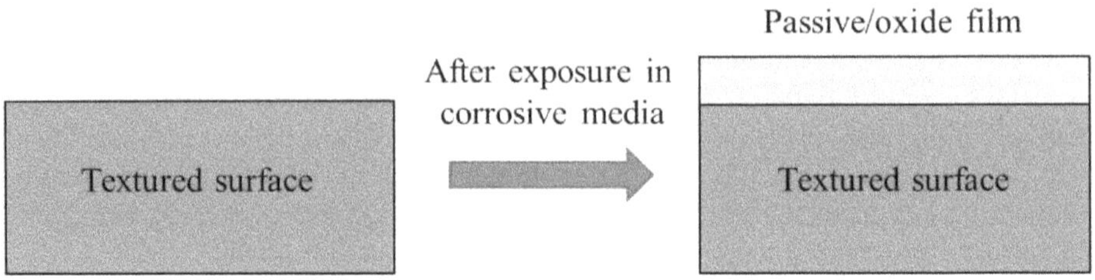

FIGURE 10.5 Passive film formation.

The stability of the passive film is just as important as its formation. A stable passive film remains intact and continues to protect the metal surface over time, even in the presence of corrosive agents such as chloride ions, which are common in marine environments. Factors that influence the stability of the passive film include its chemical composition, the presence of alloying elements, the pH of the environment, and the presence of mechanical stresses.

10.6.2 Laser Surface Texturing and Its Impact on Passive Film Formation

Laser surface texturing is a technique that involves the use of a focused laser beam to create precise micro- or nano-scale textures on a material's surface. These textures can significantly influence the surface properties of the material, including its ability to form and maintain a stable passive film [54–56]. The following sections discuss the key ways in which LST enhances passive film formation and stability.

10.6.2.1 Modification of Surface Morphology and Topography

The morphology and topography of a metal surface play crucial roles in the formation and stability of passive films. By creating micro- or nano-scale textures on the surface, LST increases the surface area and provides more nucleation sites for oxide formation. This leads to a more uniform and continuous passive film, which enhances the overall corrosion resistance of the material.

The specific patterns created by LST, such as grooves, dimples, or ridges, can improve the adhesion of the passive film to the substrate. This is particularly important because poor adhesion can lead to delamination or spalling of the passive film, exposing the underlying metal to corrosive agents. Additionally, the increased surface roughness resulting from LST can relieve tensile stresses within the passive film, which would otherwise destabilize it. This reduction in stress concentrations helps to maintain the mechanical integrity of the passive film, making it less susceptible to cracking or peeling.

10.6.2.2 Enhancement of Surface Chemistry

LST can also modify the chemical composition of the material's surface, promoting the formation of more protective oxide layers. For example, in stainless steels, LST can increase the concentration of chromium in the surface region, leading to the formation of a chromium-rich oxide layer. This chromium oxide layer is highly resistant to corrosion, and its presence in the passive film significantly improves the material's ability to resist further oxidation.

The laser-induced heating and rapid cooling cycles associated with LST can also promote the formation of other stable oxide phases, such as titanium oxide or aluminum oxide, depending on the material being treated. These oxides are known for their excellent corrosion resistance, and their presence in the passive film enhances the overall protective qualities of the film.

Moreover, LST can lead to the diffusion of alloying elements to the surface, enriching the passive film with protective species. For instance, in nickel-based alloys, LST can promote the formation of a nickel-rich oxide layer, which provides excellent resistance to high-temperature oxidation and corrosion. This surface enrichment with alloying elements can significantly enhance the long-term stability and effectiveness of the passive film.

10.6.2.3 Alteration of Electrochemical Properties

The electrochemical behavior of the passive film, including its potential and conductivity, is a critical factor in determining its stability and effectiveness. LST can modify the electrochemical properties of the passive film by altering the surface's microstructure and chemistry. For example, the textured surface created by LST may exhibit different electrochemical potentials compared to an untreated surface, leading to a more stable passive state.

The laser-induced textures can also reduce the likelihood of localized corrosion, such as pitting, by promoting a more uniform distribution of electrochemical potentials across the surface. In environments where chloride ions are present, this uniformity is particularly important because chloride ions can break down the passive film, leading to the initiation of pitting corrosion. By creating a more homogeneous and defect-free passive film, LST reduces the susceptibility of the material to such localized corrosion phenomena.

10.6.2.4 Influence on Oxide Layer Thickness and Composition

The thickness and composition of the passive film are critical factors in determining its protective capabilities. LST can influence these factors by controlling the thermal gradients and cooling rates during the texturing process. Rapid cooling rates, for instance, can lead to the formation of thinner but more dense and stable oxide layers, which are less prone to cracking or spalling.

Additionally, LST can promote the formation of mixed oxides, where the passive film is composed of multiple oxide phases. These mixed oxides can offer superior protection compared to single-phase oxides because they combine the protective qualities of different oxides. For example, a passive film composed of both chromium oxide and iron oxide may provide better resistance to a wider range of corrosive environments than a film composed of chromium oxide alone.

10.6.3 Stability of Passive Films Enhanced by Laser Surface Texturing

The long-term stability of the passive film is crucial for ensuring the corrosion resistance of the material in service [57–59]. LST enhances the stability of the passive film through several mechanisms, which are discussed next.

10.6.3.1 Mechanical Stability and Adhesion

The mechanical stability of the passive film is significantly enhanced by the textures created through LST. These textures improve the adhesion of the passive film to the substrate, making it more resistant to mechanical damage such as abrasion or impact. The compressive residual stresses introduced by LST also play a vital role in maintaining the mechanical stability of the passive film. These stresses help to prevent the formation of microcracks, which can serve as initiation sites for localized corrosion, such as pitting or crevice corrosion. Furthermore, the laser-induced textures can act as mechanical interlocks, physically anchoring the passive film to the surface. This interlocking effect reduces the likelihood of film delamination or spalling, ensuring that the passive film remains intact and protective over time.

10.6.3.2 Resistance to Environmental Degradation

The environmental stability of the passive film is another critical factor in its effectiveness. LST can enhance the environmental resistance of the passive film by promoting the formation of more stable oxide layers that are less prone to dissolution in corrosive environments. For example, the chromium-rich oxide layer formed on stainless steel through LST is more resistant to chloride ion attack, which is a common cause of pitting corrosion in marine environments. Additionally, the laser-induced surface modifications can make the passive film more resistant to temperature fluctuations, humidity, and chemical exposure. These environmental factors can accelerate the degradation of the passive film, leading to a loss of corrosion protection. By enhancing the stability of the passive film under such conditions, LST ensures that the material maintains its corrosion resistance over extended periods [60].

10.6.3.3 Self-Healing Properties

In some cases, LST can promote the self-healing properties of the passive film. This is particularly important in dynamic environments where the material is subjected to mechanical wear or chemical attack. If the passive film is damaged, the laser-textured surface may facilitate the rapid reformation of the oxide layer, restoring the material's corrosion resistance [61]. The self-healing ability of the passive film is enhanced by the presence of alloying elements that can rapidly diffuse to the surface and form new oxide layers in response to damage. For example, in titanium alloys, the presence of vanadium and aluminum can promote the rapid formation of titanium oxide in the event of film damage, effectively "healing" the passive film and maintaining corrosion protection.

10.6.3.4 Case Studies

Numerous studies have demonstrated the effectiveness of LST in enhancing passive film formation and stability across a wide range of materials and applications.

1. **Stainless steels:** LST has been shown to increase the chromium content in the passive film of stainless steels, resulting in improved resistance to pitting and crevice corrosion. This has significant implications for applications in the chemical processing and marine industries, where stainless steel components are frequently exposed to harsh environments.
2. **Titanium alloys:** In titanium alloys, LST has been effective in forming a stable titanium oxide layer, which provides excellent protection against oxidation and corrosion in both marine and aerospace environments. The enhanced corrosion resistance of LST-treated titanium alloys makes them suitable for use in aircraft components, medical implants, and offshore structures.
3. **Nickel-based alloys:** LST has been used to enhance the passive film formation on nickel-based alloys, leading to improved resistance to high-temperature oxidation and corrosion. These alloys are commonly used in gas turbines, power generation, and petrochemical industries, where they are exposed to extreme temperatures and corrosive environments.
4. **Aluminum alloys:** In aluminum alloys, LST can promote the formation of a more stable aluminum oxide layer, which enhances resistance to corrosion in automotive and aerospace applications. The ability of LST to improve the corrosion resistance of aluminum alloys is particularly valuable in lightweight structural components that are exposed to corrosive environments.

10.7 INDUSTRIAL APPLICATIONS OF LST IN CORROSION RESISTANCE

1. **Aerospace industry:** The aerospace industry requires materials that can withstand harsh environmental conditions, including exposure to corrosive agents such as high temperature humidity, and atmospheric pollutants. LST is used to enhance the corrosion resistance of critical components such as turbine blades, landing gear, and fuselage sections. By reducing surface roughness and promoting the formation of protective films, LST helps extend the service life of these components, ensuring safety and reliability [62].
2. **Automotive industry:** In the automotive sector, corrosion resistance is crucial for both structural components and aesthetic parts. LST is applied to improve the durability of body panels, chassis components, and exhaust systems. By creating micro-textures that enhance paint adhesion and reduce water accumulation, LST helps prevent rust and corrosion, particularly in vehicles exposed to harsh weather conditions and road salts.
3. **Marine industry:** Marine environments are highly corrosive due to the presence of saltwater, humidity, and biological organisms. LST has been widely adopted in the marine industry to improve the corrosion resistance of ship hulls, propellers, and offshore structures. The technique not only

enhances the protective properties of coatings but also reduces biofouling, which can further accelerate corrosion.

4. **Biomedical implants:** In the biomedical field, metallic implants such as orthopedic screws, plates, and dental implants are prone to corrosion in the human body. LST is used to improve the corrosion resistance of these implants, ensuring their longevity and biocompatibility. By creating specific surface textures, LST can enhance the osseointegration of implants, promoting better bonding with bone tissue.
5. **Electronics and microelectronics:** The miniaturization of electronic devices has led to increased demand for corrosion-resistant materials. LST is applied to improve the durability of electronic components, such as connectors and circuit boards, by enhancing their resistance to oxidation and moisture. This ensures the reliable performance of electronic devices in various environments, including humid and corrosive atmospheres.
6. **Energy sector:** In the energy industry, particularly in oil and gas exploration, materials are exposed to highly corrosive environments, including sour gas and seawater. LST is employed to enhance the corrosion resistance of drilling equipment, pipelines, and storage tanks. The technique helps reduce maintenance costs and downtime by extending the service life of critical components.

10.8 BENEFITS OF LST IN CORROSION RESISTANCE

1. **Precision and control:** LST offers unparalleled precision in modifying surface properties. The ability to control the texture at the micro- and nanoscale allows engineers to tailor surfaces for specific applications, optimizing corrosion resistance in targeted areas.
2. **Eco-friendliness:** Unlike traditional surface treatments that may involve harmful chemicals, LST is a clean and environmentally friendly process. It relies on laser energy to modify the surface, eliminating the need for hazardous materials and reducing the environmental impact.
3. **Cost-effectiveness:** While the initial setup for LST may be expensive, the long-term benefits outweigh the costs. The extended service life of components, reduced maintenance requirements, and improved performance lead to significant cost savings over time.
4. **Versatility:** LST can be applied to a wide range of materials, including metals, ceramics, and polymers. This versatility makes it a valuable tool in various industries, from aerospace to biomedical engineering [63].
5. **Improved coating adhesion:** LST can also be used to improve the adhesion of protective coatings. By creating an optimal surface texture, LST enhances the bonding between the coating and the substrate, further improving the corrosion resistance.

10.9 CHALLENGES AND FUTURE DIRECTIONS

While LST offers numerous advantages, it is not without challenges. The process requires precise control of laser parameters, and improper settings can lead to surface defects or unwanted thermal effects. Additionally, the high cost of laser equipment and the need for specialized knowledge may limit its widespread adoption.

Future research in LST for corrosion resistance is likely to focus on further refining the process to enhance efficiency and reduce costs. The development of advanced lasers with higher precision and energy control will enable more complex and effective surface textures [64]. Additionally, the integration of LST with other surface treatment techniques, such as chemical passivation and coating applications, could lead to even greater improvements in corrosion resistance.

10.10 CONCLUSION

LST stands out as an innovative and effective technique for enhancing corrosion resistance across various materials and industries. By precisely modifying surface properties at the micro- and nano-scale, LST addresses some of the fundamental mechanisms that contribute to corrosion, such as surface roughness, wettability, residual stresses, and passive film stability. The ability to create highly controlled textures not only promotes the formation of protective films but also allows for the development of superhydrophobic surfaces that can significantly reduce corrosion rates, particularly in aggressive environments. The applications of LST are broad, ranging from the aerospace and automotive industries, where durability and performance under extreme conditions are crucial, to the biomedical field, where corrosion-resistant implants are essential for patient safety and implant longevity. Additionally, the technique's environmental friendliness, precision, and potential for cost savings make it an attractive alternative to traditional corrosion prevention methods, which often involve the use of hazardous chemicals. Despite its many advantages, the widespread adoption of LST is still challenged by the high costs associated with laser equipment and the need for specialized expertise in process optimization. However, ongoing advancements in laser technology and increasing demand for high-performance, corrosion-resistant materials are likely to drive further innovation in this field. Future research and development efforts will focus on refining the process to make it more accessible and cost effective, as well as exploring new applications and material combinations. Thus, LST represents a promising solution for enhancing corrosion resistance in a wide array of applications. Its ability to tailor surface characteristics with high precision offers significant benefits in terms of material longevity, performance, and environmental impact. As the technology continues to mature, it is poised to play an increasingly important role in the development of advanced, corrosion-resistant materials and structures, contributing to safer, more durable products across multiple industries.

10.11 ACKNOWLEDGEMENT

The authors like to thank the editors of this book for presenting an opportunity to contribute this chapter.

10.12 CONFLICTS OF INTEREST

The authors declare no conflicts of interest.

REFERENCES

[1] Caraguay SJ, Pereira TS, Giacomelli RO, Cunha A, Pereira M, Xavier FA. The effect of laser surface textures on the corrosion resistance of epoxy coated steel exposed to aggressive environments for offshore applications. Surface and Coatings Technology. 2022 May 15;437:128371.

[2] Wahab JA, Ghazali MJ, Yusoff WM, Sajuri Z. Enhancing material performance through laser surface texturing: A review. Transactions of the IMF. 2016 Jul 3;94(4):193–198.

[3] Karuppasamy SS, Yang CH. Adapting the time-domain synthetic aperture focusing technique (T-SAFT) to laser ultrasonics for imaging the subsurface defects. Sensors. 2023 Sep 22;23(19):8036.

[4] Zhang L, Tan Z, Zhang C, Tang J, Yao C, You X, Hao B. Research on metal corrosion resistant bioinspired special wetting surface based on laser texturing technology: A review. Micromachines. 2022 Aug 30;13(9):1431.

[5] Natarajan J, Yang CH, Karuppasamy SS. Investigation on microstructure, nanohardness and corrosion response of laser cladded colmonoy-6 particles on 316L steel substrate. Materials. 2021 Oct 18;14(20):6183.

[6] Trdan U, Hočevar M, Gregorčič P. Transition from superhydrophilic to superhydrophobic state of laser textured stainless steel surface and its effect on corrosion resistance. Corrosion Science. 2017 Jul 15;123:21–26.

[7] Grabowski A, Sozańska M, Adamiak M, Kępińska M, Florian T. Laser surface texturing of Ti6Al4V alloy, stainless steel and aluminium silicon alloy. Applied Surface Science. 2018 Dec 15;461:117–123.

[8] Verma PK, Vasudev H, Jeyaprakash N. Thermal barrier coatings and claddings for aerospace applications in extreme environment. In Aerospace Engineering Materials 2026 (pp. 110–133). CRC Press.

[9] Conradi M, Kocijan A, Klobčar D, Godec M. Influence of laser texturing on microstructure, surface and corrosion properties of Ti-6Al-4V. Metals. 2020 Nov 11;10(11):1504.

[10] Zhang L, Lin N, Zou J, Lin X, Liu Z, Yuan S, Yu Y, Wang Z, Zeng Q, Chen W, Tian L. Super-hydrophobicity and corrosion resistance of laser surface textured AISI 304 stainless steel decorated with Hexadecyltrimethoxysilane (HDTMS). Optics & Laser Technology. 2020 Jul 1;127:106146.

[11] Karuppasamy SS, Jeyaprakash N, Yang CH. Microstructure, nanoindentation and corrosion behavior of colmonoy-5 deposition on SS410 substrate using laser cladding process. Arabian Journal for Science and Engineering. 2022 Jul;47(7):8751–8767.

[12] Cao L, Chen Y, Cui J, Li W, Lin Z, Zhang P. Corrosion wear performance of pure titanium laser texturing surface by nitrogen ion implantation. Metals. 2020 Jul 22;10(8):990.

[13] Wang Z, Song J, Wang T, Wang H, Wang Q. Laser texturing for superwetting titanium alloy and investigation of its erosion resistance. Coatings. 2021 Dec 16;11(12):1547.

[14] Jeyaprakash N, Yang CH, Karuppasamy SS, Dhineshkumar SR. Evaluation of microstructure, nanoindentation and corrosion behavior of laser cladded Stellite-6 alloy on Inconel-625 substrate. Materials Today Communications. 2022 Jun 1;31:103370.

[15] Kedia S, Bonagani SK, Majumdar AG, Kain V, Subramanian M, Maiti N, Nilaya JP. Nanosecond laser surface texturing of type 316L stainless steel for contact guidance of bone cells and superior corrosion resistance. Colloid and Interface Science Communications. 2021 May 1;42:100419.

[16] Ma Q, Tong Z, Wang W, Dong G. Fabricating robust and repairable superhydrophobic surface on carbon steel by nanosecond laser texturing for corrosion protection. Applied Surface Science. 2018 Oct 15;455:748–757.

[17] Jeyaprakash N, Yang CH, Susila P, Karuppasamy SS. Laser cladding of NiCrMoFeNbTa particles on Inconel 625 alloy: Microstructure and corrosion resistance. Transactions of the Indian Institute of Metals. 2023 Feb;76(2):599–612.

[18] Travessa DN, Guedes GV, de Oliveira AC, Cardoso KR, Roche V, Jorge Jr AM. The effect of surface laser texturing on the corrosion performance of the biocompatible β-Ti12Mo6Zr2Fe alloy. Surface and Coatings Technology. 2021 Jan 15;405:126628.
[19] Rezayat M, Morales M, Moradi M, Mateo A. Laser wobbling surface texturing of AISI 301LN steel for enhancement of the corrosion resistance at high temperature. Optics & Laser Technology. 2024 Apr 1;171:110375.
[20] Kumar V, Verma R, Kango S, Sharma VS. Recent progresses and applications in laser-based surface texturing systems. Materials Today Communications. 2021 Mar 1;26:101736.
[21] Jeyaprakash N, Kantipudi MB, Yang CH. Study on microstructure and anti-corrosion property of stainless steel particles deposition on Ti–6Al–4 V substrate using laser cladding technique. Lasers in Manufacturing and Materials Processing. 2022 Jun;9(2): 214–227.
[22] Conradi M, Sever T, Gregorčič P, Kocijan A. Short-and long-term wettability evolution and corrosion resistance of uncoated and polymer-coated laser-textured steel surface. Coatings. 2019 Sep 19;9(9):592.
[23] Shivakoti I, Kibria G, Cep R, Pradhan BB, Sharma A. Laser surface texturing for biomedical applications: A review. Coatings. 2021 Jan 22;11(2):124.
[24] Jeyaprakash N, Yang CH, Karuppasamy SS, Duraiselvam M. Stellite 6 cladding on AISI type 316L stainless steel: Microstructure, nanohardness and corrosion resistance. Transactions of the Indian Institute of Metals. 2023 Feb;76(2):491–503.
[25] Morales M, Rezayat M, Mateo A. Enhancing the corrosion resistance of 2205 duplex stainless steel in molten carbonate salts by laser-surface texturing. Journal of Energy Storage. 2024 Feb 1;78:110053.
[26] Rafieazad M, Jaffer JA, Cui C, Duan X, Nasiri A. Nanosecond laser fabrication of hydrophobic stainless steel surfaces: The impact on microstructure and corrosion resistance. Materials. 2018 Sep 1;11(9):1577.
[27] Liu H, Tong Z, Gu J, Liu C, Wang X, Ren X. Effect of laser surface texturing depth and pattern on the bond strength and corrosion performance of phosphate conversion coating on magnesium alloy. Optics & Laser Technology. 2022 Sep 1;153:108164.
[28] Gupta RK, Anandkumar B, Choubey A, George RP, Ganesh P, Upadhyaya BN, Philip J, Bindra KS, Kaul R. Antibacterial and corrosion studies on nanosecond pulse laser textured 304 L stainless steel surfaces. Lasers in Manufacturing and Materials Processing. 2019 Sep 15;6(3):332–343.
[29] Kumar V, Verma R, Bairwa HK. Fabrication of superhydrophobic surfaces by laser surface texturing and autoxidation. Journal of Electrochemical Science and Engineering. 2022 Sep 18;12(4):639–649.
[30] Boinovich LB, Emelyanenko AM, Modestov AD, Domantovsky AG, Emelyanenko KA. Synergistic effect of superhydrophobicity and oxidized layers on corrosion resistance of aluminum alloy surface textured by nanosecond laser treatment. ACS Applied Materials & Interfaces. 2015 Sep 2;7(34):19500–19508.
[31] Karuppasamy SS, Jeyaprakash N, Yang CH. Application of corrosion-resistant laser claddings. In Handbook of Laser-Based Sustainable Surface Modification and Manufacturing Techniques 2023 Jul 5 (pp. 27–56). CRC Press.
[32] Caraguay SJ, Pereira TS, Cunha A, Pereira M, Xavier FA. The effect of laser surface textures on the adhesion strength and corrosion protection of organic coatings-experimental assessment using the pull-off test and the shaft load blister test. Progress in Organic Coatings. 2023 Jul 1;180:107558.
[33] Aulakh SS, Kaushal G. Laser texturing as an alternative to grit blasting for improved coating adhesion on AZ91D magnesium alloy. Transactions of the IMF. 2019 Mar 4;97(2):100–108.

[34] Jeyaprakash N, Yang CH, Karuppasamy SS. Laser cladding of Colmonoy 6 particles on Inconel 625 substrate: Microstructure and corrosion resistance. Surface Review and Letters. 2022 Aug 15;29(8):2250102.

[35] Liu H, Tong Z, Yang Y, Zhou W, Chen J, Pan X, Ren X. Preparation of phosphate conversion coating on laser surface textured surface to improve corrosion performance of magnesium alloy. Journal of Alloys and Compounds. 2021 Jun 5;865:158701.

[36] Mao B, Siddaiah A, Liao Y, Menezes PL. Laser surface texturing and related techniques for enhancing tribological performance of engineering materials: A review. Journal of Manufacturing Processes. 2020 May 1;53:153–173.

[37] Hajam MI, Khan MM. Effect of laser surface texturing on the corrosion behaviour and biocompatibility of zinc hydroxyapatite implant material. Materials Today Communications. 2024 Mar 1;38:108110.

[38] Kumari R, Scharnweber T, Pfleging W, Besser H, Majumdar JD. Laser surface textured titanium alloy (Ti–6Al–4V)–Part II–Studies on bio-compatibility. Applied Surface Science. 2015 Dec 1;357:750–758.

[39] Cuao-Moreu CA, Campos-Silva I, Delgado-Brito AM, Garcia-Sanchez EO, Juarez-Hernandez A, Diabb-Zavala JM, Hernandez-Rodriguez MA. Effect of laser surface texturing and boriding on the tribocorrosion resistance of an ASTM F-1537 cobalt alloy. Wear. 2023 Jun 15;523:204799.

[40] Kaushik P, Prakash S. Fabrication of improved corrosion resistant Al-Mg alloy surfaces using underwater Laser surface texturing. Proceedings of the Institution of Mechanical Engineers, Part L: Journal of Materials: Design and Applications. 2025 Sep 3:14644207251374133.

[41] Lazov L, Teirumnieks E, Angelov N, Yankov E. Modification of the roughness of 304 stainless steel by laser surface texturing (LST). Laser Physics. 2023 Mar 2;33(4):046001.

[42] Karuppasamy SS, Yang CH. Laser utrasonic imaging based on synthetic aperture focusing technique algorithm. International Journal of Civil and Environmental Engineering. 2014 Jan 1;15(1):1–5.

[43] Liu Z, Niu T, Lei Y, Luo Y. Metal surface wettability modification by nanosecond laser surface texturing: A review. Biosurface and Biotribology. 2022 Jun;8(2):95–120.

[44] Ahuir-Torres JI, Arenas MA, Perrie W, Dearden G, De Damborenea J. Surface texturing of aluminium alloy AA2024-T3 by picosecond laser: Effect on wettability and corrosion properties. Surface and Coatings Technology. 2017 Jul 15;321:279–291.

[45] Jeyaprakash N, Karuppasamy SS, Yang CH. Interlayers in FSSW process: An overview. Friction Stir Spot Welding.:21–38.

[46] Seid Ahmed Y, DePaiva MJ, Amorim LF, Torres DR, de Rossi W, Veldhuis CS. Laser surface texturing and characterization of austenitic stainless steel for the improvement of its surface properties. The International Journal of Advanced Manufacturing Technology. 2021 Jul;115(5):1795–1808.

[47] Jeyaprakash N, Yang CH, Kumar DR. Laser surface modification of materials. In Practical Applications of Laser Ablation 2020 Oct 31. IntechOpen.

[48] Lou D, Liang E, Chen P, Jiang G, Wang L, Guan L, Wang Y, Liu D. The effect of dimple overlap on wettability and corrosion resistance of laser-textured stainless steel. Crystals. 2022 May 13;12(5):695.

[49] Zhao C, Ma C, Du W, Yu Z, Zhang Z. Study on the reduction of residual stress in laser cladding layers through groove texture. Scientific Reports. 2024 Jul 10;14(1):15911.

[50] Shivakoti I, Kibria G, Das S, Sharma A, Pradhan BB, Chatterjee S. Laser surface texturing on Ti-6Al-4V. Materials and Manufacturing Processes. 2021 May 19;36(7):858–867.

[51] Prakash FP, Jeyaprakash N, Duraiselvam M, Prabu G, Yang CH. Droplet spreading and wettability of laser textured C-263 based nickel superalloy. Surface and Coatings Technology. 2020 Sep 15;397:126055.

[52] Moldovan ER, Concheso Doria C, Ocaña JL, Baltes LS, Stanciu EM, Croitoru C, Pascu A, Roata IC, Tierean MH. Wettability and surface roughness analysis of laser surface texturing of AISI 430 stainless steel. Materials. 2022 Apr 18;15(8):2955.
[53] Schwibbert K, Richter AM, Krüger J, Bonse J. Laser-textured surfaces: A way to control biofilm formation? Laser & Photonics Reviews. 2024 Jan;18(1):2300753.
[54] Li H, Wang X, Zhang J, Wang B, Breisch M, Hartmaier A, Rostotskyi I, Voznyy V, Liu Y. Experimental investigation of laser surface texturing and related biocompatibility of pure titanium. The International Journal of Advanced Manufacturing Technology. 2022 Apr;119(9):5993–6005.
[55] Li S, Liu C. Preparation of the wear and corrosion-resistant coating using a composite process of laser surface texturing technology and plasma electrolytic oxidation. AIP Advances. 2023 Mar 1;13(3).
[56] Misyura SY, Feoktistov DV, Morozov VS, Orlova EG, Kuznetsov GV, Islamova AG. Effect of heat treatment on corrosion of laser-textured aluminum alloy surfaces. Journal of Materials Science. 2021 Aug;56(22):12845–12863.
[57] Cholkar A, Chatterjee S, Richards C, McCarthy É, Perumal G, Regan F, Kinahan D, Brabazon D. Biofouling and corrosion protection of aluminum alloys through ultrafast laser surface texturing for marine applications. Advanced Materials Interfaces. 2024 Feb;11(6):2300835.
[58] Kümmel D, Hamann-Schroer M, Hetzner H, Schneider J. Tribological behavior of nanosecond-laser surface textured Ti6Al4V. Wear. 2019 Mar 15;422:261–268.
[59] Lei P, Zhang P, Song S, Liu Z, Yan H, Sun T, Lu Q, Chen Y, Gromov V, Shi H. Research status of laser surface texturing on tribological and wetting properties of materials: A review. Optik. 2023 Dec 25:171581.
[60] Riveiro A, Maçon AL, del Val J, Comesaña R, Pou J. Laser surface texturing of polymers for biomedical applications. Frontiers in Physics. 2018 Feb 27;6:16.
[61] Moura CG, Carvalho O, Gonçalves LM, Cerqueira MF, Nascimento R, Silva F. Laser surface texturing of Ti-6Al-4V by nanosecond laser: Surface characterization, Ti-oxide layer analysis and its electrical insulation performance. Materials Science and Engineering: C. 2019 Nov 1;104:109901.
[62] Yuan S, Lin N, Wang W, Zhang H, Liu Z, Yu Y, Zeng Q, Wu Y. Correlation between surface textural parameter and tribological behaviour of four metal materials with laser surface texturing (LST). Applied Surface Science. 2022 May 1;583:152410.
[63] Pfleging W, Kumari R, Besser H, Scharnweber T, Majumdar JD. Laser surface textured titanium alloy (Ti–6Al–4V): Part 1–Surface characterization. Applied Surface Science. 2015 Nov 15;355:104–111.
[64] Vishnoi M, Kumar P, Murtaza Q. Surface texturing techniques to enhance tribological performance: A review. Surfaces and Interfaces. 2021 Dec 1;27:101463.

11 Applications of Laser Surface Texturing on Additively Manufactured Parts

T. Sathies, V. M. Basanthy, and Jeyaprakash Natarajan

11.1 INTRODUCTION

Additive manufacturing (AM), also known as 3D printing is extensively used in academia and industries for manufacturing 3D components from polymers, ceramics, and metallic materials [1]. AM techniques like stereolithography, selective laser sintering, material jetting, and extrusion-based 3D printing can fabricate polymer components while methods like wire arc additive manufacturing, selective laser melting, electron beam melting, friction stir additive manufacturing, laser engineered net shaping, and ultrasonic AM are employed for fabricating metallic components [2]. The accuracy and precision of the components produced by AM depend on factors like the energy source, AM technique, and various process-related parameters [1,3].

11.1.1 Issues Reported on Surface Characteristics of Parts Built by Laser Powder Bed Fusion

The metallic components fabricated by laser powder bed fusion (LPBF) techniques often exhibit high surface roughness, staircase effect, balling and spatter, and surface oxidation. The presence of partially melted particles on the surface of LPBF-built parts leads to high surface roughness, and it can affect the heat conduction during component fabrication. Poor selection of layer thickness and scan strategy results in stair stair-stepping effect and can affect the geometries. Because of unstable melt pool and the Marangoni effect, balling and spattering phenomena are noticed in LPBF parts, and it is commonly observed in over-melted regions. When the components are exposed to oxygen during selective laser melting, it may result in surface oxidation and form oxides such as TiO_2, Al_2O_3, and NiO, as a result, affecting wettability, corrosion resistance, wear resistance, and coating adhesion of LPBF-built parts. To address these issues, the LPBF-built part can be post-processed with laser surface texturing [2,4–6].

DOI:10.1201/9781003542087-11

11.1.2 Issues Reported on Surface Characteristics of Directed Energy Deposited Components

Since the directed energy deposition (DED) method involves multiple track deposition, the usage of an improper overlap ratio results in waviness and bead overlap irregularities. Because of this, components with non-uniform layer height might be produced in the DED process. The usage of an inconsistent feed rate causes droplet solidification ridges, leading to a rough and uneven surface. The DED of Al and Ti alloys in an open atmosphere usually results in surface oxidation. The incomplete fusion between beads during DED generates porosity and undercut, which affects the sealing and fatigue properties of DED-fabricated parts. The remelting of previously solidified layers during DED results in heat accumulation and affects the final shape of the printed component [7].

11.1.3 Issues Reported on Surface Characteristics of Binder-Jetted and Extrusion-Based 3D-Printed Components

The presence of issues like poor binder spreading and lower powder particle resolution in binder jetting affects the surface definition of green components and increases the surface roughness before sintering. During the sintering phase of binder jetting, shrinkage-induced distortions are common and can alter the surface characteristics and dimensional accuracy. Incomplete removal of the binder causes the powder to adhere improperly to the component and results in local rough patches [8]. In the case of material extrusion-based 3D printing, incomplete interlayer fusion leads to porosity between rasters and affects the strength and surface quality of printed components. Further, there are chances for non-uniform cooling in the FDM process, leading to warping and delamination. During composite 3D printing, frequent nozzle blockage and vibrations were observed and it creates surface waviness and ripples. These defects affect the mechanical and tribological performance [9,10].

When the surface quality of a 3D printed component is not good, it affects the mechanical strength, tribological behavior, corrosion resistance, adhesion, and coating performance and dimensional accuracy and tolerance. The surface defects act as a stress generator and reduce fatigue and tensile strength. The components with high surface roughness exhibit reduced wear resistance. The surface pores and oxides act as corrosion initiation sites and reduce the corrosion resistance of printed components. Further, the high surface roughness and oxide reduce the adhesion and coating performance of components fabricated through additive manufacturing. To improve the functional characteristics of AM, post-processing treatments such as machining, sandblasting, heat treatment, chemical treatment, laser polishing, laser surface texturing, and laser shock peening are performed on components fabricated by AM. Among the different surface treatment methods, laser surface texturing is widely used by researchers for improving the functional characteristics of components fabricated in conventional manufacturing routes. The research progress in the laser surface texturing of additively manufactured parts is discussed in the forthcoming sections.

11.2 LASER SURFACE TEXTURING

Surface texturing methods are widely used by researchers to modify the topography of a material's surface at the micro- or nano-scale for improving properties such as friction, wear resistance, adhesion, wettability, lubrication, or biological response. Surface texturing methods are generally categorized as laser-based surface texturing, mechanical based texturing (abrasive blasting, shot peening, micro milling, grinding, rolling), chemical and electrochemical texturing methods (chemical etching, electrochemical etching, anodization, plasma or ion etching), deposition-based methods (physical vapor deposition, chemical vapor deposition, thermal spray coating), and hybrid texturing methods. Among the different texturing methods, laser-based texturing methods are widely investigated because laser-based texturing operates at high speeds, reduces processing time, is environmentally friendly, and provides precise control over texture size and shapes. By modifying the energy density, lasers can effectively treat a wide range of materials, including hardened steels, ceramics, polymers, and crystalline structures. LST modifies the material's surface properties by changing its texture and roughness [11,12]. The different types of laser-based texturing and their corresponding advantage and applications are provided in the Table 11.1. The components fabricated by AM have several limitations. The forthcoming section discusses the different properties of laser surface-textured additively manufactured components.

TABLE 11.1
Different Types of Laser-Based Texturing and Their Applications

Method	Description	Advantages	Typical Applications
Laser surface texturing	Creates micro-dimples, grooves, or grids using pulsed laser ablation.	High precision, no tool wear, controllable patterns.	Tribology (reducing friction/wear), sealing, lubrication control
Laser-induced periodic surface structures	Uses interference between incident and scattered laser light to create nano-scale ripples.	Nano-structuring, scalable, improves wettability and antibacterial properties.	Biomedical implants, sensors, optics
Direct Laser interference patterning	Interference of multiple laser beams forms periodic patterns (1D, 2D).	Uniform large-area patterning, controlled periodicity.	Functional surfaces, photonics, tribology
Femtosecond/ picosecond laser texturing	Ultra-short pulses minimize heat-affected zones.	High precision, minimal melting, applicable to sensitive materials.	Microelectronics, aerospace, medical devices
Laser shock texturing	High-energy laser pulses induce micro-deformation (shock waves).	Improves hardness, fatigue strength.	Turbine blades, engine components

11.3 APPLICATIONS OF LASER-BASED SURFACE TEXTURING IN ADDITIVE MANUFACTURING

11.3.1 Effect of Laser Surface Texturing on the Bonding Strength of AM Components

Laser surface texturing is used as a post-processing technique by researchers for improving the surface quality and functionality of FDM-fabricated polymer composites. In a recent study, nanosecond fibre laser texturing was applied to FDM-printed PAHT CF15, a high-temperature polyamide reinforced with 15% carbon fibre. The effects of laser power (20–50 W), scanning speed (2000–4000 mm/s), and texturing direction (longitudinal and transverse) on surface roughness, wettability, and moisture absorption were examined. A defocused laser beam induced controlled surface melting (~29 μm depth) without altering the bulk tensile strength (~61 MPa). The optimal condition (20 W, 4000 mm/s) achieved a surface roughness of 0.742 μm, while the most hydrophobic surface (contact angle ≈ 92°) occurred under 30 W and 4000 mm/s. Transverse texturing improved smoothness, whereas longitudinal direction enhanced hydrophobicity. The tensile strength of the samples was not changed during texturing, which confirms that the shallow melted layer did not affect bulk mechanical properties. Further, laser-textured surfaces absorbed less moisture, and this will enable them to be suitable for humid or marine environments. The study demonstrated that nanosecond laser texturing can enhance surface finish and reduce moisture sensitivity in FDM-printed polyamide composites, which could be beneficial for aerospace and marine applications. The authors utilized linear raster textures, and further work could be carried out by using complex texturing patterns such as grids, dimples, and circular motifs to improve the tribological and functional surface behavior of FDM composites. Additionally, laser texturing could be combined with other post-processing methods like surface coatings and plasma treatments to enhance the performance [13].

Researchers are extensively investigating the additive manufacturing (AM) of conductive layers on polymeric substrates for producing conformal antennas, flexible circuits, and integrated sensors. Among various polymers used for substrate, polyether ether ketone (PEEK) is widely investigated due to its thermal stability, fatigue resistance, and chemical inertness. However, PEEK has low surface energy and exhibits poor interfacial adhesion with metallic inks, leading to delamination and peeling of printed conductive layers from the PEEK substrate. To address this limitation, laser surface modification is used as a non-contact technique by researchers for enhancing the interfacial bonding strength between PEEK substrates and printed conductive layers. Liu et al. utilized a 355 nm nanosecond UV laser to treat PEEK surfaces under varying scanning spacings and spot overlap rates for achieving surface morphology and chemistry with improved adhesion performance. The development of periodic microgrooves and textured features on the PEEK surface through laser texturing can increase the surface roughness and enable mechanical interlocking between the textured substrate and deposited silver particles. The optimized laser parameters (scanning spacing of 100 μm and overlap rate of 50%) identified in the study achieved uniform microstructure and it promoted better ink spreading and anchoring during the 3D printing of silver inks on PEEK substrate. X-ray photoelectron spectroscopy (XPS) analysis of the textured PEEK substrate revealed an increase in the oxygen-to-carbon (O/C) ratio indicating the occurrence of partial surface oxidation and the generation of oxygen-containing functional groups such as carbonyl and hydroxyl species during laser texturing. These chemical modifications resulted in the formation of hydrogen bonds at the PEEK–Ag interface and helped in improving the chemical bonding in addition to the physical interlocking. Figure 11.1 shows the bonding mechanism behind the textured polymer sample and the conductive ink. As a result, the interfacial bond strength between the polymer substrate and inkjet-printed silver ink improved significantly. The adhesion force improved from 0.37 MPa for untreated PEEK to 1.9 MPa for laser-treated substrates, and it represents a 413% enhancement. Additionally, laser-modified surfaces exhibited

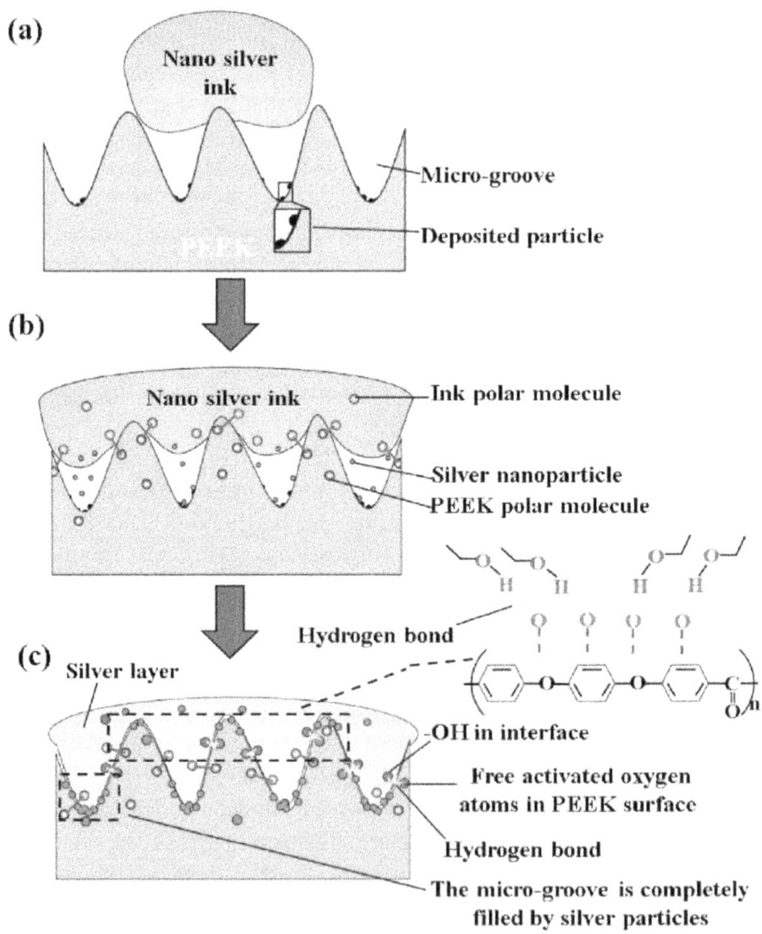

FIGURE 11.1 Bonding mechanism behind textured PEEK sample and conductive ink [14].

better wettability, and the contact angle of the textured surface decreased from 21° to 12–13°, because of the enhancement in surface free energy. The enhancement in the wettability of the textured surface helped in uniform silver nanoparticle distribution within the laser-formed grooves, and improved the bonding. The findings confirm that the laser texturing promoted mechanical bonding through microgrooves and chemical bonding through functional group activation. However, the authors did not evaluate the electrical properties of the silver films and further investigation is required for developing high-frequency or flexible electronic systems through 3D printing [14].

Additionally, Xu et al. introduced a method for enhancing the adhesion of conductive layers on FDM printed PEEK dielectric substrates. Poor interfacial bonding between conductive layers and the FDM-printed dielectric layers reduce device durability and it limits the usage of 3D printing for such application. To address this issue, two surface modification techniques namely concentrated sulfuric acid etching and ultraviolet (UV) laser treatment were investigated by the authors. The proposed post-processing treatments altered the surface morphology, roughness, wettability, and chemical composition, and improved adhesion between conductive layer and 3D printed polymer substrate. The untreated PEEK exhibited a binding force of 0.37 MPa, and it increased to 1.99 MPa after sulfuric acid treatment and 2.21 MPa after UV laser treatment. When both treatments were combined with optimized sequencing of laser texturing followed by acid treatment, a highest binding force of 2.77 MPa was achieved. Improvements in the binding force were attributed to increased surface oxygen content (O/C ratio), hydrophilicity, and surface free energy, and it promotes the conductive ink spreading and mechanical interlocking between dielectric layer and conductive layer. The study demonstrated that controlled surface modification can enhance interlayer adhesion and reliability in additively manufactured electronic devices [15].

Atahan et al. analysed the effect of bio-inspired surface textures developed based on gecko's foot and toe structures (hexagonal, lamellae, and mixed), on the mechanical behavior of single-lap adhesive joints fabricated via LPBF. LPBF was employed to fabricate metal adherends directly with considered designed surface textures, and adhesive joints were prepared and tested as per ASTM D1002–10. Bio-mimicked surfaces exhibited super-hydrophilic behavior and possessed better adhesive wetting and spreading compared to as-printed surfaces. All textured joints showed improved lap shear strength, and the mixed texture achieved maximum improvement due to the combined mechanical interlocking effects of hexagonal and lamellae-like geometries. The as-printed joints failed adhesively, while bio-mimicked joints exhibited mixed (adhesive–cohesive) failure mechanisms [16].

The surface characteristics of polymeric materials produced by AM determine their adhesion and bonding performance, particularly in structural applications where adhesive joints are used. In the case of polyamide 12 (PA12) produced via multi jet fusion (MJF), the as-printed surfaces contain partially fused powder particles and limited surface energy, which can affect the mechanical strength of bonded joints. To address this issue, Wojdat investigated the influence of laser micro-texturing on the surface properties and adhesive joint strength of PA12 polyamide components fabricated via MJF additive manufacturing. By optimizing laser parameters such as power (12–14 W), scan speed (200–300 mm/s), and repetition rate (100–400 kHz, the author was able to produce controlled micro-grooved textures with depths

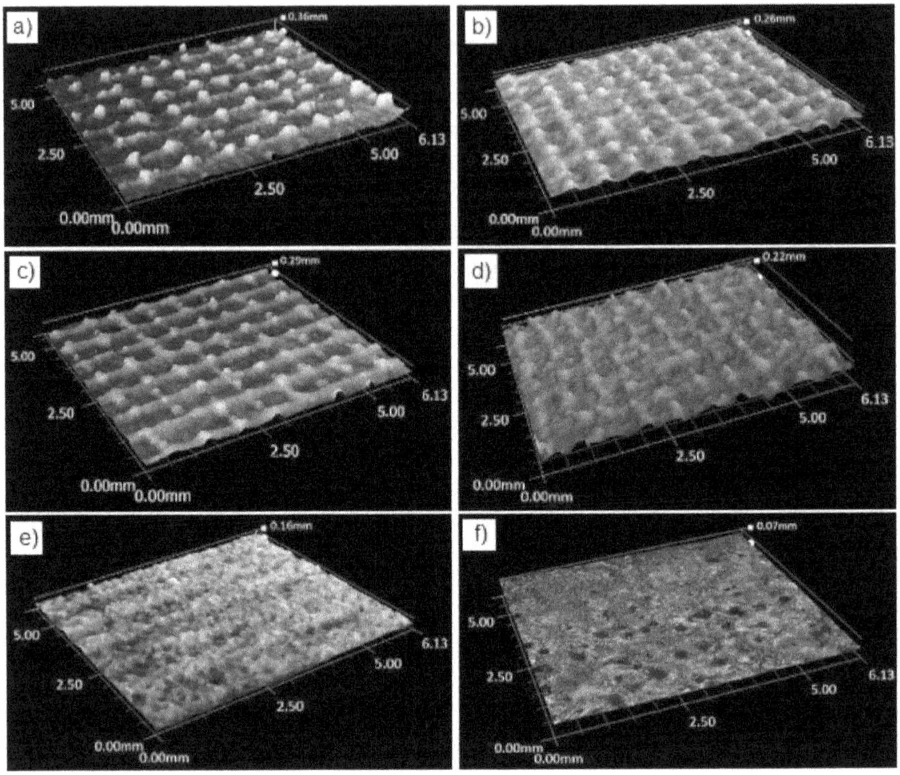

FIGURE 11.2 Three-dimensional surface morphology obtained from optical profilometry for samples subjected to different processing conditions [17]: (a–f) represent the surface topographies of the respective treated specimens, showing changes in peak height, surface roughness, and texture uniformity.

ranging from 150 to 200 μm and alter the surface morphology and energy. The laser micro-texturing removes or melts residual powder particles from the printed surface and generates a well-developed hierarchical structure that promotes mechanical interlocking and improved wettability. Figure 11.2 shows the 3D images of different textures developed on PA 12 samples. A grid-type texture with groove depths of 150–200 μm enhanced surface free energy (SFE) up to 49% (from 37.0 to 55.1 mJ/m²) and improved adhesive joint shear strength (40 MPa), which is double that of untreated samples. The improved SFE and wettability resulted in higher peel strength when cyanoacrylate adhesive was used. However, excessive laser melting and surface smoothing reduced SFE (~27.0 mJ/m²) and bond strength up to 25% [17].

11.3.2 Effect of Laser Surface Texturing on Biological Properties of Additively Manufactured Components

LPBF can produce intricate geometries and porous structures, and it can be used for the development of patient-specific biomedical implants. But the osseointegration, tissue integration, and implant life depend on the cell–material interactions, and it is influenced by the surface topography of the printed implant. Numerous post-processing methods are suggested by researchers for achieving the desired surface topography on additively manufactured implants. Among various surface modification methods, laser surface texturing (LST) is found to enhance osteoblast activity significantly. Sartori et al. studied the biological response of Ti6Al4V ELI (extra low interstitial) substrates fabricated via LPBF by subjecting the printed components to Corundum-sandblasting (SB-C), Zirconia-sandblasting (SB-Z), and Femtosecond LST. Samples were fabricated in XY and XZ orientations to examine the effect of build orientation on cell interaction. Human osteoblasts were cultured directly on the post-processed surfaces to assess morphology, metabolic activity, and differentiation potential. Gene expression (TGF-β1, BMP2, ALPL, COL1A1, SPP1) and protein synthesis (collagen type I, osteonectin, osteocalcin, IL-1, IL-6) were evaluated to understand the biological response of topographically modified surfaces. Micro-structured topographies with enhanced microroughness were developed by laser texturing compared to sandblasting and as-built surfaces. Further, uniform microfeatures required for cellular attachment and guidance were developed by laser texturing, and it was absent in sandblasting. Samples built in XZ-oriented surfaces exhibited slightly enhanced biological performance in comparison to XY orientation. Morphometric analysis revealed that osteoblasts on LST surfaces were smaller and more elongated, with branched morphologies. It indicates the active cellular engagement on the laser-textured surface. Elongation and contact guidance were enhanced because of the reduction in roundness values, and improvement in aspect ratios was increased. Circularity indices obtained in the study suggested increased cytoplasmic protrusions. Further, LST surfaces exhibited higher metabolic activity over time ($p < 0.001$) compared to SB surfaces. Microtopography-induced mechano-transduction pathways were generated on LST surfaces, which help in cell proliferation and differentiation. LST surfaces also maintained stable expression of osteogenic markers (ALPL, COL1A1, SPP1), which indicates the presence of sustained osteoblastic phenotype. As testing duration increases, collagen type I production increased steadily on LST surfaces, and it promotes extracellular matrix deposition. Osteocalcin increased on LST surfaces compared to sandblasted samples, but osteonectin was lower on LST surfaces. Similarly, interleukin-1 and IL-6 secretion was also lower on LST surfaces, suggesting a reduced inflammatory response. These findings confirms that laser texturing promotes cell attachment, metabolic activity, and osteogenic signalling [18].

Hariharan et al. performed surface functionalization of additively manufactured Ti-13Nb-13Zr implants using picosecond (ps) and nanosecond (ns) direct laser interference patterning (DLIP) to improve cell adhesion, proliferation, and osteogenic differentiation. The ns-DLIP generated single-scale line patterns, while ps-DLIP produced multi-scale hierarchical textures with combined micro- and nano-topographies. Both ns- and ps-DLIP textures enhanced human bone marrow stromal cell (hBMSC) adhesion and proliferation compared to as-built samples. But the Ps-DLIP achieved faster and large-area cell spreading due to nano-topographies, while ns-DLIP helped in cytoskeletal stretching along grooves. Higher alkaline phosphatase (TNAP) activity and osteoblast differentiation were observed on ns-DLIP because of the presence of surface hydroxyl groups and improved wettability. Groove width, periodicity, and nano topographies guided cell alignment and focal adhesion formation. Cell behavior was influenced by surface roughness, energy, and the chemical state of the textured surface. The investigation concluded that Ps-DLIP was suitable for rapid cell coverage and osseointegration, while ns-DLIP helped to enhance osteogenic differentiation and corrosion resistance [19].

Villapun et al. investigated the effect of the bioinspired laser surface texturing on the physicochemical and biological properties of LPBF-fabricated Ti-6Al-4V (Ti64) substrates. The micro- and nano-patterns (including grid, dual, channel, triangle, and shark-skin textures) developed by LST altered surface roughness, kurtosis, and time-dependent contact angle behavior. All the textured surfaces exhibited improved early-stage cell adhesion, and the cell adhesion with micro-patterns was better than nano-textures. Triangle and shark-skin textures supported higher DNA content and achieved better proliferation. Channel, grid, and dual textures demonstrated higher osteogenic differentiation. Long-term mineralization (21 days) was enhanced in grid, dual, triangle, and shark-skin textures, and it indicates improved osteoconductivity and bone-forming potential. A combination of roughness kurtosis and glycerol contact angle accounted for 82% of the variation in long-term mineralization [20].

The antibacterial performance of biomaterial surfaces depends on the nanoscale topography and surface wettability, because it governs the bacterial adhesion, colonization, and biofilm formation. Among various laser-based surface engineering techniques, laser-induced periodic surface structures (LIPSSs) are widely investigated by researchers for inducing nanoscale ripples on biomaterials so that bacterial behavior can be modulated through mechanical and physicochemical interactions. These periodic nanostructures with sub-micron spacing can damage bacterial cell membranes or alter surface energy and wettability, and minimize bacterial attachment. Recent developments in metal AM methods, especially laser powder bed fusion (LPBF or PBF-LB), enabled the fabrication of metallic biomedical components such as implants and surgical tools through AM. However, the as-built LPBF parts are rough and porous, and they promote bacterial adhesion, posing a challenge for medical applications. To address this issue, researchers have proposed a hybrid surface modification approach by integrating LPBF and LIPSS formation. In this approach, Ti6Al4V components fabricated using LPBF were irradiated by femtosecond or picosecond lasers at varying wavelengths (1064, 532, 355, and 266 nm). The laser treatment generates ordered LIPSS patterns on the complex geometries and irregular surfaces of LPBF substrates without affecting their functional properties. These multiscale surfaces comprising microscale LPBF roughness and nanoscale LIPSS textures can exhibit enhanced antibacterial functionality. Experimental studies revealed that PBF-LB + LIPSS-treated Ti and Ti6Al4V surfaces effectively inhibited *Escherichia coli* and *Staphylococcus aureus* adhesion, compared to either polished or pure LIPSS surfaces. The improvement was attributed to the interplay between the micro-porous LPBF morphology, residual powder particles, and the periodic nanostructures generated by laser irradiation. This multilevel texture generates a surface environment unfavourable for bacterial colonization through (1) physical membrane rupture, (2) limiting bacterial growing area and (3) modification of local surface energy and wettability. The combination of increased surface roughness, altered surface energy, and nanostructure-induced mechanical stress enhances an antibacterial efficacy which cannot be achieved through either LPBF or LIPSS alone. Hence next generation implantable devices, surgical tools, and contact killing bio-materials with tunable antibacterial properties can be produced through hybrid approach [21].

11.3.3 Development of a Hydrophobic Surface on AM Components through LST

Hydrophobic and superhydrophobic surfaces find application in anti-corrosion coating, anti-icing, self-cleaning, anti-fouling surface, drag reduction, oil water separation, printing, and electronic and optical protection. But it is difficult to produce such a surface directly through AM. Xu et al. reported the fabrication of superhydrophobic surfaces on selective-laser-melted Ti–6Al–4V (TC4) alloy by combining LST, micro-arc oxidation (MAO), and chemical modification with perfluorooctyltrichlorosilane (PFOCTS). The developed surface exhibited super hydrophobicity with a water contact angle (WCA) of 156.8°. The MAO coating produced a dense, porous oxide layer, and it helped in enhancing adhesion, wear resistance, and protected the micro-structured surface during the wear test. Under abrasion tests using 600# sandpaper at 2.9 kPa over 1020 cm, the superhydrophobic surface retained its hydrophobic behavior, and it indicates the mechanical durability of the proposed approach. Wear test revealed a lower and more stable coefficient of friction (COF) and narrower wear tracks compared to untreated or LST-only surfaces. The abrasion mechanism observed on the textured surface is shown in Figure 11.3. The study confirmed that the MAO layer delayed the Cassie–Wenzel transition and maintained the surface hydrophobicity under mechanical stress. The combined LST–MAO–chemical modification approach was found to develop wear-resistant, superhydrophobic coatings that can be used in harsh industrial environments [22].

Huang et al. utilized chemical immersion treatment (CIT) and hybrid surface nano-structuring (HSN), comprising laser surface texturing and chemical treatment to achieve superhydrophobic surfaces on AlSi10Mg and Ti6Al4V produced by laser powder bed fusion (L-PBF). The CIT approach imparted a rose petal effect on the metal surface, and it was characterized by high water contact angles (θw) and high-water adhesion. The HSN approach achieved the lotus leaf effect with high water contact angles and low water adhesion. As-built LPBF parts were hydrophilic but became more hydrophobic (θw increases by 5–45°) after a month of air exposure

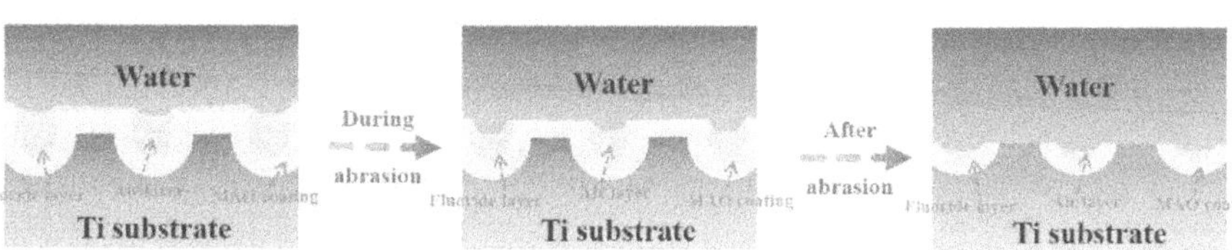

FIGURE 11.3 Abrasion mechanism observed on textured titanium substrate [22].

due to organic molecule adsorption. Rose petal surfaces created by CIT have smaller Spc (~2/μm) and low feature curvature ($r < 4$), while the HSN approach induces lotus surfaces that exhibit larger Spc (>4/μm) with sharp microstructures. CIT-treated surfaces have higher water adhesion due to lower fluorine functionalization. But the fluorosilane reagents reacted with laser-textured surfaces during HSN, and it enhanced fluorocarbon coverage and lowered surface energy. As a result, water adhesion on HSN-treated surfaces was reduced. Hence, the combined surface chemistry and laser-induced topography approach can be used to modify the wettability on LPBF-built metal samples [23].

Kusat et al. utilized nanosecond direct laser interference patterning (ns-DLIP) for surface modification of AM Ti-6Al-4V and Scalmalloy® (Al-Mg-Sc) alloys. The ns-DLIP reduced the surface roughness of Ti-6Al-4V from 94.1 ± 23.3 μm to 67.7 ± 7.2 μm (≈28% reduction), while Scalmalloy® exhibited a more pronounced reduction from 83.1 ± 21.8 μm to 47.3 ± 3.6 μm (≈43% reduction). DLIP enabled the fabrication of periodic line-like micro-patterns with structure depths of up to 3.9 ± 0.6 μm (Ti-6Al-4V) and 11.4 ± 2.8 μm (Scalmalloy®). The surface texturing significantly altered wettability characteristics from hydrophilic to hydrophobic. For Ti-6Al-4V, the static water contact angle increased from 25.9° to 93.4°, while for Scalmalloy®, it increased from 16.3° to 131.4°. The study demonstrates that ns-DLIP can reduce roughness, improve uniformity, and introduce controlled microstructures on additively manufactured metallic alloys. By enabling tailored wettability and topographical control, DLIP can enhance the applications of LPBF-built parts in biomedical applications (enhanced biocompatibility) and aerospace components [24].

11.3.4 Effect of Laser Surface Texturing on Mechanical, Tribological, and Corrosion Properties

The mechanical, tribological, and corrosion properties of AM components depend on the printing conditions and the resulting surface characteristics. FDM-printed iron-polylactic acid (Ir-PLA) composites are susceptible to corrosion, and their limited mechanical properties limits their usage in various applications. To address these challenges, Rezayat et al. utilized LST as a post-processing technique for tailoring the surface morphology, wettability, and mechanical properties of 3D-printed Ir-PLA composite. Using a design of experiments (DoE) approach, the study analysed the effects of laser power and scanning speed on surface roughness, contact angle, tensile properties, and corrosion resistance. LST generated controlled surface patterns and microstructures, and it results in decreased roughness and better wettability. Surface roughness (Ra) decreased with increasing laser power and scanning speed, and the optimal conditions were identified at 50 W and 3.6 mm/s. Contact angle measurements revealed that laser power negatively influenced wettability while scanning speed had a positive effect, and an optimal contact angle of ~47–50° was achieved, which enhanced hydrophobicity and corrosion resistance. Further, LST-treated Ir-PLA samples exhibited improved tensile strength (up to 25%) and ductility (15%). The observed results were attributed to surface-induced microstructural modifications and the formation of patterned textures that enhanced load distribution and deformation resistance. Corrosion tests in saline environments demonstrated a reduction in degradation. Laser-treated surfaces exhibited minimal rust formation, while untreated samples showed severe corrosion. The usage of laser parameters 60 W and 6 mm/s provided maximum corrosion resistance and mechanical performance. Hence, it is concluded that LST can serve as a non-contact post-processing technique for improving the mechanical reliability, surface integrity, and corrosion resistance of 3D-printed Ir-PLA [25].

The corrosion and stress corrosion cracking (SCC) resistance is an important property for additively manufactured metallic materials, which will be exposed to aggressive environments such as marine or chloride-rich industrial systems. Femtosecond laser texturing (FLT) is a promising surface engineering approach that could be utilized to enhance the durability of AM stainless by creating hierarchical micro/nano-scale surface architectures. In a recent investigation by Xiong et al., FLT was applied to binder jet printed (BJP) 316L stainless steel in both its as-printed and hot isostatic pressed (HIP) conditions to evaluate its impact on wetting, corrosion, and SCC behaviors. Hierarchical micro/nano-textures with low-, medium-, and high-depth structures were generated through 200, 400, and 800 laser passes. These laser parameters generated crossed-line micro-protrusion arrays decorated with LIPSSs, yielding well-defined hierarchical morphologies. Surface wettability of the developed structure was dependent on texture depth. The 400- and 800-pass conditions induced Cassie–Baxter-type superhydrophobicity with static contact angles above 150° and contact angle hysteresis below 10°, because of which the liquid adhesion and penetration were reduced. The 200-pass condition produced a high-adhesion, "rose-petal"-type hydrophobic regime. Superhydrophobic surface was created on BJP 316L, but only a hydrophobic surface was generated on HIP 316L. Electrochemical tests provided the influence of FLT on localized corrosion. Cyclic potentiodynamic polarization results showed that FLT improved the localized corrosion resistance of as-printed BJP 316L, but FLT diminished the corrosion resistance of HIP-treated 316L sample. The deterioration in HIP samples was attributed to their smoother surfaces and reduced porosity, because of which the aggressive ion is trapped at the deep textures, and it suppresses the benefits offered by developed hydrophobicity. Importantly, corrosion rates in both material states were found to increase with surface roughness, following the order 200 < 400 < 800 laser passes. In the untreated condition, porous BJP 316L exhibited higher SCC susceptibility due to high defect density and δ-ferrite formation, and it promotes galvanic coupling. However, medium-depth FLT (400 passes) reduced SCC initiation and propagation, and no visible cracks were detected on the surface. Post-FLT, the SCC resistance of BJP 316L and HIP-treated becomes closer. For HIP samples, FLT neither enhanced nor degraded SCC resistance, and it indicates a neutral effect. The underlying mechanisms for this behavior are discussed further. The superhydrophobic surface created by FLT reduces the residence time and contact area of corrosive electrolytes and limits chloride ion penetration and passive film breakdown. The hierarchical micro/nanostructure enhances mechanical barrier effects, reduces crack initiation sites, and distributes stresses evenly. However, excessive roughness (as in 800-pass textures) locally concentrates stress and increases corrosion rates. The optimized medium-depth FLT condition (400 passes) identified in the investigation achieved a balance between superhydrophobicity, corrosion protection, and SCC resistance. Further, the work demonstrates that surface engineering at micro- and nanoscale levels can mitigate drawbacks associated with porous AM stainless steels components. Since FLT is a maskless, single-step post-processing technique, it can be easily integrated with AM to modify the substrate surface without altering bulk microstructure or introducing foreign elements. The process simultaneously imparts functional superhydrophobicity, corrosion mitigation, and SCC resistance enhancement, which is required for marine, offshore, desalination, and chemical processing components exposed to chloride-rich environments [26].

11.3.5 Effect of Laser Remelting on Mechanical, Tribological, and Corrosion Properties

LPBF of NiTi intermetallics often results in poor surface finish and near-surface porosity, limiting their mechanical performance and practical application. To address these challenges, Obeidi et al. investigated laser polishing as a non-contact post-processing method for LPBF-fabricated NiTi parts. Laser polishing creates localized remelting and redistributes material from surface peaks to valleys with minimal material loss. Because of this, the surface gets smoothened and pores get sealed and connected. The authors evaluated the performance of continuous-wave CO_2 and femtosecond lasers. The CO_2 laser-based polishing achieved a 91% reduction in surface roughness (from 38 μm to 3 μm) at a power of 75 W, scanning speed of 11 mm/s, and energy density of 34 J/mm^2. XRD analysis indicated a shift in phase transformation temperatures, but the austenitic crystal structure was retained on the surface. In contrast, femtosecond laser polishing caused Ni evaporation and TiO_2 formation, leading to limited improvement in surface quality [27].

Hot cracking in the centres of deep molten pools is a major limitation in LPBF of nickel-based superalloys such as Haynes 230. Hodgir et al. employed an in situ remelting strategy to eliminate cracks and improve the mechanical properties of LPBF-built Haynes 230 alloy. By controlling the remelting depth to 70–80% of the original layer, liquid metal backfilled cracks reduced crack density up to 80% and produced crack-free specimens. The process also alleviated elemental segregation at grain boundaries (W, Cr, Mo, C, O) and promoted the precipitation of high-melting-point M23C6 carbides, and reduced residual stresses and crack propagation. Laser remelting modified the dendrite structure, increased grain boundary serration, and enhanced grain coarsening, because of which residual stresses were reduced. Further, the remelted samples exhibited improvements in strength and plasticity, comparable to ceramic-particle-reinforced Haynes 230, due to increased dislocation density and favourable grain orientation and enhanced dislocation strengthening [28].

Laser-based directed energy deposition (L-DED) enables the fabrication of complex stainless steel components at high deposition rates but often results in surface defects and porosity, and affects their performance. Hodgir et al. used laser remelting to enhance surface integrity, wear resistance, and corrosion performance of L-DED 316L stainless steel. By integrating periodic remelting during L-DED of SS316L, Hodgir et al. achieved a deposition rate of 25 g/min, which is twice the deposition rate achievable with conventional L-DED. Laser remelting eliminated solute segregation and reduced δ-ferrite formation along grain boundaries. The optimal results were obtained at high laser power (2000 W) and low scan speed (400 mm/min), which increased melt pool depth and reduced porosity. Lower power and higher scan speeds promoted finer grains and higher hardness. The mechanism behind pore mitigation is shown in Figure 11.4 [29]. The remelting process promotes grain refinement through rapid solidification, leading to a 34% increase in microhardness compared to the as-deposited condition. The specific wear rate of the remelted surface was decreased by 66%. Electrochemical impedance spectroscopy tests in 3.5% NaCl solution further confirmed the improvement in corrosion resistance after laser remelting, and it was attributed to the refined microstructure and smoother surface. Parametric analysis indicated that lower laser power and higher scan speed favour reduced wear and higher hardness due to higher cooling rates [30].

Thanumoorthy et al. aimed to improve the surface quality of laser powder directed energy deposited (LP-DED) Inconel 718 (IN718) components through laser surface remelting (LSR). Owing to the rough surface and poor machinability of DED-fabricated IN718, conventional post-processing methods are not efficient. With the help of the same deposition system, the authors examined the influence of remelting parameters on surface morphology, roughness, microhardness, and residual stress. The results demonstrated that LSR can produce a remelted zone with refined grains and improve the microhardness. The remelting process melts surface asperities and reduces surface roughness up to 80%. Residual stress was evaluated through X-ray diffraction, and it was also significantly reduced after remelting. A regression analysis revealed that laser power was the most influential parameter and positively affected surface finish and depth of the remelted zone. But the laser

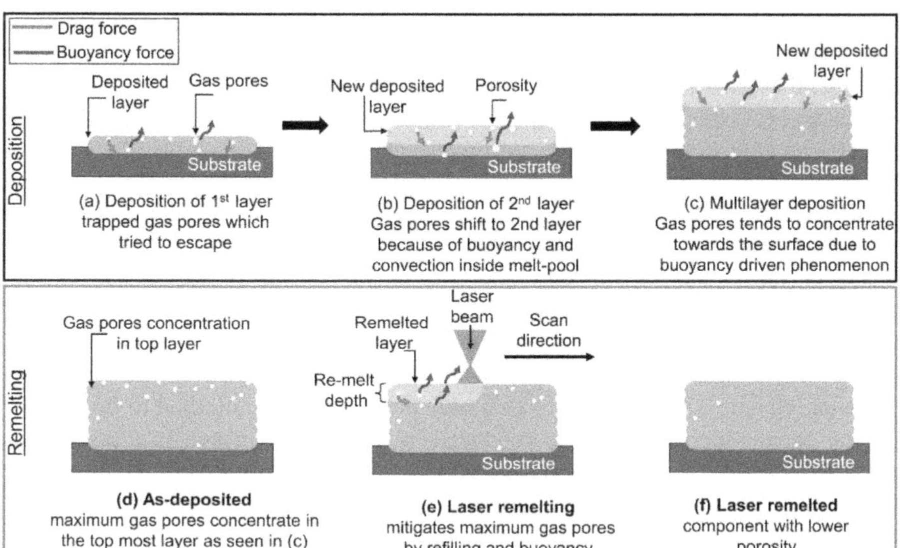

FIGURE 11.4 Effect of laser remelting on pore mitigation [29].

inversely influenced residual stress and hardness. Optimal surface conditions were identified as laser power of 150 W and scan speed of 1140 mm/min to achieve a balance between surface smoothness, microhardness, and residual stress [31].

11.4 SUMMARY AND CONCLUSION

The microscale grooves, dimples, LIPSS/nano-structures, and remelted tracks developed by laser surface texturing change the surface topography, oxides, and near-surface microstructure of the additively manufactured parts. These effects can influence the corrosion, wear, bonding, biological characteristics, and mechanical behavior of the additively manufactured components positively.

LST can improve the corrosion resistance of AM components by (1) closing open surface pores developed in AM through localized remelting, (2) producing a uniform passive oxide layer by modifying surface chemistry, and (3) enabling the proper adherence of protective coatings. However, usage of improper laser texturing parameters (excessive energy) can create microcracks or brittle oxides on the additively manufactured components and reduce the corrosion resistance.

The surface textures developed by LST on AM components improve the wear resistance by (1) trapping debris in dimples, (2) acting as lubricant reservoirs, (3) creating micro-hydrodynamic effects, and (4) hardening the near-surface through rapid solidification. The textured pattern geometry, like dimple size, spacing, depth, and orientation relative to the building direction and wear testing sliding direction, determines the mass loss during the wear test. The production of wrong patterns on AM components may concentrate stress and increase the wear rate. Overall, tribology studies on LPBF and other AM metallic components showed improvements in wear life after LST [32]. Laser texturing increases the mechanical interlocking and improves the bonding shear strength by improving the wettability for coatings/adhesives. In the case of additively manufactured components, LST removes loosely bound powder particles and creates a defect-free metallurgical surface for improving diffusion bonding or coating adhesion. However, excessive surface melting or development of recast layers during LST can reduce adhesion [33]. For the additively manufactured implants (Ti, stainless steel, porous scaffolds), LST is an efficient method for modifying the biological responses. The micro- and nano-textures developed on the implant can promote osteoblast adhesion and osseointegration. Specific patterns produced by LST can reduce bacterial colonization, and improve the antibacterial properties. By controlling the roughness of implants, tissue ingrowth in porous implants can be tailored according to the requirement. The development of specific textures through LST enables researchers to adjust wettability and protein adsorption to get desirable cell behavior. Hence, LST can act as a desirable post-processing method to functionalize the patient-specific implants produced by AM [34]. Further, it is possible to perform texturing or remelting during additive manufacturing itself. However, the number of research works available on LST of AM components is minimal and extensive research can be performed to assess the benefits and limitations associated with LST of AM components.

REFERENCES

[1] Sathies, T., & Senthil, P. (2021). A review on milestones achieved in the additive manufacturing of functional components. Advanced Materials and Manufacturing Processes, 135–154.

[2] Chandra, S., Radhakrishnan, J., Huang, S., Wei, S., & Ramamurty, U. (2025). Solidification in metal additive manufacturing: Challenges, solutions, and opportunities. Progress in Materials Science, 148, 101361.

[3] Sathies, T., Senthil, P., & Anoop, M. S. (2020). A review on advancements in applications of fused deposition modelling process. Rapid Prototyping Journal, 26(4), 669–687.

[4] Raghuraman, V. (2024). Influence of in-situ process parameters, post heat treatment effects on microstructure and defects of additively manufactured maraging steel by laser powder bed fusion: A comprehensive review. Physica Scripta, 99(5), 052001.

[5] Prabu, G., Yang, C. H., Alnaser, I. A., & Jeyaprakash, N. (2025). Analysis of tribological behavior of Laser Powder Bed Fusion (LPBF) CuCrZr alloy through reciprocating loading on the columnar structure. Tribology International, 202, 110301.

[6] Prabu, G., Jeyaprakash, N., Yang, C. H., & Alnaser, I. A. (2024). Evaluation of friction and wear depth during the cyclic loading on partly melted LPBF particles of LPBF Cu alloy. Materials Characterization, 217, 114386.

[7] Liu, M., Kumar, A., Bukkapatnam, S., & Kuttolamadom, M. (2021). A review of the anomalies in directed energy deposition (DED) processes & potential solutions-part quality & defects. Procedia Manufacturing, 53, 507–518.

[8] Yu, K., Zhang, Y., Cai, J., Zhang, B., & Qu, X. (2025). Research progress on densification behavior of binder jetting additive manufacturing. Review of Materials Research, 100073.

[9] Wickramasinghe, S., Do, T., & Tran, P. (2020). FDM-based 3D printing of polymer and associated composite: A review on mechanical properties, defects and treatments. Polymers, 12(7), 1529.

[10] Dhakal, N., Wang, X., Espejo, C., Morina, A., & Emami, N. (2023). Impact of processing defects on microstructure, surface quality, and tribological performance in 3D printed polymers. Journal of Materials Research and Technology, 23, 1252–1272.

[11] Lei, P., Zhang, P., Song, S., Liu, Z., Yan, H., Sun, T., . . . & Shi, H. (2024). Research status of laser surface texturing on tribological and wetting properties of materials: A review. Optik, 298, 171581.

[12] Ince, F. D., & Özel, T. (2025). Laser surface texturing of materials for surface functionalization: A holistic review. Surface and Coatings Technology, 131818.

[13] Prasad, S., & Prakash, S. (2025). Experimental investigation of laser texturing on surface roughness and wettability of PAHT CF15 fabricated by fused deposition modeling. Polymer Composites, 46(7), 6443–6457.

[14] Liu, Q., Wei, D., Lv, J., Wang, Z., Xu, Z., Yang, P., . . . & Li, C. (2024). Enhancing interfacial bond strength between PEEK and inkjet-printed silver film through laser surface modification for additive manufacturing of electronics. Journal of Materials Research and Technology, 30, 6724–6736.

[15] Xu, Z., Hui, J., Lv, J., Wei, D., Yan, Z., Zhang, H., & Wang, J. (2024). An investigation of methods to enhance adhesion of conductive layer and dielectric substrate for additive manufacturing of electronics. Scientific Reports, 14(1), 10351.

[16] Atahan, M. G., Maskery, I., Ashcroft, I., Apalak, M. K., & Pappas, A. (2025). Effect of bio-mimicked surface texturing on the shear strength of additively manufactured metal single-lap joints: An innovative approach. Engineering Failure Analysis, 174, 109460.

[17] Wojdat, T. (2025). Effect of laser micro-texturing MJF 3D printed polymers on surface properties and mechanical strength of adhesive joints. Precision Engineering, 94, 236–250.
[18] Sartori, M., Bregoli, C., Carniato, M., Cavazza, L., Maglio, M., Giavaresi, G., . . . & Fini, M. (2025). Biological characterization of Ti6Al4V additively manufactured surfaces: Comparison between ultrashort laser texturing and conventional post-processing. Advanced Healthcare Materials, 14(4), 2402873.
[19] Hariharan, A., Goldberg, P., Schell, F., Hempel, U., Striggow, F., Hantusch, M., . . . & Gebert, A. (2024). Single-and multiscale laser patterning of 3D printed biomedical titanium alloy: Toward an enhanced adhesion and early differentiation of human bone marrow stromal cells. Advanced Functional Materials, 34(8), 2310607.
[20] Villapun, V. M., Man, K., Carter, L., Penchev, P., Dimov, S., & Cox, S. (2023). Laser texturing of additively manufactured implants: A tool to programme biological response. Biomaterials Advances, 153, 213574.
[21] Chen, L., Kanetaka, H., Furuya, M., Yokota, K., Ueda, K., Takano, N., & Mizutani, M. (2025). Laser-beam powder bed fusion followed by laser-induced periodic surface structuring leads to antibacterial properties. Materials & Design, 114078.
[22] Xu, G., Liu, P., Liu, A., Zhang, C., Zhang, G., Wu, Q., & Lu, H. (2024). Mechanically stable superhydrophobic surfaces constructed by laser surface texturing and micro-arc oxidation of TC4 alloy prepared based on SLM. Materials Chemistry and Physics, 325, 129791.
[23] Huang, W., Nelson, B., Tian, S., Ordikhani-Seyedlar, R., Auyeung, R. C., Samanta, A., . . . & Ding, H. (2022). Superhydrophobic surface processing for metal 3D printed parts. Applied Materials Today, 29, 101630.
[24] Kuisat, F., Ränke, F., Lasagni, F., & Lasagni, A. F. (2021). Simultaneous micro-structuring and surface smoothing of additive manufactured parts using DLIP technique and its influence on the wetting behavior. Materials, 14(10), 2563.
[25] Rezayat, M., Karamimoghadam, M., Dezaki, M. L., Zolfagharian, A., Casalino, G., Mateo, A., & Bodaghi, M. (2024). Enhancing corrosion resistance and mechanical strength of 3D-printed iron polylactic acid for marine applications via laser surface texturing. Advanced Engineering Materials, 2402591.
[26] Xiong, Y., Zhou, S., Zhao, T., Shi, W., Suzuki, A., Kobashi, M., & Ramamurty, U. (2025). Effects of femtosecond laser texturing on the corrosion and stress corrosion cracking resistances of the 316L austenitic stainless steel additively manufactured by binder jet printing. Acta Materialia, 121615.
[27] Obeidi, M. A., Al-Hamaoy, A. R., Cholkar, A., Agarwal, N., & Brabazon, D. (2024). Comparing the surface characteristics of additively manufactured nitinol parts polished by femto-second and CO2 laser. Applied Surface Science Advances, 24, 100637.
[28] Xi, X., Lin, D., Song, X., Wei, H., Yan, M., Luo, W., . . . & Minami, F. (2025). In-situ remelting induced healing of cracks and strength-ductility synergy in additively manufactured Haynes 230 alloy. Additive Manufacturing, 98, 104638.
[29] Hodgir, R., Singh, R., & Mujumdar, S. (2025). Simultaneous enhancement of build rate and microstructure in SS316L additive manufacturing through in-situ laser remelting. Journal of Manufacturing Processes, 133, 865–878.
[30] Hodgir, R., Singh, R. K., & Mujumdar, S. (2024). Improved wear and corrosion resistance of additively manufactured SS316L by laser remelting process. Manufacturing Letters, 41, 938–946.
[31] Thanumoorthy, R. S., Jadhav, S. V., Oyyaravelu, R., Bontha, S., & Ass, B. (2024). Effect of surface remelting on the characteristics of IN718 components fabricated using laser powder directed energy deposition. Engineering Research Express, 6(3), 035548.

[32] Saravana Kumar, M., Jeyaprakash, N., & Yang, C. H. (2024). Influence of re-melting rate on the dry sliding wear characteristics of LPBFed Cu–Cr–Zr alloy. Proceedings of the Institution of Mechanical Engineers, Part E: Journal of Process Mechanical Engineering, 09544089241289748.
[33] Stribick, S., & Alhadad, B. (2025). Laser surface texturing for surface preparation of adhesive bonding of polypropylene. The International Journal of Advanced Manufacturing Technology, 1–16.
[34] Shabir, S., Khan, N. Z., & Siddiquee, A. N. (2025). Laser-induced surface modifications in biomedical materials: A comprehensive review. Surface Topography: Metrology and Properties, 13(2), 023002.

12 Laser Surface Texturing for Biomaterial Applications

Kannan Ganesa Balamurugan, K. Kamalakannan, and Jeyaprakash Natarajan

12.1 INTRODUCTION

In recent times, studies into the biological interface of materials have been greatly on the rise as a result of growing interdisciplinary demand in biomaterials and regenerative medicine [1,2]. Biomaterials are materials designed to interact with biological systems for therapeutic or diagnostic interventions [3]. These developments have facilitated applications in tissue engineering, controlled drug delivery, implantable medical devices, and biosensing technologies [4,5], driving innovation in contemporary healthcare solutions. Thus, the function of biomaterials has become a vital facilitator for the replacement and regeneration of injured tissues and organs [6]. This field of research combines multidisciplinary disciplines like material science, chemistry, biology, and mechanical engineering [7].

Biomaterials can either be natural or be synthetic, and the most important functional need of biomaterials is biocompatibility with the ability to safely interact with living tissue [3,8]. As indicated in Figure 12.1, biomaterials can be classified broadly into metallic, ceramic, polymeric, and composite categories [9]. Stainless steel and titanium alloys are the metals of choice for load-bearing applications because they have a high strength-to-weight ratio and mechanical stability [10]. Ceramics have better corrosion and wear resistance, while biopolymers like chitosan and alginate possess biodegradability and flexibility for soft tissue engineering [11]. Composite materials with various phases provide customized structural and biofunctional characteristics [12].

For successful biomedical implant performance, surface interaction with biological tissues becomes more important than bulk characteristics [13]. Thus, surface modification methods—like thin film coatings, plasma processing, self-assembled monolayers, laser patterning through additive manufacturing-based approaches, and laser surface treatment—are extensively utilized to customize interfacial reactions [14,15]. The development of laser-based fabrication, for instance, laser powder bed fusion (LPBF), has shown immense control over microstructural fining and wear properties in alloys like Inconel-718 and CuCrZr [16,17], pointing to the significance of laser–material interaction physics, which is the basis for laser surface texturing (LST) on biomaterials (Figure 12.2).

 DOI:10.1201/9781003542087-12

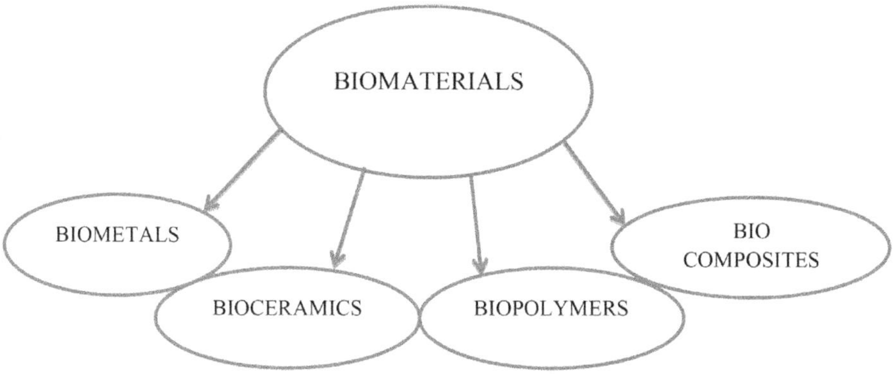

FIGURE 12.1 Biomaterials classification.

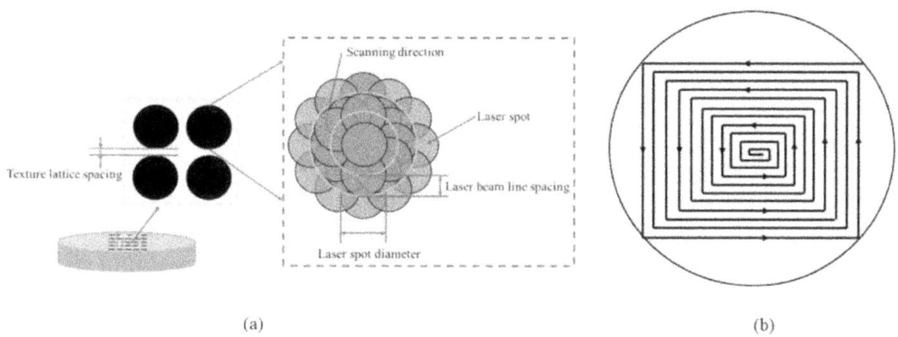

FIGURE 12.2 (a) Laser-surface-textured surface, (b) texture hatch design.

TABLE 12.1
Applications and Benefits of LST

Application	Benefits
Biomedical implants	Enhanced biocompatibility and tissue integration
Automotive components	Reduced friction and wear resistance
Aerospace parts	Improved fatigue resistance and durability
Food processing equipment	Enhanced hygiene and easy cleaning
Marine applications	Corrosion resistance and surface protection
Manufacturing tools	Increased tool lifespan and performance

Table 12.1 presents the applications and benefits of Laser Surface Texturing (LST) across various industries. Among the several laser modification techniques, LST is a highly effective approach owing to its ability to produce micro/nano-scale features like dimples, grooves, and laser-induced periodic surface structures (LIPSS),

which greatly promote surface wettability, protein adsorption, and cellular response [18–20]. In contrast with chemical etching and sandblasting, LST facilitates greater reproducibility, contamination-free processing, and adjustable texture morphology [19]. Thus, this chapter critically discusses the influence of LST on different biometallic materials and its significance in biological performance, corrosion properties, tribological characteristics, and clinical conversion [20].

12.2 LASER SURFACE TEXTURING ON BIO-METALS

12.2.1 LST on Titanium Alloys

Ti-6Al-4V is a familiar titanium alloy known for its strength-to-weight ratio. It has been gaining much attention in critical engineering applications. It is unique because of the combination of its mechanical properties, corrosion resistance, and biocompatibility, which makes it a good material for aerospace components, marine environments, and medical implants. Adopting optimized LST on the Ti6Al4V alloys significantly improves their tribological characteristic. Presence of micro-textures entraps the lubrication in their cavities to form films, which reduce the contact between the metallic surfaces. In LST, the laser beam interacts with the substrate by the interaction of photons with the electrons of the substrate, which causes the heat generation due to the electronic transitions. The generated heat energy results in ablation of the material surface. The high energy density and pulse length are the significant factors governing the ablation. These modifications influence biocompatibility through three major routes. Topographical guidance, wettability changes, and antibacterial effects are those major routes. In topographical guidance, the cell adhesion and spreading could be achieved by mimicking the extracellular matrix by the microgrooves and dimples. Wettability changes promote hydrophilic surfaces, which facilitates the protein adsorption. Creation of LIPSSs by LST disrupts the bacterial membranes and reducing the pre implant infection risk.

12.2.1.1 Effects on Osteoblast Behavior and Osseointegration

Initial osseointegration begins with protein adsorption, and then adhesion of osteoblast. LST creates microgrooves, pits, and hierarchical micro–nano textures, which serve as sites of anchorage, enhancing cell attachment. Osteoblasts and BMSCs grown on laser-textured Ti-6Al-4V have greater areas of spreading, increased focal adhesion sites, and improved cytoskeleton organization when compared to polished or grit-blasted ones. Wu et al. [21] demonstrated that femtosecond laser-fabricated microtopographies facilitated strong adhesion and spreading of BMSCs, ensuring cytocompatibility.

Osteoblast proliferation and differentiation over a period are essential for de novo bone formation. Micron-scale grooves and hierarchical micro/nanostructures generated by LST promote greater osteoblast proliferation than Ti-6Al-4V with smooth surface. Kedia et al. [22] showed picosecond laser-induced hybrid grooves to promote enhanced extracellular matrix deposition and mineralization with greater expression of osteogenic markers like alkaline phosphatase (ALP) and osteocalcin.

Anisotropic microstructures like laser-generated grooves direct osteoblast orientation and extracellular matrix (ECM) assembly. Such alignment replicates natural bone lamellae, enhancing the strength of bone–implant interfaces and stress transfer throughout the junction. Biomimetic ECM deposition under such conditions improves implant stability.

Early osseointegration could be achieved in animal models by LST. Increased removal torque values and greater BIC percentages are usually found at 2–4 weeks after implantation. A recent review by Kolarovszki et al. [23] pointed out that laser-modified Ti dental implants exhibited higher early-stage BIC and more rapid osseointegration in preclinical models, while maintaining the necessity of soft-tissue integration to provide bacterial barrier function.

Substrate surface could be tailored to superhydrophilic to enhance the adsorption of proteins like fibronectin. While some surfaces become hydrophobically recovered over time, this is reversible by storage in aqueous conditions or chemical functionalization.

12.2.2 LST on 316L-Stainless Steel

Austenitic stainless steel 316L is still one of the most common biomaterials used in surgical instruments, orthopedic implants, and cardiovascular devices because it is not only very good mechanically, corrosion-resistant, and cost effective [24], but also relatively bioinert, by which it restricts tissue integration and may allow bacterial colonization [25]. To mitigate such problems, LST has proven to be a capable, non-contact method of designing surface topography, chemistry, and wettability at micro- and nano-metric scales without the introduction of extrinsic coatings [26].

LST allows for accurate and reproducible alteration of 316L surfaces, promoting cellular adhesion, osseointegration, and antibacterial performance with the substrate's mechanical integrity [27]. In contrast to typical surface treatments, including sandblasting, etching, or coatings, LST provides better spatial control and reproducibility [28].

12.2.2.1 LST Process Parameters

LST can be conducted with femtosecond (fs), picosecond (ps), or nanosecond (ns) pulsed lasers, each with different micro/nano features. Pulse duration, fluence, wavelength, and scanning velocity have significant effects on the resultant morphology and oxide layer properties [29].

Femtosecond lasers (100–300 fs, $\lambda \approx 800$–1030 nm) create LIPSSs or ripples, with negligible thermal influence. Highly ordered nano-ripples on 316L surfaces were created by Martínez-Calderón et al. [30] using fs lasers, which enhanced osteoblast alignment and migration considerably. Because of the extremely short pulse duration, fs-LST reduces the heat-affected zone (HAZ) and preserves bulk properties [31].

Picosecond lasers (10–50 ps) produce uniform microgrooves and micro-pits with enhanced surface wettability and resistance against corrosion [33]. Neves et al. [33] showed that 316L treated with a ps-laser had increased passive Cr_2O_3 layer, lowering corrosion current density in simulated body fluid (SBF).

Nanosecond lasers (10–100 ns) are less expensive and commercially available for industrial uses. They create bigger features like dimples and crosshatch textures (20–80 μm diameter). Jang et al. [34] found that ns-LST decreased *Staphylococcus aureus* and *Escherichia coli* adhesion by 50–70% without loss of mechanical strength. Nonetheless, high energy input may lead to re-melting and oxide heterogeneity, reducing corrosion resistance slightly [35].

12.2.2.2 Effect of LST on the Surface Properties

Laser processing alters the chemical composition, wettability, and surface roughness (Ra), all of which determine biocompatibility. Femtosecond and picosecond lasers produce nanoscale roughness (Ra ≈ 0.1–0.8 μm) and ns lasers form micro-textures (Ra ≈ 1–5 μm) [30, 33].

Increased Cr enrichment, as shown by Neves et al. [33], enhances passivation and minimizes pitting susceptibility under saline conditions.

12.2.2.3 Effect of LST on Biological and Corrosion Performance

Laser-textured 316L has clear biological benefits. Micro- and nano-patterns expand protein adsorption sites (e.g., fibronectin, vitronectin), enhancing osteoblast adhesion and differentiation [30]. Xu et al. [36] reported that femtosecond-textured stainless steel surfaces triggered aligned cytoskeletal structure and enhanced osteogenic markers like ALP and Runx2.

Corrosion research has established that optimized LST enhances the passive film stability. EIS tests indicate greater polarization resistance and lower corrosion current density for laser-treated samples [33, 35]. These enhancements result from homogeneous oxide formation and decreased defect density.

Concurrently, antibacterial functionality is increased by optimizing surface microgeometry. Nano-ripples and hierarchical structures disrupt bacterial membranes, reducing adhesion forces [34]. For instance, Jang et al. [34] showed that nano-patterned 316L produced a 60% decrease in *S. aureus* attachment compared to polished controls. Table 12.2 presents the LST parameters used by various researchers.

12.2.3 LST on Co–Cr Alloy

Cobalt–chromium (Co–Cr, commonly CoCrMo) alloys are widely used for orthopedic and dental implants because of their high strength, wear resistance, and corrosion resistance. Implant performance depends strongly on surface topography, chemistry, roughness, and residual stress; these factors influence friction/wear, corrosion, protein adsorption, bacterial colonization, and bone cell response. LST is a controllable tool to tailor such surface properties at micro- and nano-scales without adding coatings, making it attractive for Co–Cr implant surfaces [26,37].

Pulse duration, pulse energy (fluence), repetition rate, spot size, and scan strategy determine feature depth, geometry, and the heat-affected zone. Ultrashort pulses (fs/ps) produce high-resolution features with minimal thermal damage, while longer pulses (ns) remove material faster but produce larger melt/recast zones and oxide formation [38,39]. Reports on Co–Cr indicate that LST can be used to precisely tune roughness, wettability, oxide chemistry, and local residual stresses by selecting appropriate laser regimes [37,40]. Table 12.3 shows the process parameters used by the researchers to investigate their effect on Co–Cr alloy.

TABLE 12.2
LST Parameters and Their Effect on 316L Stainless Steel

Laser Type	Key Parameters	Surface Morphology/ Microstructure	Surface Chemistry and Oxide Composition	Wettability (Contact Angle)	Corrosion/Biological Effects	Reference
Femtosecond (fs) laser	Pulse: 120–300 fs; λ = 1030 nm; fluence = 2.7 J/cm²; scan speed = 1 mm/s	Periodic uniform nano-ripples (LIPSS), period between 200–500 nm; micro and nano hierarchical topography	Uniform formation of Cr_2O_3 + Fe_2O_3; less re-melting; very little HAZ	50–65° (hydrophilic) immediately after texturing; becomes 110–130° after aging	Improved osteoblast adhesion and migration; increased corrosion resistance in PBS due to dense passive film	Martínez-Calderón et al. (2016) [[30]
Nanosecond (ns) laser	λ = 1064 nm; Power = 7 W 50 kHz	Microgrooves (3–10 μm) and nano-rough overlay; smooth transitions	Richer Cr_2O_3/Fe_3O_4 ratio; mild oxidative	45–70°; stable hydrophilicity for several days	Enhanced fibroblast proliferation; reduced corrosion current density by ~35% in SBF	Neves et al. (2024) [33]
Nanosecond (ns) laser — fiber/Nd:YAG	Pulse: 50–100 ns; λ = 1064 nm; fluence = 2–8 J/cm²; scan speed = 5–15 mm/s	Micro-dimples (20–80 μm diameter), grooves or grids; remelted layer more deeply	Thick Fe_3O_4 + Cr_2O_3 oxide layer; moderate HAZ (1–2 μm)	70–110°, depending on feature depth and oxidation	Reduced *S. aureus* and *E. coli* adhesion (~50–70%); passive film stability improved if fluence < 5 J/cm²	Jang et al. (2018) [34]

TABLE 12.3
Process Parameters and Their Effect on Co–Cr Alloy

Laser Type	Pulse Duration	Laser Power / Energy (Fluence)	Scan Speed (mm/s)	Spot Size (μm)	Hatch/ Overlap (%)	Atmosphere	Observed Surface Features/Effects	Resulting Property Changes	Reference
Nanosecond fiber laser (1064 nm)	70 ns	25W	50–150	45	-	Air	Micro-dimples shallow melt pool, recast oxide layer	↑ Hardness (by 10–20%), ↑ surface roughness (Ra ~2–5 μm), moderate increase in wettability	Sandoval-Robles et al. (2020) [37]
Nanosecond pulsed fibre laser (1064 nm)	–250 ns	50 W	100–500	30	50%	Air	Melted dimples, larger HAZ	↑ Hardness (by rapid solidification), possible tensile stress	Vidyasagar et al. (2024) [39]

12.2.4 LST on Magnesium Alloys

Magnesium (Mg) and magnesium alloys also have been of enormous interest as biodegradable implant materials due to their superior blend of biocompatibility, biodegradability, low density, and elastomodulus of approximately the same order as for natural bone [41]. Their accelerated corrosion in body environments leads to premature mechanical failure, widespread hydrogen evolution, and local degradation, limiting their clinical prospect [42,43].

In order to solve such a problem, surface modification techniques have been extensively studied to modulate the degradation rate as well as optimize the biological performance. Among them, LST has been identified as a controlled, precise, and contactless technique that can be used to modify magnesium surface topology, chemistry, and wettability [26]. Laser processing offers the capability to create micro/nano-scale structures with potential to influence corrosion kinetics, cell adhesion, and osseointegration [44,45].

Laser treatment of Mg is normally associated with local melting and fast solidification, resulting in grain refinement and homogenization of the remelted layer [42]. Pereira et al. [46] showed that Nd:YAG laser surface texturing of pure Mg formed orderly craters and re-solidified microstructure, which altered the corrosion channels. Along the same lines, Chen et al. [47] have stated that pulsed laser surface texturing of Mg alloys enhanced the microstructure and encouraged the development of a dense and protective oxide layer, enhancing corrosion stability.

12.2.4.1 Corrosion Behavior of LST Magnesium

Mitigation of corrosion remains an enduring imperative within the scope of magnesium surface engineering. The enhanced electrochemical reactivity coupled with the fleeting nature of passivation layers associated with magnesium translates into intensified corrosion within biological settings. Researches conducted by LST expose the possibility of modifying surface chemistry and morphology to enhance corrosion resistance [47]. Pereira et al. [46], too, showed that laser-treated Mg displayed depressed corrosion current density and enhanced corrosion morphology uniformity in SBF, associated with smaller grain sizes coupled with the formation of an oxide layer denser than before. Analogously, Conradi et al. made a comparison between laser-textured and sandblasted surfaces of Mg, concluding that laser surface treatment greatly reduced corrosion rates while fostering improved surface uniformity. Zhang et al. [42] noted that even though laser processing offered the possibility of enhanced surface finish, excessive inputting of energy had the possibility of causing the development of microcracks, consequently propagating localized corrosion. As such, optimization of scanning speed coupled with pulse energy is fundamentally necessary to optimize both structural integrity and corrosion resistance.

12.2.4.2 Effect of LST on Biocompatibility and Osseointegration

Surface topology is also the prime determinant for cellular response and osseointegration. Laser-textured surfaces have the capability to guide osteoblast directionality and increase adhesion by simulating intrinsic extracellular matrix structures. The laser texturing of the Mg–Gd–Ca alloy led to enhanced MC3T3-E1 pre-osteoblast

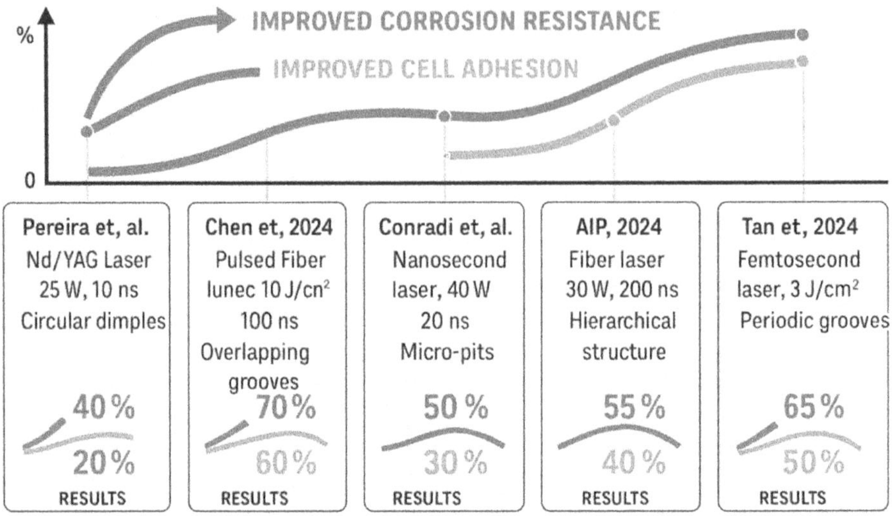

FIGURE 12.3 The effect of LST on corrosion and cell adhesion on magnesium.

cell growth and viability compared to control surfaces, confirming the bioactive potential of LST [46]. Conradi et al. [44] demonstrated that enhanced bioactivity and wettability boosted by LST surfaces led to improved adsorption of proteins and cell adhesion compared to sandblasting. The enhancements are attributed to the combined effects from the micro/nano-scale roughness and boosted hydrophilicity correlated to the laser-textured surfaces, improving the initial cell-implant interactions. However, limited in vivo studies are available, and the long-term biocompatibility under degradation conditions remains to be demonstrated. Figure 12.3 clearly indicates that LST significantly improves the corrosion resistance and cell adhesion.

Table 12.4 summarizes the effect of LST parameters on the magnesium and its alloys by various researchers.

12.2.5 LST on Zinc

Zinc and zinc alloys have attracted increasing interest in the last few years because of their extensive industrial and biomedical applications. Zinc has previously been utilized as a sacrificial coating for steel and an integral material in galvanization and corrosion protection. It has emerged more recently as a leading candidate as a biodegradable metal for temporary implants, batteries, and microelectronic devices [49,50].

LST is a cutting-edge, high-precision, green technology for the controlled alteration of surface morphology, wettability, and chemical composition. It improves the functional properties of zinc by its corrosion resistance, surface energy, and biological interactions tailored to required specifications [51–53].

TABLE 12.4
Summary of Laser Processing Parameters

Study / Reference	Material	Laser Type/ Wavelength	Power/Energy Density	Pulse Duration	Scanning Speed (mm/s)	Texture Type/ Morphology	Key Observations	Corrosion Behavior	Biocompatibility Effects
Pereira et al. [46]	Pure Mg	Nd:YAG (1064 nm)	25–50 W	10 ns	100–500	Circular dimples, shallow craters	Smooth and uniform re-solidified layer	Reduced corrosion rate due to oxide film formation	Enhanced wettability; improved protein adsorption
Chen et al. [47]	Rare-earth WE43 magnesium alloys	Pulsed fiber laser (1064 nm)	5–15 J/cm²	100 ns	200	Overlapping microgrooves	Fine-grain formation, homogeneous layer	Reduced Icorr; more stable passive layer	Promoted osteoblast adhesion and proliferation
Conradi et al. [44]	99.9%Pure Mg	Nanosecond laser	5 W	0.5 ms		Uniform micro-pits	Reduced surface roughness variation	Reduced corrosion current density; uniform attack	Higher bioactivity and wettability than sandblasting
J. Zhang [48]	Mg–Gd–Ca alloy	Fiber laser (1064 nm)	2.04 × 106W/cm		60	Hierarchical micro–nano structure	Re-solidified zone with refined grains	Stable degradation profile; reduced corrosion rate	Improved MC3T3-E1 cell adhesion and proliferation
Zhang et al. [42]	AZ31 Mg alloy	CO_2 laser (10.6 μm)	50–100 W	Continuous	50–300	Melt pool with ridge textures	Improved microhardness; small HAZ	Reduced corrosion with moderate fluence; microcracks at high power	No cytotoxic effects observed

LST induces extreme transformations in the morphology and chemical composition of the zinc surface depending on the laser pulse duration, energy density, and scanning parameters. The zinc surface, when treated with high-energy laser pulses, exhibits rapid melting, vaporization, and resolidification, which collectively form micro- and nano-scale structures such as grooves, dimples, ridges, and LIPSSs [52].

12.2.5.1 Corrosion and Wettability Behavior

The corrosion behavior of zinc has been governed by its surface oxides (ZnO, $Zn(OH)_2$). LST on zinc substrate alters the oxide layer thickness and defect density. Hajam et al. [51] performed a nanosecond LST at 1.2 J/cm² and 300 mm/s scan speed experiment on the zinc substrate. Results indicate that the corrosion current density was reduced by 60%. Improved passivity had been observed through the positive corrosion potential. However, high fluence or slow scanning speed could lead to oxide spallation or microcracking, which leads to the formation of galvanic cells that accelerate the localized corrosion. Optimizing the fluence window is challenging to balance the oxide formation and surface integrity [52].

Femtosecond (fs) laser texturing creates ultra-fine nanostructures and flat LIPSS with little or no thermal damage. Nanostructures facilitate the formation of dense, adherent ZnO films upon laser treatment or post-oxidation.

The improved corrosion performance of LST-Zn is mainly due to the following:

- Development of a stable passive oxide layer ($ZnO/Zn(OH)_2$) that retards anodic dissolution
- Decrease in active area exposure with micro-dimple geometries that encapsulate air and restrict electrolyte contact
- Refined microstructure within the remelted zone minimizing grain boundary corrosion pathways (Balamurugan & Ramesh, 2024)

Nevertheless, some of the studies also observe that the rate of corrosion may enhance under certain texture geometries (deep grooves, porosity) or when oxide uniformity is not good. The direction of change (improvement or worsening) therefore relies on the interaction among texture morphology, oxide compactness, and surface chemistry [51,52].

12.2.5.2 Wettability Behavior

Surface structures induced by laser and changes in oxide chemistry have a profound effect on zinc surface wettability. Wettability, quantified using the contact angle (θ), dictates the nature of interaction of water and biological fluids with the material surface, a crucial consideration for biomedical and coating application areas [54].

Immediately following laser texturing, the majority of zinc surfaces are superhydrophilic ($\theta < 10°$) owing to the presence of newly created hydrophilic ZnO and $Zn(OH)_2$ phases and enhanced surface energy. With time, surface aging or contamination results in hydrophobic recovery and contact angles up to 140–150°, especially if hierarchical micro/nano-structures are present [43]. Balamurugan [54] investigated the impact of laser fluence on pure zinc wettability systematically and concluded the following:

- Low fluence (0.8 J/cm^2) → moderate hydrophilicity ($\theta \approx 65°$) because of restrained oxidation and shallow relief.
- Medium fluence (1.2 J/cm^2) → transition zone ($\theta \approx 100°$) because micro-dimples and oxide thickening take place.
- At high fluence (1.6 J/cm^2) → high hydrophobicity ($\theta \approx 135$–$145°$) due to hierarchical roughness and air pockets trapped in keeping with the Cassie–Baxter theory.

Laser-induced LIPSSs also increase water repellency by creating fine-scale anisotropic textures that control droplet spreading. Such control is beneficial for biomedical implants, where surface wetting controls protein adsorption and osteoblast adhesion [49,50].

Zinc has moderate corrosion rate, biological metabolic function, and nontoxicity in the physiological concentration range [49,50]. However, controlling degradation rate and inducing favorable cell response are key issues for clinical applications. Laser surface texturing provides a clean and direct approach to resolving these limitations by modulating surface topography and oxide content without chemical coatings.

12.2.5.3 Controlled Corrosion and Degradation Behavior

For biodegradable implants, the zinc degradation rate must be associated with tissue healing. Excessive corrosion results in premature mechanical failure, while slow degradation results in bioresorption delay. Hajam et al.'s [51] and Balamurugan's [54] research demonstrated that LST permits fine-tuning of zinc corrosion kinetics by varying the oxide layer thickness and roughness. Laser-textured surfaces with micro-dimpled surface topology exhibited a 60% reduction in corrosion rate, providing a stable passive ZnO layer that dissolves progressively under physiological conditions.

Further, ultrashort-pulse (femtosecond) laser treatment produces nanoscale surface textures that stabilize oxide layers further and minimize localized pitting, a serious problem for biodegradable implants [52]. This allows for more reliable degradation rates and renders LST an appealing route toward developing next-generation zinc-based orthopedic and cardiovascular implants.

12.2.5.4 Improved Cell Adhesion and Biointegration

Surface roughness and wettability are very important in protein adsorption, osteoblast adhesion, and osseointegration thereafter. A relatively hydrophilic level (contact angle ~70–90°) is most favorable for cell adhesion [50]. The laser-textured zinc surfaces may be made to possess controllable wettability, and therefore the surface–cell interaction will be controlled. Moreover, thin, adherent ZnO film development improves not only corrosion control but also antibacterial properties by releasing Zn^{2+} ions, thus reducing the risk of implant-related infection [43]. Such combined effects render zinc surfaces modified with LST particularly well suited for applications in biodegradable stents, bone fixation implants, and antimicrobial coatings. Figure 12.4 elaborates the application of LST on zinc.

12.2.5.5 A Case Study in LST on Zinc

Kannan Ganesa Balamurugan and Kamalakannan K. (authors) have demonstrated a LST experiment on the pure zinc substrate using a 40 W femtosecond laser system with 1035 nm wavelength, 50 μJ@800kHz pulse energy, and beam diameter of 3 mm. The authors attempted to create dimple textures (circular) on the zinc substrate. The process parameters are shown in Table 12.5. The frequency and the power demand are the varied parameters. Figure 12.5 shows the micrograph of the dimple texture.

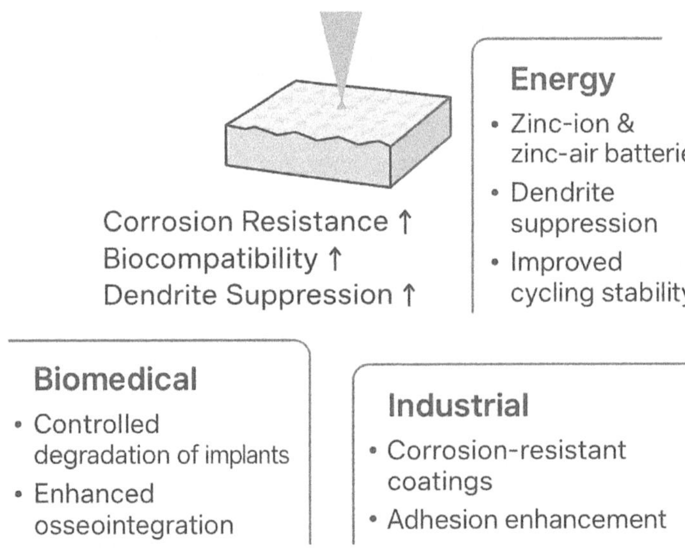

FIGURE 12.4 LST application on zinc.

TABLE 12.5
LST Process Parameters on Pure Zinc

Sample Number	Dimple Diameter (μm)	Number of Loops	Scan Speed (mm/s)	Power Demand (%)	Frequency (KHz)
1	250	50	500	50	10
2	250	50	500	50	5
3	250	50	500	55	10
4	250	50	500	55	5

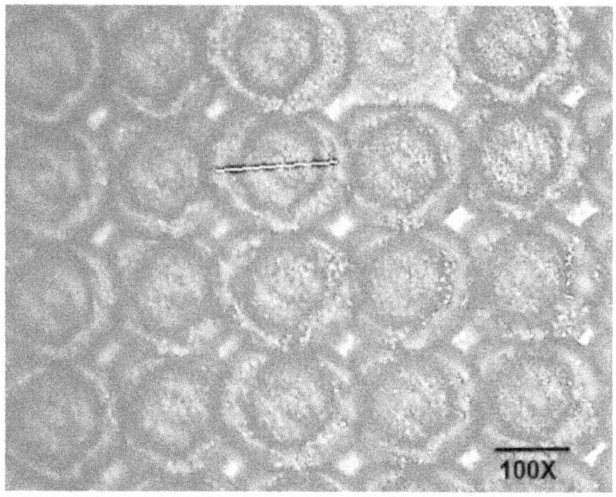

FIGURE 12.5 Optical micrograph of the dimple textures on the zinc substrate.

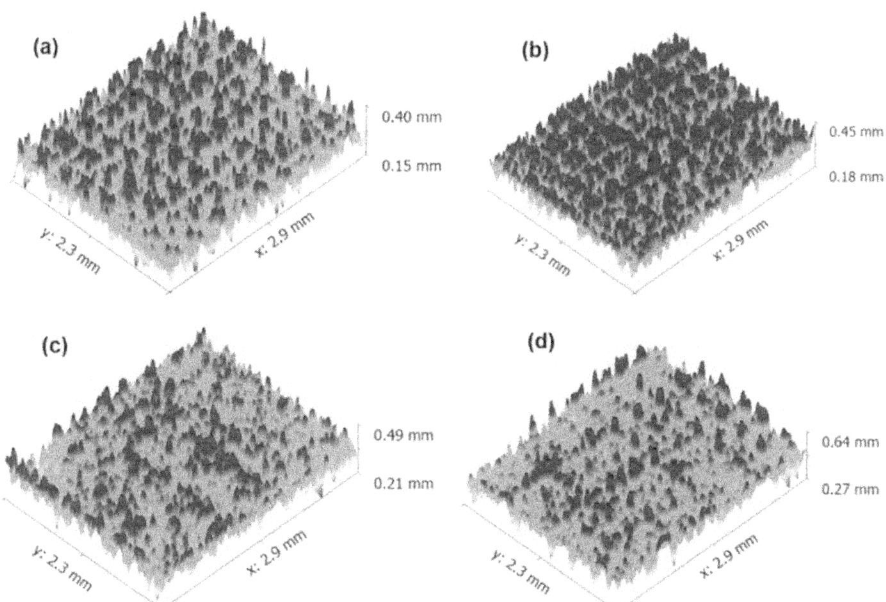

FIGURE 12.6 3D topographies of the textures zinc surfaces (a) Sample-1, (b) Sample-2, (c), Sample-3, (d) Sample-4.

The 3D topographies (Figure 12.6) show the roughness variations on the textured surface of zinc. Table 12.6 indicates the surface roughness Ra and texture depth of LST zinc substrates.

This experimental work revealed that higher laser frequency and power demand percentage produced higher depth textures. Higher laser power and frequency

TABLE 12.6
Surface Roughness Ra and Texture Depth Values of LST Zinc

Sample Number	Ra (μm)	Texture Depth (μm)
1	33.5	100
2	36.6	88
3	33.1	122
4	41.1	114

increases the laser fluence, which results in more material vaporization. However, there are no significant Ra changes in the textured surface except in sample-4, which was textured with 55% power demand and 5 KHz laser frequency. This combination might produce more re-solidification materials around the textures. Re-solidification materials formation around and inside the textures might increase the surface roughness. Further SEM and EDS analysis may reveal the clear texture formation and chemical changes due to laser interactions. Moreover, the effect of laser parameters on the corrosion and cell growth behavior should be analyzed as a future work.

12.3 FUTURE PERSPECTIVES OF LST ON BIOMETALLIC MATERIALS

12.3.1 Titanium Alloys

LST of Ti and Ti-6Al-4V will be transformed from osseointegration roughness to intentional, multifunctional, patient-specific interfaces. Subsequent research will tackle hierarchical micro–nano motifs (e.g., grooves + LIPSS) formed by ultrafast/burst and beam-shaping technologies to overcome cell-scale guidance from nanoscale protein adsorption and the addition of in situ chemistry (manipulated phases of TiO_2, dopants like Ag/Cu/Zn, or calcium-phosphate nanoislands) to offer simultaneous osteoinduction, angiogenesis, and long-term antibiofilm action and pursue closed-loop AI-enhanced process control that maps laser parameters to biologic endpoints via Bayesian/ML optimization and digital twins of patient bone quality, enabling textures controlling mechanotransduction (integrins, YAP/TAZ) for target indications (accelerated dental loading, osteoporotic fixation). LST will also be co-designed with additive-manufactured lattices texturing strut exteriors and pore walls to enable early fixation and fluid transport. Safety and longevity will be guaranteed: tribocorrosion-resistant fretting contact coatings, shelf- and sterilization-life-stand antimicrobial efficacy, and long-term in vivo confirmation in aging and large-animal models. Scalable manufacture (5-axis galvo/robotics for curved geometries, in-line metrology) and hybrid stacks (LST + PEO/ALD or bioactive films) will drive commercialization, with texture descriptor standards and biological tests (ISO 10993-compliant). Last but not least, intelligent manufacturable, regulatory approved, affordable drug-eluting, and immunomodulatory textures will define the decade of LST for titanium biomedical implants.

12.3.2 L Stainless Steel

The future of biomedical stainless steel parts is direct manufacturing + functional surface engineering. In situ LST during layer deposition of AM could make post-processing obsolete. Researchers have already shown initial success by using ns-LST on SLM-constructed 316L, enhancing surface roughness and bone cell adhesion (patient-specific implants with predesigned bioactive surface zones directly programmed into the build process).

12.3.3 Co–Cr Alloy

Future investigations of LST of Co–Cr alloys for bioapplications will increasingly move away from rudimentary surface modification in the direction of mechanism-based, multifunctional surface tailoring to the needs of targeted implant environments. Future studies need to concentrate on comprehending the ways in which highly engineered micro- and nano-structures alter protein adsorption kinetics, macrophage polarization, and downstream osteogenic or endothelial responses. Rather than just enhancing wettability, LST needs to be applied to engineer cell-instructive surfaces that direct biological processes in a sequentially programmed manner from protein corona assembly to tissue integration. Meanwhile, long-term tribocorrosion resistance under physiologically relevant conditions is an unsolved challenge. There needs to be future work to determine the behavior of LST-modified surfaces under cyclic loading, micro-motion, third-body wear (debris), and protein-containing synovial or blood-like conditions, while ensuring that increased cell adhesion is not at the expense of higher ion release or corrosion-assisted fatigue.

12.3.4 Magnesium Alloys

The fabrication of hierarchical micro–nano textures that replicate the natural bone morphology provides an appealing pathway to improve osteointegration, corrosion resistance, and antibacterial performance together. The integration of LST with laser interference lithography or ultrafast femtosecond laser processing can create multifunctional surfaces that are tunable in terms of wettability and protein adsorption. Gradient texturing, in which texture depth and density spatially change to design local degradation and tissue response, needs to be investigated in future designs as well.

The majority of existing research is restricted to in vitro testing involving model body fluids and short terms. But there is a need for in vivo studies involving physiological mechanical loading and dynamic biochemical environments to ensure long-term biocompatibility, degradation rates, and osseointegration. Immune response, inflammation behavior, and long-term tissue–implant interface development should also be evaluated in future work. Standardized animal testing protocols for textured Mg implants would speed up clinical translation.

12.3.5 Zinc

Through combined finite element modeling (FEM) and electrochemical data, subsequent work might forecast implant degradation behavior under physiological loads

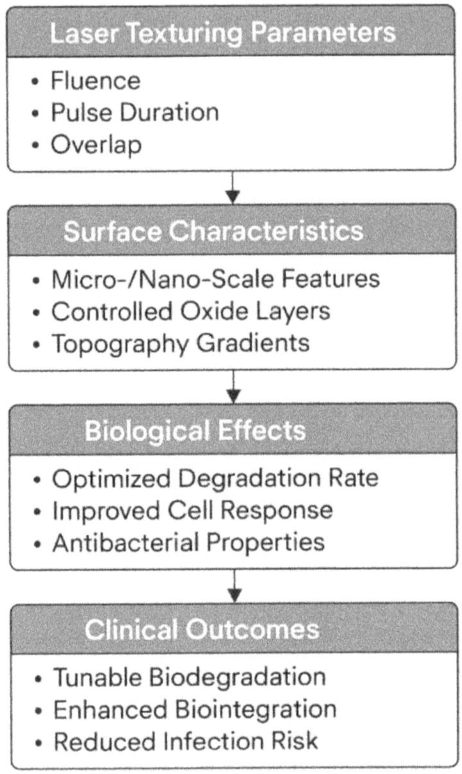

FIGURE 12.7 Future perspective road map of LST on zinc for biomedical applications.

and direct texture design for individual anatomical locations. Combining with biophysical tools like AFM-based force mapping and real-time confocal microscopy will allow for improved measurement of how Zn surfaces micromodulated cytoskeletal arrangement and adhesion kinetics. Figure 12.7 shows the road map for the future perspective of LST on zinc for the biomedical applications.

12.4 CONCLUSION

Laser surface texturing has shown to have a transformative role in engineering biometallic implant surfaces with functionalized properties beyond mere roughness enhancement. By facilitating controlled micro- and nano-scale feature formation, LST fundamentally enhances cell adhesion, protein adsorption, early-stage osseointegration, and corrosion resistance while also limiting bacterial attachment and tribological wear. Titanium and stainless steel derive significantly improved bioactivity and hydrophilicity following femtosecond and picosecond laser structuring, whereas Co–Cr alloys display pulse duration and fluence-dependent frictional and biological responses. Magnesium- and zinc-based biodegradable implants record excellent improvements in controlled degradation as well as cytocompatibility once laser parameters are optimized judiciously to balance oxide layer growth and microstructural refinement.

Yet, with rigorous laboratory-scale demonstration, clinical translation is still hampered by limitations in large-area texturing, parameter reproducibility, long-term in vivo function under dynamic loading, and regulatory standardization. Future progress will likely address AI-aided process control, incorporation of LST into additive manufacturing protocols, and creation of multifunctional textures that can provide simultaneous osteoinduction, stimulation of angiogenesis, and antibacterial protection. With the alignment of digital manufacturing, surface engineering, and biointerface science, LST is set to progress from a laboratory method into a clinically viable technology that makes possible next-generation intelligent implant surfaces optimized for patient-specific therapeutic processes.

REFERENCES

[1] Ratner, Buddy D., Allan S. Hoffman, Frederick J. Schoen, and Jack E. Lemons. *Biomaterials science: an introduction to materials in medicine.* Elsevier, 2004.

[2] Williams, David F. "On the nature of biomaterials." *Biomaterials* 30, no. 30 (2009): 5897–5909.

[3] Hench, Larry L. "Bioceramics: from concept to clinic." *Journal of the american ceramic society* 74, no. 7 (1991): 1487–1510.

[4] Langer, Robert, and Nicholas A. Peppas. "Advances in biomaterials, drug delivery, and bionanotechnology." *AIChE Journal* 49, no. 12 (2003): 2990–3006.

[5] X. Zhang, H. Xie, Z. Liu, C. Tan, Z. Luo, H. Li, J. Lin, L. Sun, W. Chen, Z. Xu, L. Xie, W. Huang and H. Zhang, *Angew. Chem., Int. Ed.*, 2015, 54, 3653–3657

[6] O'brien, Fergal J. "Biomaterials & scaffolds for tissue engineering." *Materials today* 14, no. 3 (2011): 88–95.

[7] Caldon J Esdaille, Kenyatta S Washington, Cato T Laurencin, Regenerative engineering: a review of recent advances and future directions. Regenerative Medicine, Volume 16, 2021 , 495–512

[8] Thangaraju, Pugazhenthan, and Shoban Babu Varthya. "ISO 10993: biological evaluation of medical devices." In *Medical device guidelines and regulations handbook*, pp. 163-187. Cham: Springer International Publishing, 2022.

[9] Narayan R. Biomedical Materials. Springer; 2021.

[10] Geetha, Manivasagam, Ashok K. Singh, Rajamanickam Asokamani, and Ashok K. Gogia. "Ti based biomaterials, the ultimate choice for orthopaedic implants–A review." *Progress in materials science* 54, no. 3 (2009): 397–425.

[11] Rinaudo, M. 2006. "Chitosan: Properties and Applications." *Progress in Polymer Science* 31 (7): 603–632.

[12] Ramakrishna, S., Joerg Mayer, Erich Wintermantel, and Kam W. Leong. "Biomedical applications of polymer-composite materials: a review." *Composites science and technology* 61, no. 9 (2001): 1189–1224.

[13] Liu, Xuanyong, Paul K. Chu, and Chuanxian Ding. "Surface modification of titanium, titanium alloys, and related materials for biomedical applications." *Materials Science and Engineering: R: Reports* 47, no. 3-4 (2004): 49–121.

[14] Xiandui Dong, Juntao Lu, Chuansin Cha. Self-Assembled Monolayers: Preparation, Properties and Application[J]. *Journal of Electrochemistry*, 1995, 1(3): 248–258.

[15] Vallet-Regí, M., and J. M. González-Calbet. 2004. "Calcium Phosphates as Substitution of Bone Tissues." *Progress in Solid State Chemistry* 32 (1–2): 1–31.

[16] Jeyaprakash, N., Che-Hua Yang, and K. R. Ramkumar. "Microstructural, mechanical and wear behaviour of Inconel-718 produced through laser-powder bed-fused additive manufacturing." *Materials Science and Technology* 37, no. 3 (2021): 326–337.

[17] Prabu, G., Che-Hua Yang, Ibrahim A. Alnaser, and N. Jeyaprakash. "Analysis of tribological behavior of Laser Powder Bed Fusion (LPBF) CuCrZr alloy through reciprocating loading on the columnar structure." *Tribology International* 202 (2025): 110301.

[18] Cunha, Alexandre, Omar Farouk Zouani, Laurent Plawinski, Ana Maria Botelho do Rego, Amélia Almeida, Rui Vilar, and Marie-Christine Durrieu. "Human mesenchymal stem cell behavior on femtosecond laser-textured Ti-6Al-4V surfaces." *Nanomedicine* 10, no. 5 (2015): 725–739.

[19] Kedia S, Singh R, Dave K, et al. Picosecond laser-induced micro/nano hybrid surface for enhanced cell adhesion. J Biomed Mater Res B Appl Biomater. 2023;111(5):1230–1242.

[20] Kolarovszki B, Ficsor S, Frank D, Katona K, Soos B, Turzo K. Unlocking the potential: laser surface modifications for titanium dental implants. Lasers Med Sci. 2024 Jun 24;39(1):162.

[21] Wu, Xinhui, Haiyong Ao, Zhaoru He, Qun Wang, and Zhaoxiang Peng. "Surface modification of titanium by femtosecond laser in reducing bacterial colonization." *Coatings* 12, no. 3 (2022): 414.

[22] Kedia, Sunita, Rahul Checker, Santosh Kumar Sandur, and J. Padma Nilaya. "Picosecond laser-induced hybrid groove structures on Ti-6Al-4V bio-alloy to accelerate osseointegration." *Journal of Biomedical Materials Research Part B: Applied Biomaterials* 111, no. 10 (2023): 1775–1784.

[23] Kolarovszki, Bela, Szabolcs Ficsor, Dorottya Frank, Krisztian Katona, Balazs Soos, and Kinga Turzo. "Unlocking the potential: laser surface modifications for titanium dental implants." *Lasers in Medical Science* 39, no. 1 (2024): 162.

[24] A. C., Elias. "Principles and Presentation of Corrosion - D.A. Jones." ISBN 0 -13-359 993-0, 1996.

[25] Long, Marc, and H. J. Rack. "Titanium alloys in total joint replacement—a materials science perspective." *Biomaterials* 19, no. 18 (1998): 1621–1639.

[26] Shivakoti, Ishwer, Golam Kibria, Robert Cep, Bal Bahadur Pradhan, and Ashis Sharma. "Laser surface texturing for biomedical applications: a review." *Coatings* 11, no. 2 (2021): 124.

[27] Gnecco, Enrico, and Ernst Meyer, eds. *Fundamentals of Friction and Wear on the Nanoscale*. Cham: Springer, 2015.

[28] Lawrence, Jonathan R., and David Waugh, eds. *Laser surface engineering: Processes and applications*. Elsevier, 2014.

[29] Sedao, M., A. Rudenko, J. Colin, et al. 2019. "Ultrafast Laser-Induced Surface Structuring of Metals." *Applied Surface Science* 471: 983–999.

[30] Martínez-Calderon, Miguel, Miguel Manso-Silván, A. Rodríguez, M. Gómez-Aranzadi, J. P. García-Ruiz, S. M. Olaizola, and R. J. Martín-Palma. "Surface micro- and nano-texturing of stainless steel by femtosecond laser for the control of cell migration." *Scientific reports* 6, no. 1 (2016): 36296.

[31] Bonse, J., S. Hohm, S. Kirner, A. Rosenfeld, and J. Krüger. 2017. "Laser-Induced Periodic Surface Structures: A Scientific Evergreen." *IEEE Journal of Selected Topics in Quantum Electronics* 23 (3): 9000615.

[32] Crouse, T., H. Li, and J. Xie. 2019. "Picosecond Laser Texturing of Stainless Steel for Wettability Control." *Applied Physics A* 125 (4): 1–10

[33] Neves, V. D., F. Q. Correa, M. O. Ferreira, A. R. Rodrigues, W. Wolf, R. Galo, F. M. Yasuoka, and J. A. Moreto. 2024. "Nanosecond Pulsed Fiber Laser Treatment to Improve Corrosion Resistance of 316L Stainless Steel Surgical Devices." *Materials* 17 (24): 6178.

[34] Jang, Yeongseon, Won Tae Choi, Christopher T. Johnson, Andrés J. García, Preet M. Singh, Victor Breedveld, Dennis W. Hess, and Julie A. Champion. "Inhibition of bacterial adhesion on nanotextured stainless steel 316L by electrochemical etching." *ACS biomaterials science & engineering* 4, no. 1 (2018): 90–97.
[35] Xu, Xiaofeng, Laifei Cheng, Xiaojiao Zhao, Jing Wang, and Xinyi Chen. "Micro/nano periodic surface structures and performance of stainless steel machined using femtosecond lasers." *Micromachines* 13, no. 6 (2022): 976.
[36] Sandoval-Robles, Jesús A., Ciro A. Rodríguez, and Erika García-López. "Laser surface texturing and electropolishing of CoCr and Ti6Al4V-ELI alloys for biomedical applications." *Materials* 13, no. 22 (2020): 5203.
[37] Filipov, Emil, Liliya Angelova, Sanjana Vig, Maria Helena Fernandes, Gerard Moreau, Marie Lasgorceix, Ivan Buchvarov, and Albena Daskalova. "Investigating potential effects of ultra-short laser-textured porous poly-ε-caprolactone scaffolds on bacterial adhesion and bone cell metabolism." *Polymers* 14, no. 12 (2022): 2382.
[38] Vidyasagar, KE Ch, and Dinesh Kalyanasundaram. "Enhanced corrosion resistance of CoCrMo by laser-based surface modification." *Surface Engineering* 40, no. 1 (2024): 100-111.
[39] Man, Kun, Sangram Mazumder, Narendra B. Dahotre, and Yong Yang. "Surface nanostructures enhanced biocompatibility and osteoinductivity of laser-additively manufactured CoCrMo alloys." *ACS omega* 8, no. 50 (2023): 47658–47666.
[40] Aikin, Mykyta, Vadim Shalomeev, Volodymyr Kukhar, Andrii Kostryzhev, Ihor Kuziev, Viktoriia Kulynych, Oleksandr Dykha et al. "Recent advances in biodegradable magnesium alloys for medical implants: evolution, innovations, and clinical translation." *Crystals* 15, no. 8 (2025): 671.
[41] Zhang, Shiliang, Jing Jiang, Xianrui Zou, Ning Liu, Hongshui Wang, Lei Yang, Huan Zhou, and Chunyong Liang. "Progress of laser surface treatment on magnesium alloy." *Frontiers in Chemistry* 10 (2022): 999630.
[42] Kumar, K., R. S. Gill, and U. Batra. "Challenges and opportunities for biodegradable magnesium alloy implants." *Materials technology* 33, no. 2 (2018): 153–172.
[43] Shivakoti, I., G. Kibria, R. Cep, B. B. Pradhan, and A. Sharma. "Laser surface texturing for biomedical applications: a review. Coatings. 2021; 11: 124." *Metal Surfaces* (2021): 5.
[44] Conradi, Marjetka, Aleksandra Kocijan, and Bojan Podgornik. "Enhancing magnesium bioactivity for biomedical applications: effects of laser texturing and sandblasting on surface properties." *Materials* 17, no. 20 (2024): 4978.
[45] Tan, Chee Ying, Cuie Wen, and Hua Qian Ang. "Influence of laser parameters on the microstructures and surface properties in laser surface modification of biomedical magnesium alloys." *Journal of Magnesium and Alloys* 12, no. 1 (2024): 72–97.
[46] Pereira, H., O. Carvalho, G. Miranda, and F. S. Silva. "Pure magnesium laser surface modification using Nd: YAG laser." *Materials Technology* 36, no. 13 (2021): 811–815.
[47] Chen, Rui, Yansheng Yao, Jiale Yong, Suwan Zhu, Xinpeng Xu, and Nianwei Dai. "Pulsed laser surface texturing enhancing corrosion resistance of rare-earth WE43 magnesium alloys in simulated body fluid environment." *Journal of Alloys and Compounds* 1005 (2024): 176197.
[48] Zhang, Jiaru, Wenting Lin, Yingchun Guan, and Xuenan Gu. "Biocompatibility enhancement of Mg-Gd-Ca alloy by laser surface modification." *Journal of Laser Applications* 31, no. 2 (2019).
[49] Kabir, Humayun, Khurram Munir, Cuie Wen, and Yuncang Li. "Recent research and progress of biodegradable zinc alloys and composites for biomedical applications: Biomechanical and biocorrosion perspectives." *Bioactive materials* 6, no. 3 (2021): 836–879.
[50] Rao, Jiahui, Hairui Gao, Jiwei Sun, Ran Yu, Danlei Zhao, and Yumei Ding. "A critical review of biodegradable zinc alloys toward clinical applications." *ACS Biomaterials Science & Engineering* 10, no. 9 (2024): 5454–5473.

[51] Hajam, Mohammad Irfan, and Mohammad Mohsin Khan. "Effect of laser surface texturing on the corrosion behaviour and biocompatibility of zinc hydroxyapatite implant material." *Materials Today Communications* 38 (2024): 108110.
[52] Mustafa H. Ultrashort pulsed laser surface texturing of zinc: Mechanisms and morphology control. Ph.D. Thesis, University of Twente, The Netherlands; 2019.
[53] Kang, Peng, Yi Yuan, Funian Mo, and Haibo Hu. "Surface laser texturing and alloying: front-end design optimization of zinc metal anode for dendrite-free deposition." *ACS nano* 19, no. 16 (2025): 15994–16010.
[54] Balamurugan, Kannan Ganesa. "Investigation on the effect of laser surface texturing on the wettability of pure zinc substrate." *Journal of Engineering and Applied Science* 71, no. 1 (2024): 142.

Index

L

M

N

O

P

R

S

T

U

W

For Product Safety Concerns and Information please contact our EU
representative GPSR@taylorandfrancis.com
Taylor & Francis Verlag GmbH, Kaufingerstraße 24, 80331 München, Germany

www.ingramcontent.com/pod-product-compliance
Ingram Content Group UK Ltd.
Pitfield, Milton Keynes, MK11 3LW, UK
UKHW020157260926
475702UK00047B/497
* 9 7 8 1 0 3 2 8 9 2 7 8 8 *